NUCLEIC ACID CHEMISTRY
PART ONE

R. STUART TIPSON

LEROY B. TOWNSEND

NUCLEIC ACID CHEMISTRY

IMPROVED AND NEW SYNTHETIC PROCEDURES, METHODS AND TECHNIQUES

PART ONE

Edited by

Dr. LEROY B. TOWNSEND

The University of Utah
Salt Lake City, Utah

Dr. R. STUART TIPSON

(Retired)

A WILEY-INTERSCIENCE PUBLICATION

JOHN WILEY & SONS, New York • Chichester • Brisbane • Toronto

Library of Congress Cataloging in Publication Data:

Main entry under title:

Nucleic acid chemistry.

"A Wiley-Interscience publication."
"Successor to volume 1 of Synthetic Procedures in nucleic acid chemistry, edited by W. W. Zorbach and R. S. Tipson."
Includes indexes.
1. Chemistry, Organic—Synthesis. 2. Nucleic acids. I. Townsend, Leroy B. II. Tipson, R. Stuart.

QD262.N8 547'.596 77-22816
ISBN 0-471-88090-6 (part 1)
ISBN 0-471-04680-9 (part 2)

Printed in the United States of America

10 9 8 7 6 5 4 3 2 1

PREFACE

The present two volumes contain a collection of new or improved synthetic procedures, methods, and techniques in the field of nucleic acid chemistry and are subdivided into discussions of eight main topics: in Part 1, I, Heterocyclic Compounds; II, Carbohydrates; and III, Nucleosides; in Part 2, IV, Nucleotides and Polynucleotides; V, Isotopically Labeled Compounds; VI, Chemical and Enzymic Syntheses; VII, Reagents, Intermediates, and Miscellaneous Compounds; and VIII, Instrumental or Analytical Techniques and Applications. They provide an up-to-date source of information on all the important aspects of the subject. Each contribution was written by experienced research workers to guide the reader by giving representative descriptions with ample details, so that even a novice will be able to apply the information.

The present collection was intended to be a successor to Volume 1 of *Synthetic Procedures in Nucleic Acid Chemistry,* edited by W. W. Zorbach and R. S. Tipson, but an enlarged scope and different format necessitated both a title change and an extension to two volumes.

Although intended primarily for use by organic chemists, these books should prove valuable to medicinal chemists and biochemists, because the rapid expansion of these fields has produced an urgent need for a compilation of reliable methods. The extensive literature now makes it difficult, even for the expert in the field, to select a suitable procedure, but the detailed information given here exemplifies the most modern approaches to the various problems encountered. Most of the authors of the articles were investigators who had either originated these methods or acquired detailed knowledge of them through extensive use in the laboratory. We thank all of them for their enthusiasm and their gratifying responses to our request for contributions. We also thank Ms. M. E. Siedenstrang for typing camera ready copy of both volumes, compiling and typing subject and author indices, and handling all correspondence related to the publication of these two volumes. We also thank Ms. S. Mason and Dr. D. S. Wise for their help in the preparation and proofing of reaction schemes. The reception accorded Volumes 1 and 2 will determine whether this series will be continued.

LEROY B. TOWNSEND
R. STUART TIPSON

Salt Lake City, Utah
Kensington, Maryland
April 1978

CONTENTS

HETEROCYCLIC COMPOUNDS

Imidazole

Purines

Pyrimidines

Others

CARBOHYDRATES

NUCLEOSIDES

Monocyclic Aglycone

Imidazoles

Pyrimidines

Contents: Part 2

NUCLEOSIDES

Bicyclic Aglycone

Purines

Others

Monocyclic and Bicyclic Aglycones

NUCLEOTIDES AND POLYNUCLEOTIDES

Monocyclic Aglycone

Bicyclic Aglycone

Monocyclic and Bicyclic Aglycones

Others

ISOTOPICALLY LABELED COMPOUNDS

CHEMICAL AND ENZYMIC SYNTHESES

REAGENTS, INTERMEDIATES, AND MISCELLANEOUS COMPOUNDS

NUCLEIC ACID CHEMISTRY
PART ONE

[1] 4(OR 5)-(3,3-DIMETHYL-1-TRIAZENO)IMIDAZOLE-5(OR 4) CARBOXAMIDE[a]

An Improved Procedure Adaptable to Large-Scale Production

Arthur J. Zambito, Edward Hendrzak, and Robert D. Babson

Merck & Co., Inc., Rahway, New Jersey 07065

INTRODUCTION

The synthesis of 4(or 5)-(3,3-dimethyl-1-triazeno)imidazole-5(or 4)carboxamide (DIC) (*3*) via the coupling reaction of 5(or 4)-diazoimidazole-4(or 5)carboxamide (*1*) with dimethylamine was first reported by Shealy and co-workers.[1-3] As this compound has consistently shown promise as an antitumor agent, particularly in malignant melanomas, it was necessary to develop a procedure suitable for the preparation of the large quantities needed in extensive clinical-evaluation programs. To achieve this the literature procedure required considerable modification. By the original coupling reaction *2*, is isolated and then added as a solid to the amine solution, an operation that poses no serious problems on a laboratory scale. On a large scale, however, this procedure

---

[a]This investigation was supported by Contract NIH-72-2002 from the Division of Cancer Treatment, National Cancer Institute, National Institutes of Health, Department of Health, Education, and Welfare Bethesda, Maryland 20014.

is unsafe, owing to the ease with which the diazonium salt *2* explodes on impact when dry.

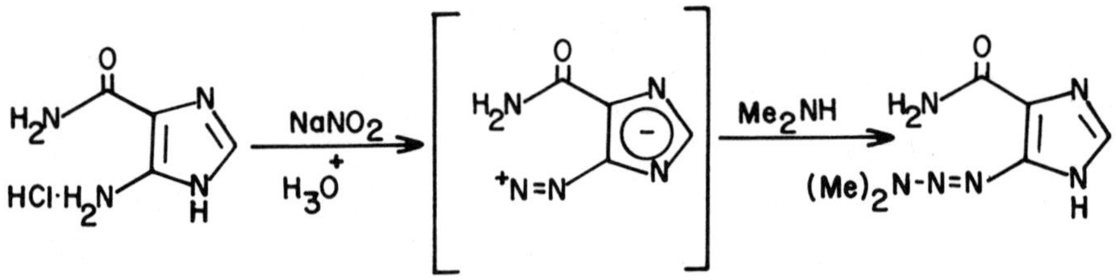

1 (162.6) 2 (137.1) 3 (182.2)

Under optimal conditions *3* is obtained as a cream-colored solid, although an orange-to-red product may occasionally result. A more serious problem is that on formulation of *3* in aqueous D-mannitol citric acid, a pink or red precipitate develops on standing, even when it is protected from light. The cause or nature of the impurity has not been elucidated, but it is believed to be caused by the competing side-reaction coupling of *1* with *2* during the diazotization. Also, as a result of the dissociation of solutions of *3* to *2* in the presence of light or acid, the regenerated *2* subsequently couples with any unreacted *1* still present in the final mixture. Modification of the diazotization step and heating of the reaction mixture to maximize the formation of color and the coupling of *1* with *2* have both consistently produced

DIC of satisfactory quality. Described here is a procedure that: (a) uses a slurry of *2*, thus avoiding its isolation, (b) is adaptable to large-scale production, and (c) affords pure DIC.

PROCEDURE

Pilot Plant Operation[b]

Purification of 4(or 5)-Aminoimidazole-5(or 4)carboxamide Hydrochloride (1). 4(or 5)-Aminoimidazole-5(or 4)carboxamide hydrochloride (*1*)[c] (10 kg, 61.5 mol) (dry basis) is dissolved in water (16 gal) at 40° to 45° in a 30-gal, agitated, glass-lined reactor. The dark solution is treated with Darco KB (500 g), the suspension is filtered through an 8-in Hastelloy Sparkler filter, and the reactor and charcoal cake are washed with water (2.5 gal). The filtrate and wash are combined and stored overnight in a suitable glass-lined kettle (or used directly in the diazotization step), and immediately prior to diazotization, concentrated hydrochloric acid (6.35 l) is added with stirring.

---

[b]The pilot plant work was supervised by A. D. Epstein, Chemical Engineering Research and Development Department; the purity specifications were established by J. P. Gilbert, B. Singleton, and G. V. Downing, Analytical and Physical Chemistry Department, Merck & Co., Inc., Rahway, NJ.

[c]Commercially available from Heinrich Mack Co., Chem. Pharmazeutische Fabrik, Iltertissen, Bavaria, West Germany; and from Calbiochem, Los Angeles, Calif.

4(or 5)-Diazoimidazole-5(or 4)carboxamide (2).

NOTE: Before starting the diazotization of *1*, the lights are turned off, and care is taken to exclude most of the light until the final product (*3*) is placed in a drying oven, as solutions of *3* are light-sensitive.

Sodium nitrite[d] (5.5 kg, 77.3 mol) is dissolved in water (20 gal) in a 50-gal, agitated, glass-lined reactor. The tan, acidified solution of *1* is pumped into the nitrite solution at -5° to 0°, nitrogen gas being simultaneously admitted into the reactor at the rate of 3 to 5 $ft^3 \cdot min^{-1}$ via a tube concentric with, and larger than, the tube through which the solution of *1* is pumped. [A localized excess of the amino carboxamide solution (*1*) results in a red by-product. To avoid this, the addition is so performed that the solution of *1* falls directly into the rapidly stirred mixture while a constant stream of nitrogen displaces the nitrous fumes.] The reactor is cooled by circulating a (low-temperature) methanol-dry-ice solution through the jacket. When approximately three-quarters of *1* has been charged, a solution of sodium nitrite (1.075 kg, 15.1 mol) in water (2.5 l) is added to the mixture, and after complete addition of *1* the mixture is stirred for an additional 5 min at 0° to 5°.

---

[d]Merck reagent, 97% minimum assay.

4(or 5)-(3,3-Dimethyl-1-triazeno)imidazole-5(or 4)carboxamide (3). The foregoing aged batch of diazonium intermediate (*2*) is added during a 2-hr period to dimethylamine (50 gal) in a 100-gal, glass-lined, agitated reactor.[e] The jacket of the reactor is also supplied with the low-temperature coolant, thus permitting the addition to be conducted below 5°. The diazotization vessel is washed with water (3 gal), and the wash is added to the reaction mixture. The pink to salmon-colored batch is aged for 1 hr at 0° to 5°, warmed to 25° during 30 min, heated with free steam to 55° in ≈30 min, and finally kept for 5 min at 55°. Concentration is begun immediately, the vacuum being applied cautiously (to prevent bumping), and the volume of the deep-red reaction mixture is diminished to 43 gal during 2 hr. The rapid evaporation of dimethylamine cools the mixture from 55° to 30°. At this point, high vacuum (28.5 in) is used at a temperature below 35° to concentrate to a volume of ≈30 gal during ≈12 hr. The batch is finally cooled for 30 min at 0° to 5° and filtered through polyethylene cloth[f] on a 30-in pot. Small amounts of product adhering to the reactor are transferred to the filtered cake with the mother liquors, and the cake is washed with cold water

---

[e]For convenience, the dimethylamine is collected the day before, and is kept cold until ready for use. It is charged, as a liquid, directly from the cylinder into the reactor.

[f]Linear polyethylene sheet, 1/16-in thickness, 45-μm porosity, F-1255, Bel-Art Products, Pequannock, NJ.

(3 × 5 gal) followed by cold acetone (3 × 3 gal). The solid is tray-dried at 50°/29 in for 15 hr; yield 9.0 kg, 214° to 215° (dec.).

The tan, crude product (4.5 kg[g]) is dissolved in 40% aqueous dimethylamine (76 gal,[h] 570 lb) in a 100-gal, agitated, glass-lined reactor. The solution is treated with Darco KB (900 g), and the mixture is heated to 55° using hot water in the jacket. When this temperature is reached, the reactor is pressurized with nitrogen, and the batch is blown through an 8-in, Sparkler filter. The reactor is washed with 40% aqueous dimethylamine (9 gal, 68 lb) and then with water (5 gal). The filtrate and wash are combined and concentrated to 36 gal in a 100-gal, agitated, glass-lined reactor under vacuum, starting with hot water in the jacket, and later using free steam. Initially, a vacuum of ≃100 torr prevails, but toward the end of the concentration it improves to ≃25 mm. During the concentration (3 to 4 hr), the temperature never exceeds 34°. (At higher temperatures, a dark product results.) The concentrated batch is cooled to 0° to 5° and filtered through polyethylene cloth on a 32-in filter pot. The reactor is rinsed with water (25 gal), which, in turn, is used to slurry-wash the cake four times. The solid is subsequently

---

[g]The entire 9.0 kg of crude product cannot be processed in a single operation due to equipment limitations.

[h]A smaller volume may be used.

washed with water (5 gal) and then with acetone (3 × 5 gal). After it has been dried at 50°/29 in for 15 hr, the yield of slightly pink product is 3.0 kg; m.p. 208° to 211° (dec.). Similarly, another 3 kg is obtained from the remaining 4.5 kg of crude product, and, of this combined 6 kg, 5.56 kg is recrystallized[i] as already described, using 40% aqueous dimethylamine (94 gal, 703 lb) and Darco KB (1.11 kg). The charcoal is removed by filtration, and the filtrate and washings are concentrated to 46 gal, giving 5.215 kg (50.2%)[j] of pure, light-cream, crystalline 4(or 5)dimethyltriazenoimidazole(5 or 4)carboxamide (*3*); m.p. 212° to 213° (dec.).

The following specifications apply:

Test	Limits of Specifications
1. *Elemental analysis*	C, 39.25 to 39.85
$C_6H_{10}N_6O$ (182.20)	H, 5.23 to 5.83
	N, 45.83 to 45.43
2. *Equivalent weight*	180.4 to 184.0
($HClO_4$ titrant)	
3. *Color of solution*	$A\% \times 10^3$ at 495 nm = <25

A solution of DIC (1.6 g) in 100 ml of stock solution of citric acid—D-mannitol is kept for 20 to 24 hr in the dark,

---

[i]A second recrystallization is not always necessary.

[j]Yields as high as 56% have been obtained.

filtered, and the optical absorbance is measured in a suitable spectrophotometer (10-cm cell). Stock solution: 17.5 g of citric acid monohydrate and 4.0 g of D-mannitol dissolved in 987 ml of water.

4. *Thin-layer chromatography* Analtech Silica Gel G.F., benzene, acetone, 1-butanol, water, glacial acetic acid, (1 part each); detected with a solution of 6% of vanillin and 1% (vol) of sulfuric acid in 1:1 water—methanol; also detected by ultraviolet (UV) light. — Single spot ($RF \simeq 0.5$)

5. *Infrared spectrum* (Nujol) — Conforms to proposed structure

6. *Ultraviolet spectrum* (0.1*M* hydrochloric acid) — 322 nm (ε_{mM} 19.26 to 20.44); 277 nm (ε_{mM} 7.69 to 8.08); 224 nm (ε_{mM} 7.41 to 7.89)

Laboratory Scale

The procedure is similar to the foregoing. A solution of *1* (41 g, 0.25 mol) in water (250 ml) is decolorized with Darco KB (2 g), and the charcoal cake is washed with water (2 × 25 ml). To the filtrate and washings is added concentrated hydrochloric acid (51 ml), and the resulting solution is added to sodium nitrite (22 g, 0.32 mol) in water (300 ml) at 0° to 5° as previously described. After 20 min, a solution of sodium nitrite (4.5 g,

65 mmol) in water (10 ml) is added, followed by four more such additions at 10-min intervals. Stirring is continued 20 min, and then the cold slurry[k] of *2* is added from a jacketed dropping funnel to dimethylamine (700 ml) at -10°. The turbid orange solution is allowed to warm to room temperature, stirred for 1.5 hr, and heated under reflux for 10 to 15 min at 55° under a cold-finger condenser (dry-ice—acetone). The solution is concentrated to ≈500 ml, the mixture is stirred for 30 min at 5°, and filtered, and the yellow-orange solid is washed twice each with cold water and acetone. The air-dried, crude DIC [32.2 g, 71%; m.p. 207° to 208° (dec.)] is boiled under reflux in 40% aqueous dimethylamine (1.55 l) as previously described, and the pink solution is treated with Darco KB (3.2 g) and filtered. Concentration of the combined filtrate and charcoal water-wash to ≈450 ml and isolation in the usual way afford 23.4 g (51.3%) of pure, light cream-colored DIC; m.p. 210° to 211° (dec.).

REFERENCES

1. Y. F. Shealy, R. F. Struck, L. B. Holum, and J. A. Montgomery, *J. Org. Chem.*, 26, 2396 (1961).

2. Y. F. Shealy, C. A. Krauth, and J. A. Montgomery, *J. Org. Chem.*, 27, 2150 (1962).

---

[k]Alternatively, the reaction slurry may be filtered and then washed twice with a minimum of cold water. The moist solid is added, in small portions, to 700 ml of dimethylamine and 100 ml of water at -10°.

3. Y. F. Shealy, C. A. Krauth, S. J. Clayton, A. T. Shortnacy, and W. R. Laster, Jr., *J. Pharm. Sci.*, 57, 1562 (1968).

[2] 2-ACETAMIDOPURIN-6-ONE (N^2-ACETYLGUANINE)

Acetylation of an Amino Group on a Purine Ring with Acetic Anhydride in an Aprotic, Dipolar Solvent

H. Hřebabecký and J. Farkaš

Institute of Organic Chemistry and Biochemistry, Czechoslovak Academy of Sciences, 16610 Prague, Czechoslovakia

INTRODUCTION

N^2-Acetylguanine (*2*) is a useful intermediate in the synthesis of guanine nucleosides. In the present procedure, acetylation of an amino group on the purine ring with acetic anhydride in an aprotic, dipolar solvent is exemplified as a modification of the acetylation of 2-amino-6-chloropurine.[1] The phosphoric acid-catalyzed acetylation of 2-amino-6-chloropurine affords a mixture of acetyl derivatives from which the required 2-acetamido-6-chloropurine is obtained by partial hydrolysis.[2] The previously

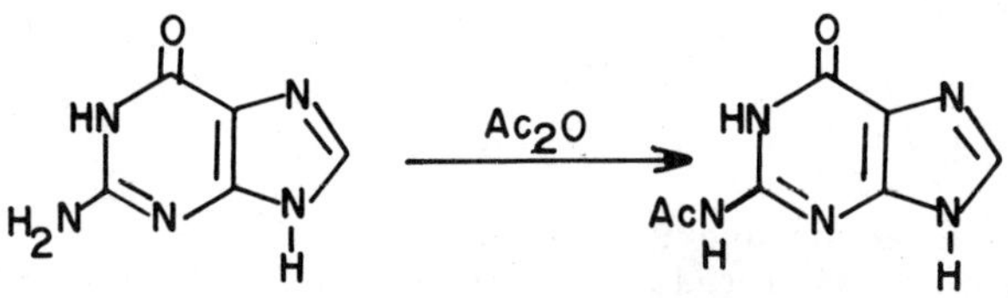

1 (151.1) 2 (193.2)

reported preparation of compound *2* by acetylation of compound *1* in acidic media requires a considerable excess of acetic anhydride and is not suitable for large-scale preparations.[3]

PROCEDURE

N^2-*Acetylguanine* (*2*)

Acetic anhydride (25 ml) is added to a stirred suspension of finely ground guanine (*1*) (15.1 g, 0.1 mol) in 1-methyl-2-pyrrolidinone[a] (150 ml), and the mixture is heated for 2 hr at 150°. The resulting solution is kept for 12 hr at room temperature, to deposit crystals that are collected with suction. The crude product (14.5 g) is crystallized from 50% aqueous ethanol to afford 11.0 g (57%) of compound *2*, m.p. 258° to 260°; λ_{max} in nm (ε_{mM}), in 0.1M HCl, 261 (15.93); at pH 7.0, 260 (15.29) and 274 (sh) (≃12.00); at pH 9.1, 264 (11.80); ^{1}H-NMR data (100 MHz; in Me_2SO-d_6, hexamethyldisiloxane as internal standard): δ 2.28 (s, CH_3) and 7.95 (s, H-8).

REFERENCES

1. W. A. Bowles, F. H. Schneider, L. R. Lewis, and R. K. Robins, *J. Med. Chem.*, 6, 471 (1963).

2. E. M. Acton and R. H. Iwamoto, in W. W. Zorbach and R. S. Tipson (eds.), *Synthetic Procedures in Nucleic Acid Chemistry*, 1, 25 (1968), Wiley, New York.

---

[a]Almost the same yield of compound *2* is obtained when 1-methyl-2-pyrrolidinone is used instead of *N*,*N*-dimethylacetamide.[4]

3. Z. A. Shabarova, Z. P. Polyakova, and M. A. Prokof'ev, *Zh. Obshch. Khim.*, 29, 215 (1959).

4. J. J. K. Novák and F. Šorm, *Collect. Czech. Chem. Commun.*, 38, 1173 (1973).

[3] 6-AMINO-9-(4-DIMETHYLAMINOBUTYL)-9*H*-PURINE

Alkylation of Adenine with Cyclic, Quaternary Ammonium Hydroxides

Lowell D. Zeleznick

Alcon Laboratories, Fort Worth, Texas 76101

INTRODUCTION

9-(ω-Dialkylaminoalkyl)purines have been prepared by a multistep synthesis involving cyclization of an aminopyrimidine.[1] Tetraalkylammonium hydroxides have been used to prepare 9-methyl-, 9-ethyl-, and 9-propyl-purine derivatives,[2] and as a modification of this method 1,1-dimethylpyrrolidinium hydroxide was used as the quaternary ammonium hydroxide. The reaction between 1,1-dimethylpyrrolidinium hydroxide (*2*) and adenine (*1*) ruptures the pyrrolidine ring, with the formation of 6-amino-9-(4-dimethyl aminobutyl)-9*H*-purine (*3*) in 38% yield. This reaction can be used to prepare other dialkylaminobutylpurines.

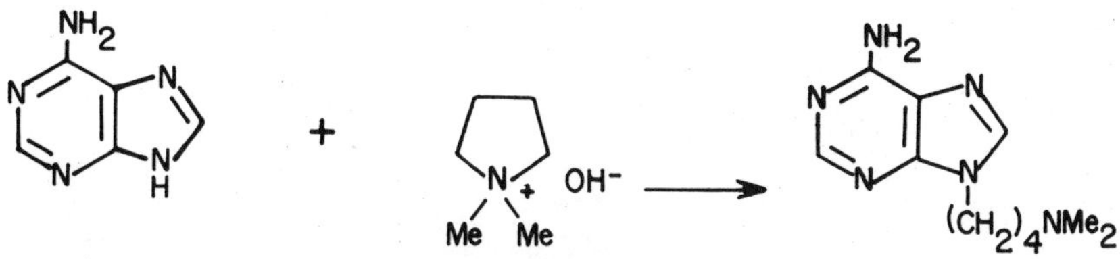

1 (135.1) 2 (117.2) 3 (234.3)

PROCEDURE

6-Amino-9-(4-dimethylaminobutyl)-9H-purine (3)

An aqueous solution of 1,1-dimethylpyrrolidinium hydroxide (5.8 g, 0.05 mol), prepared from the iodide (11.3 g),[a] is used to dissolve adenine (*1*) (6.7 g, 0.05 mol), and the resulting solution is evaporated under vacuum to a clear syrup. A dry solid is obtained by dissolving the syrup in isopropyl alcohol and removing the solvent by boiling. The residue is heated *in vacuo* in a sublimation apparatus to 200° to 210°, and sublimate is collected until no sublimand remains in the sublimation tube. The crude product (7.2 g) is extracted with benzene in a Soxhlet apparatus for 24 hr. On evaporation of the extract under diminished pressure, *3* is obtained as a white solid; yield 4.5 g, m.p. 130° to 131.5°. The benzene-insoluble material is unreacted adenine.

REFERENCES

1. J. H. Lister and G. M. Timmis, *J. Chem. Soc.*, 326 (1960).

2. T. C. Myers and L. Zeleznick, *J. Org. Chem.*, 28, 2087 (1963).

---

[a]To prepare the hydroxide, the quaternary iodide is treated with an excess of moist silver oxide.

[4] 6-CHLOROPURINE

6-Chloropurine via Oxidation of the 6-Hydrazino Derivative[a]

A. Giner-Sorolla and J. T. Segarra

Drug Development Laboratory, Memorial Sloan-Kettering Cancer Center, New York, New York 10021

INTRODUCTION

6-Chloropurine[1] (*2*) exerts a marked antitumor effect and is a key compound for the preparation of purine derivatives. The first reported synthesis of *2* consisted in the treatment of hypoxanthine with phosphorus pentachloride[1]; other methods include the ring closure of 4,5-diamino-6-chloropyrimidine[2] and the chlorination of 6-purinethiol.[3,4] The procedure described here may be considered to be a Sandmeyer type of reaction, wherein 6-hydrazinopurine[5] (*1*) is oxidized with ferric chloride to a diazonium salt that on nitrogen displacement, yields 6-chloropurine[6] (*2*).

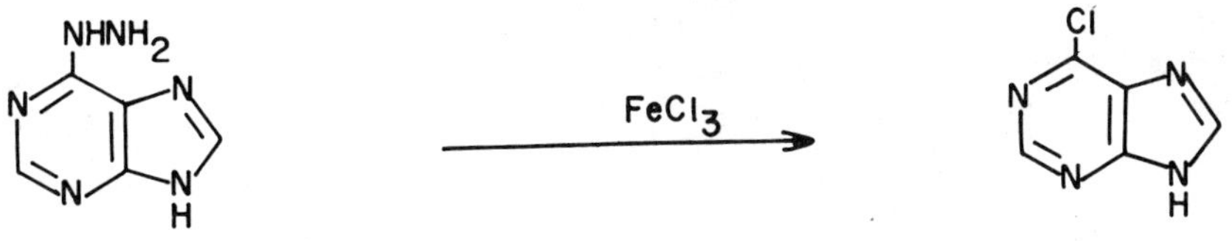

1 (150.1) 2 (154.6)

---

[a]This work was funded in part by the National Cancer Institute (Grant CA-08748).

PROCEDURE

6-Chloropurine (2)

Finely ground 6-hydrazinopurine[5,b] (*1*, 5 g, 30 mmol) is slowly added to a freshly prepared 30% aqueous solution of ferric chloride[c] (125 ml) at 25°, with efficient stirring. The mixture is stirred until the strong evolution of gas subsides, and the resulting suspension is extracted with ether in a liquid extractor for 18 hr.[d] The extracts are evaporated to dryness to yield 3.4 g (67%)[e,f] of *2*.

REFERENCES

1. A. Bendich, P. J. Russell, Jr., and J. J. Fox, *J. Am. Chem. Soc.*, 76, 6073 (1954).

2. J. A. Montgomery, *J. Am. Chem. Soc.*, 78, 1928 (1956).

---

[b]Prepared from 6-(methylmercapto)purine and hydrazine[5] in ethanolic solution and recrystallized from ethanol. Compound *1* can also be obtained from 6-chloropurine plus hydrazine.

[c]Ferric chloride having 6 mol water of crystallization per mole is used.

[d]The conversion of *1* into *2* is very rapid; when the strong evolution of nitrogen has ceased (a few minutes), the reaction is practically complete. It is, however, advantageous to continue stirring during the extraction.

[e]This reaction yields an analytically pure material. A lower yield of *2* is obtained by conducting the reaction at 5°, or above 25°, or in a more acidic solution (2M HCl). The previously reported synthesis[6] using this method gave a poor yield (6%), because of the use of a more dilute solution of ferric chloride and a more concentrated solution of acid.

[f]We have applied the reaction here described to 8-hydrazinopurine, which thereby yields 8-chloropurine.

3. Wellcome Foundation, Ltd., British Patent 767,216 (1957); C. A., 51, 19,796 (1957).

4. A. G. Beaman and R. K. Robins, *J. Appl. Chem.*, 12, 432 (1962).

5. J. A. Montgomery and J. B. Holum, *J. Am. Chem. Soc.*, 79, 2185 (1957).

6. A. Giner-Sorolla and A. Bendich, *J. Am. Chem. Soc.*, 80, 3032 (1958).

[5] 2,6-DI(HYDROXYAMINO)PURINE[a]

Stepwise, Nucleophilic Replacement of the Halogen Atoms of 6-Chloro-2-fluoropurine with Hydroxylamine

A. Giner-Sorolla, J. Longley-Cook, and J. T. Segarra

Drug Development Laboratory, Memorial Sloan-Kettering Cancer Center, New York, New York 10021

INTRODUCTION

Several mouse leukemias are markedly inhibited by low doses of (*3*) and, to a lesser extent, by 2,6-di(hydroxyamino)purine[1-3] 2-fluoro-N^6-hydroxyadenine[2,3] (*2*). Treatment of 6-chloro-2-fluoropurine[4] (*1*) with ethanolic hydroxylamine at 5° leads to 2-fluoro-N^6-hydroxyadenine[2] (*2*), whereas at the reflux temperature compound *1* yields 2,6-di(hydroxyamino)purine[1,2] (*3*). Compound *3* is also prepared by prolonged refluxing of *2* or purine 2,6-disulfonate in ethanolic hydroxylamine.[2]

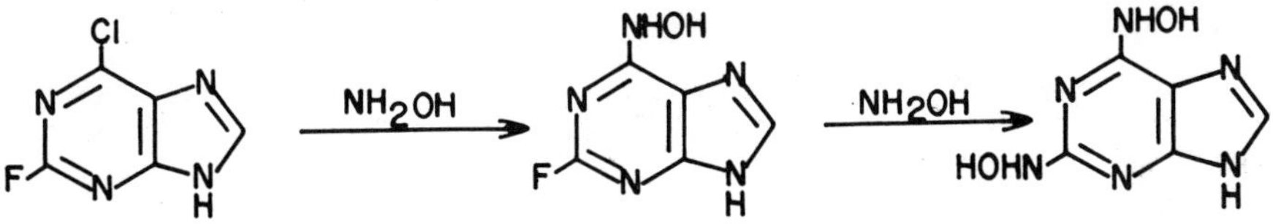

1 (137.1) | 2 (169.1) | 3 (182.1)

---

[a]This work was supported in part by the National Cancer Institute (Grant CA-08748.

PROCEDURE

2-Fluoro-N^6-hydroxyadenine (2)

A solution of 6-chloro-2-fluoropurine[b] (*1*, 4.0 g, 23 mmol) in 0.6M ethanolic hydroxylamine[c] (*1*, 500 ml) is kept for 48 hr at 5°. The small precipitate of 2,6-di(hydroxyamino)purine (*3*) that separates (0.2 g) is removed by filtration. The filtrate is concentrated[d] *in vacuo* below 25° to ≈50 ml, and the solid is collected by filtration and washed with cold water (3 × 40 ml) and ethanol (2 × 40 ml); yield, 3.26 g (83%), m.p. 270° (dec., when inserted at 260°); $\lambda_{max}^{pH\ 6.0}$ 271 nm (ε_{mM} 14.90); pK_a 2.42 (±0.07).[e]

2,6-Di(hydroxyamino)purine (3)

A solution of *1* (3.64 g, 20 mmol) in 0.6M ethanolic hydroxylamine (1.2 l) containing 0.5 g of hydroxylamine hydrochloride is boiled under reflux in a water bath for 8 hr[f] and cooled. The resulting

---

[b]The yield of pure *1* prepared by the method of Montgomery and Hewson[4] from batches of 15 g of 2-amino-6-chloropurine was increased to 30% by using 25% less tetrafluoroboric acid, prolonged extraction with ethyl ether (3 days), and repeated treatment with charcoal during recrystallization.

[c]The preparation of the ethanolic hydroxylamine used here has been described[5]; it contains water of neutralization and chloride ion; its pH is 6.8.

[d]A rotary, flash evaporator is used.

[e]Instability of the compound at other pH values does not permit further UV determinations.

[f]The abundant precipitate that forms at first is mainly *2*, which is slowly transformed into *3*. The total conversion is checked by the ultraviolet spectrum and thin-layer chromatography (TLC).

precipitate[g] is removed by filtration and washed successively with water (3 × 40 ml) and ethanol (40 ml) to yield white crystalline *3*, 2.90 g (75%), m.p. 260° (explodes when inserted at 250°); λ_{max} in nm (ε_{mM}), at pH 7.0, 279 (8.0), λ_{min} 261 (6.4).

REFERENCES

1. A. Giner-Sorolla, C. Nanos, J. H. Burchenal, M. Dollinger, and A. Bendich, *J. Med. Chem.*, 11, 52 (1968).

2. A. Giner-Sorolla and J. H. Burchenal, *J. Med. Chem.*, 14, 816 (1971).

3. A. Giner-Sorolla, J. H. Burchenal, and A. Bendich, in E. D. Bergmann and B. Pullman (eds.), *The Purines, Theory and Experiment*, The Israel Academy of Sciences, Jerusalem, 1972, pp. 478-503.

4. J. A. Montgomery and K. Hewson, *J. Am. Chem. Soc.*, 82, 463 (1960).

5. A. Giner-Sorolla and A. Bendich, *J. Am. Chem. Soc.*, 80, 3932 (1958).

---

[g] Leaving the reaction mixture overnight without filtering results in a darkening of the final product, presumably due to oxidation (formation of nitroso derivative). Once dried, compound *3* is stable; it should be kept in amber bottles, as it darkens when exposed to light.

[6] 9-FURFURYL-6-(FURFURYLAMINO)-9*H*-PURINE

Alkylation of Kinetin with Furfuryltrimethylammonium Hydroxide

Lowell D. Zeleznick

Alcon Laboratories, Fort Worth, Texas 76101

INTRODUCTION

Such analogs of *N*⁶-furfuryladenine (kinetin, *1*) as 9-furfuryl-kinetin (*3*) have been prepared by cyclization of the substituted aminopyrimidine.[1,2] The facile alkylation of adenine by tetra-alkylammonium hydroxides[3] led to the use of furfuryltrimethyl-ammonium hydroxide (*2*) in a similar alkylation. The reaction of this quaternary ammonium hydroxide with 6-(furfurylamino)-9*H*-purine (*1*) in a one-step reaction produces 9-furfuryl-6-(furfuryl-amino)-9*H*-purine (*3*) (6-furfurylkinetin).

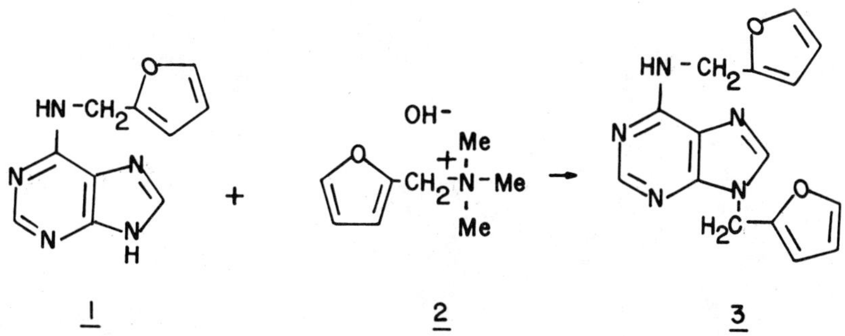

1 (215.2) 2 (157.2) 3 (295.3)

PROCEDURE

9-Furfuryl-6-(furfurylamino)-9H-purine (3)

Kinetin (*1*; 0.89 g, 4.1 mmol) is dissolved in 2.5 ml of a solution of furfuryltrimethylammonium hydroxide (*2*) prepared from furfuryltrimethylammonium iodide[a] (1 g, 3.7 mmol). The aqueous solution is evaporated *in vacuo* to a syrup which, on heating at 100°, gives an amorphous solid. This solid is then heated at 150° *in vacuo* in a sublimation apparatus. The crude sublimate obtained after several hours is recrystallized from water to yield long, silky needles of *3* in 50% yield; m.p. 129.5° to 130.5°, λ max 267 nm, λ min 237 nm.

REFERENCES

1. C. L. Leese and G. M. Timmis, *J. Chem. Soc.*, 4107 (1958).
2. R. Hull, *J. Chem. Soc.*, 481 (1959).
3. T. C. Myers and L. Zeleznick, *J. Org. Chem.*, 28, 2087 (1963).

[a]To prepare the hydroxide, the quaternary iodide is treated with an excess of moist silver oxide.

[7] METHYLATION OF CARBON ATOMS OF PURINES

Conversion of Caffeine into 8-Methylcaffeine

Mona F. Zady and John L. Wong

Department of Chemistry, University of Louisville, Louisville, Kentucky 40208

INTRODUCTION

The synthesis of 8-methylcaffeine (*2*) has been accomplished by *N*-methylation of 8-methylpurines such as 8-methyltheophylline,[1] prepared by condensation of acetic acid with 4,5-diamino-1,3-dimethyluracil, followed by ring closure in aqueous alkali,[2] and the final step, N-7 methylation, was achieved with diazomethane in a basic medium.[1a] Another route is a primary synthesis involving 4-(ethylamino)-1,3-dimethyl-5-nitrosouracil as an intermediate.[3] A more convenient method of preparing *2* is by direct methylation

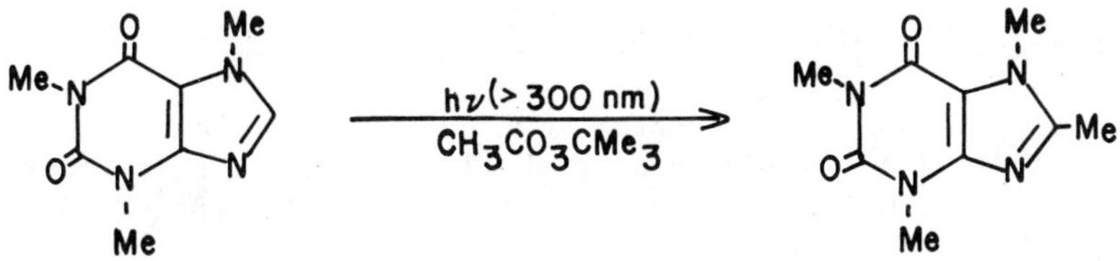

1 (194.2) 2 (208.2)

---

[a]Obtained from Eastman Organic Chemicals.

of C-8 of the readily available caffeine *1*; no such procedure has hitherto been reported. We used photodecomposition of *tert*-butyl peracetate[4] as a source of methyl radical for selective methylation at C-8 of caffeine. The same method appears to be applicable to methylation of carbon atoms of other purine compounds.

PROCEDURE

8-Methylcaffeine (2)

Caffeine[a] (*1*) (2 g, 10.3 mmol), *tert*-butyl peracetate[b] (2.64 g, 20 mmol), and water (150 ml) are placed in a 200-ml Pyrex vessel equipped with a water-cooled inner jacket. The mixture is irradiated for 6 hr[c] under nitrogen with a 1.2-kW ultraviolet source at a distance of 2.5 cm. The solution is then evaporated *in vacuo* to yield 1.14 g (53%) of *2*. Recrystallization from water affords white needles, m.p. 215° to 218°, NMR data ($CDCl_3$): δ 2.44 (s,3), 3.24 (s,3), 3.42 (s,3), and 3.87 (s,3).

REFERENCES

1. (a) A. Rybar and K. Antos, *Collect. Czech. Chem. Commun.*, 35, 1415 (1970); (b) H. Bredereck, I. Hennig, W. Pfleiderer, and G. Weber, *Chem. Ber.*, 86, 333 (1953); (c) H. Bader and J. D. Downer, *J. Chem. Soc.*, 1641 (1953);

---

[b]Obtained from K and K Division of ICN Pharmaceuticals as a 75% solution in benzene.

[c]The progress of the reaction may be monitored by PMR-spectral analysis of the residual H-8 observed at δ 7.60 in $CDCl_3$, as well as the concomitant increase in the Me-8 observed at δ 2.44 in $CDCl_3$.

1. (d) E. S. Golovchinskaya, *Zh. Obshch. Khim.*, 24, 146 (1954).

2. H. Furst and I. Ebert, *Chem. Ber.*, 93, 99 (1960).

3. H. Goldner, G. Dietz, and E. Carstens, *Tetrahedron Lett.*, 2701 (1965).

4. W. A. Pryor, D. L. Fuller, and J. P. Stanley, *J. Am. Chem. Soc.*, 94, 1632 (1972).

[8] PURINE 3-OXIDE[a]

Synthesis of an Unsubstituted Purine N-Oxide

A. Giner-Sorolla, Peter S. Fitt,[b] and M. McCravey

Drug Development Laboratory, Memorial Sloan-Kettering Cancer Center, New York, New York 10021

INTRODUCTION

Certain substituted purine *N*-oxides are potent carcinogens.[1] The unsubstituted purine 3-oxide exerts unusual carcinogenic effects in animals.[2] Its synthesis consists in the oxidation of 6-purinethiol 3-oxide[3,4] (*1*) with active manganese dioxide to purine-6-sulfinic acid 3-oxide (*2*) which, on treatment with formic acid, yields purine 3-oxide[5] (*3*). Another method for the synthesis of *3* is the oxidation of 6-cyanopurine to 6-cyanopurine 3-oxide

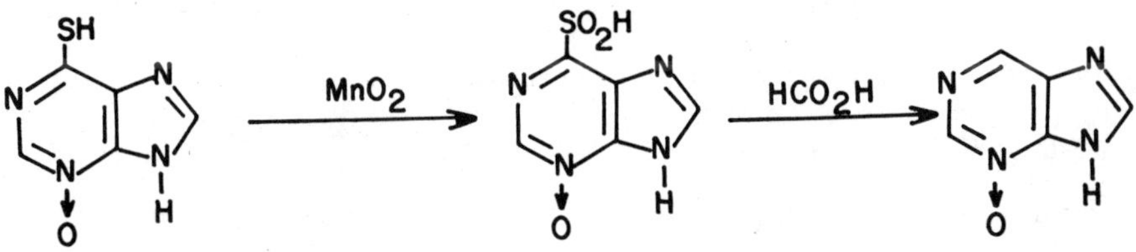

1 (168.2) 2 (200.2) 3 (136.1)

[a]This work was supported in part by the National Cancer Institute (Grant CA-08748).

[b]Present address: Department of Biochemistry, Faculty of Medicine University of Ottawa, Ottawa, Canada.

which, on alkaline hydrolysis, gives 6-carboxypurine 3-oxide; decarboxylation of this yields purine 3-oxide, although in lower yield.[5]

PROCEDURE

Purine-6-sulfinic Acid 3-Oxide (2), Sodium Salt

A suspension of 6-purinethiol 3-oxide[3,4] (*1*, 1.0 g, 6 mmol) in water (100 ml) containing sodium hydrogen carbonate (2.0 g) is heated at 50° until dissolution occurs. The solution is then cooled to 25°, and active manganese dioxide[c] (4.0 g) is added; the suspension is stirred for 5 hr and filtered, and the manganese dioxide is washed with hot water (2 × 30 ml). The filtrates are combined, the pH is adjusted to 5 with glacial acetic acid, the solution is treated with charcoal and filtered, and the filtrate is evaporated to dryness *in vacuo*. The residue is suspended in 70% aqueous ethanol (30 ml) and the solid is collected by filtration; yield 1.1 g (92%) of pale-yellow needles.[d]

Purine 3-Oxide[5] *(3)*

A solution of *2* (3.0 g, 14.9 mmol) in 90% formic acid (30 ml) is heated for 30 min at 80° and treated with charcoal; the suspension

---

[c]This reagent is available from Winthrop Laboratories, New York, N. Y.; it may be prepared according to the method of Attenburrow, *et al.*[6]

[d]Purine-6-sulfinic acid was first prepared by oxidation of compound *1* with iodine.[7] The procedure described here is much simpler, and consistently gives a purer product.

is filtered, and the filtrate is evaporated to dryness *in vacuo*.[e] The residue is suspended in 70% aqueous ethanol (50 ml), and the suspension is filtered to yield 1.3 g (64%) of pink, short prisms, m.p. 282° (dec.); λ_{max} in nm ($\varepsilon_{\underline{mM}}$), at pH 3.0, 224 (17.60), and 295 (8.40); at pH 14.0, 225 (25.40) and 300 (7.10); pK_a 6.36 (±0.02).

REFERENCES

1. G. B. Brown, *Prog. Nucleic Acid Res. Mol. Biol.*, 8, 209 (1968).

2. M. N. Teller, A. Giner-Sorolla, J. M. Budinger, and G. B. Brown, *Proc. Amer. Assoc. Cancer Res.*, 17, 588 (1976).

3. G. B. Brown, G. Levin, S. Murphy, A. Sele, H. C. Reilly, G. S. Tarnowski, F. A. Schmid, M. N. Teller, and C. C. Stock, *J. Med. Chem.*, 8, 190 (1965).

4. A. Giner-Sorolla, C. Gryte, A. Bendich, and G. B. Brown, *J. Org. Chem.*, 34, 2157 (1969).

5. A. Giner-Sorolla, C. Gryte, M. Cox, and J. C. Parham, *J. Org. Chem.*, 36, 1228 (1971).

6. J. Attenburrow, A. F. B. Cameron, J. H. Chapman, R. M. Evans, B. A. Hems, A. B. A. Jansen, and T. Walker, *J. Chem. Soc.*, 1094 (1952).

7. I. Scheinfeld, J. C. Parham, S. Murphy, and G. B. Brown, *J. Org. Chem.*, 34, 2153 (1969).

---

[e] The final point of the reaction is determined both by UV spectroscopy and TLC; if some starting material (*2*) is still present, the heating should be continued until its complete conversion occurs.

[9] ISOMERIZATION OF 6-(TRIMETHYLAMMONIO)PURINIDES

Formation of 6-(Dimethylamino)-9-methyl- and 6-(Dimethylamino)-3-methylpurines by Thermal Rearrangement

John H. Lister

The John Curtin School of Medical Research, Canberra, Australia

and John Kiburis

Department of Biological Chemistry, University of Athens, Greece

INTRODUCTION

When equimolar amounts of 6-chloropurines and trimethylamine interact, the 6-trimethylpurinium chlorides result. On being heated, these derivatives undergo the expected Hofmann degradation with formation of the 6-(dimethylamino)purine analogs. If,

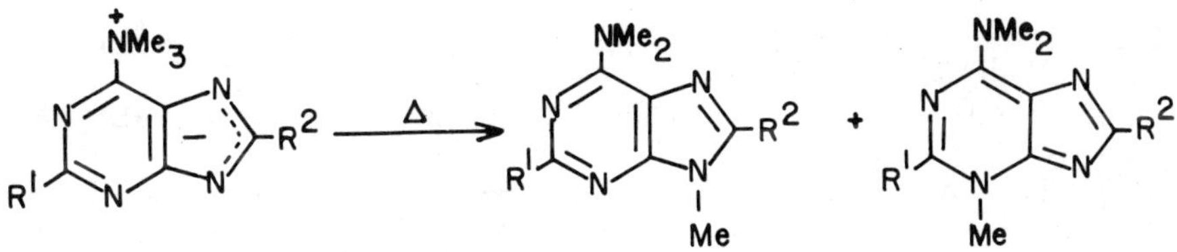

1 2 3

(a) R^1 = Cl, R^2 = H (211.7)

(b) R^1 = H, R^2 = Cl (211.7)

(c) R^1 = R^2 = Cl (246.1)

however, the trimethylamine is in excess, the (trimethylammonio)-purinides (*1*) are obtained. On thermal treatment, these betaines are not degraded, but isomerize to a mixture of the corresponding 6-(dimethylamino)-9-methyl- (*2*) and 6-(dimethylamino)-3-methyl-purine (*3*), in which the former is usually the major component.[1]

Purines having a quaternary group at C-6 show antigrowth activity against a variety of experimental tumors[2-4] and some human neoplasms.[5]

PROCEDURE

2-Chloro-6-(trimethylammonio)purinide (1a)

To a stirred solution of 2,6-dichloropurine (1 g, 5.3 mmol) in 1,2-dimethoxyethane (10 ml) is added dropwise trimethylamine (10 ml), and the mixture is refrigerated for 70 hr. The crystalline precipitate (0.8 g, 90%) is removed by filtration, washed with ether, triturated with a small volume of cold water, and dried at 40° *in vacuo* over phosphorus pentaoxide; m.p. 191° to 192°; NMR data (Me_2SO): δ 8.75 (s, H-8) and 3.89 (s, $\overset{+}{N}Me_3$ protons).

8-Chloro-6-(trimethylammonio)purinide (1b)

A solution of 6,8-dichloropurine (450 mg, 2.4 mmol) in cold 1,2-dimethoxyethane (10 ml) is treated with trimethylamine (3 ml), as in the previous preparation, and kept for 30 hr. After being dried, the resulting, crystalline purinide (260 mg, 58%) has m.p.

205° to 206°; NMR data (Me_2SO): δ 9.05 (s, H-2) and 3.87 (s, $\overset{+}{N}Me_3$ protons).

2,8-Dichloro-6-(trimethylammonio)purinide (1c)

2,6,8-Trichloropurine (600 mg, 2.6 mmol) is treated as in the foregoing reaction for 30 hr to give 515 mg (85%) of the purinide, m.p.[a] > 200°; NMR datum (Me_2SO): δ 3.87 (s, $\overset{+}{N}Me_3$ protons).

8-Chloro-6-(dimethylamino)-9-methyl- (2b) and 8-Chloro-6-(dimethyl amino)-3-methylpurine (3b)

8-Chloro-6-(trimethylammonio)purinide (300 mg, 1.5 mmol) is placed in a cold-finger sublimation apparatus and for 10 min heated at 185° (oil bath) at 15 torr. The sublimate and residue are combined and taken up in a small volume of chloroform; the suspension is filtered, silica gel (1.5 g) is added, and the mixture is evaporated to dryness under diminished pressure. The residue of silica gel containing the adsorbed product is then placed on top of a column[b] (10 × 1 cm) of silica gel (Merck Kieselgel 7734), and the column is eluted with ether. Evaporation of the eluate gives 113 mg (37%) of *2b*, m.p. 89° to 91° (after recrystallization from ether—heptane); $\lambda_{max}^{pH\ 1}$ 272 nm ($\varepsilon_{\underline{mM}}$ 20.10), $\lambda_{max}^{pH\ 7}$ 280 nm ($\varepsilon_{\underline{mM}}$ 19.80), $\lambda_{max}^{pH\ 13}$ 280 nm ($\varepsilon_{\underline{mM}}$ 19.90); NMR data (Me_2SO): δ 8.29

---

[a]The melting point is indefinite, as the compound starts to isomerize at this temperature.

[b]The column is prepared from a slurry of silica gel and the eluant. The column is allowed to settle and is then thoroughly washed with the solvent prior to use.

(s, H-2), 3.72 (s, N-Me protons), and 3.47 (s, NMe_2 protons).

Further elution of the column with 4:1 ether—ethyl acetate gives the 3-methyl isomer (115 mg, 38%), m.p. 223° to 225° (after recrystallization from benzene—heptane; $\lambda_{max}^{pH\ 1}$ 294 nm ($\varepsilon_{\underline{mM}}$ 23.50), $\lambda_{max}^{pH\ 7}$ 296 nm ($\varepsilon_{\underline{mM}}$ 20.00), $\lambda_{max}^{pH\ 13}$ 295 nm ($\varepsilon_{\underline{mM}}$ 19.50); NMR data (Me_2SO): δ 7.90 (s, H-2) and 3.94 (s, N-Me protons).[c]

2,8-Dichloro-6-(dimethylamino)-9-methyl- (2c) and 2,8-Dichloro-6-(dimethylamino)-3-methylpurine (3c)

Sublimation of 2,8-dichloro-6-(trimethylammonio)purinide (210 mg, 0.85 mmol) is performed for 45 min at 205°/15 torr by the techniques already described. The two isomers are separated on a column[b] (10 × 2 cm) of silica gel by using 1:1 petroleum ether (b.p. 40° to 60°)—ether to elute the 9-methyl isomer (112 mg, 53%), m.p. 208° to 209°, $\lambda_{max}^{0.1\underline{M}\ HCl}$ 281 nm ($\varepsilon_{\underline{mM}}$ 20.10), $\lambda_{max}^{pH\ 7}$ 282 nm ($\varepsilon_{\underline{mM}}$ 20.00), $\lambda_{max}^{0.1\underline{M}\ NaOH}$ 282 nm ($\varepsilon_{\underline{mM}}$ 19.90); NMR data (Me_2SO): δ 3.67 (s, N-Me protons) and 3.45 (s, NMe_2 protons).

Ether is then employed as the eluant, giving 26 mg (12%) of the 3-methyl isomer, m.p. 238° to 239° (after crystallization from benzene—heptane); $\lambda_{max}^{0.1\underline{M}\ HCl}$ 296 nm ($\varepsilon_{\underline{mM}}$ 22.70), $\lambda_{max}^{pH\ 7}$ 301 nm ($\varepsilon_{\underline{mM}}$ 18.40), $\lambda_{max}^{0.1\underline{M}\ NaOH}$ 290 nm ($\varepsilon_{\underline{mM}}$ 24.60); NMR data (Me_2SO): δ 4.01 (s, N-Me protons), 3.81, and 3.31 (broad doublet, NMe_2 protons).

---

[c]Because of rotational effects, the signal given by the protons of the dimethylamino group is too broad for definition.

REFERENCES

1. J. Kiburis and J. H. Lister, *J. Chem. Soc.*, (C), 1587 (1971).

2. J. P. Horwitz and V. K. Vaitkevicius, *Experientia*, 17, 552 (1961).

3. L. R. Lewis, F. H. Schneider, and R. K. Robins, *J. Org. Chem.*, 26, 3837 (1961).

4. W. A. Bowles, F. H. Schneider, L. R. Lewis, and R. K. Robins, *J. Med. Chem.*, 6, 471 (1963).

5. V. K. Vaitkevicius and M. L. Reed, *Proc. Am. Assoc. Cancer Res.*, 7, 72 (1966).

[10] 5-FLUOROCYTOSINE

Sequential Replacement of Two Chlorine Atoms

Robert Duschinsky

Swiss Institute for Experimental Cancer Research, 1011 Lausanne, Switzerland

INTRODUCTION

In contrast to its nucleosides and to uracil and thymine, cytosine is minimally metabolized[a] by rats[1] and undergoes deamination by yeast.[2] 5-Fluorocytosine (*3*) shows similar behavior in man[3] and acts as an antimetabolite versus fungi,[4] thus providing a clinically useful drug for the treatment of mycoses.[5] The present synthesis of *3*, starting from commercially available 5-fluorouracil[6] via[b] *1*, offers technical advantages over the original procedure.[8] In contrast to the behavior of 2,4-dichloropyrimidine, which gives a mixture of isomers,[9] the fluorinated analog *1* is preferentially aminated at C-4.

PROCEDURE

4-Amino-2-chloro-5-fluoropyrimidine (*2*)

---

[a]The formation of metabolites can possibly be attributed to microbes present in the intestinal tract.[3]

[b]The first step was very adequately described.[7]

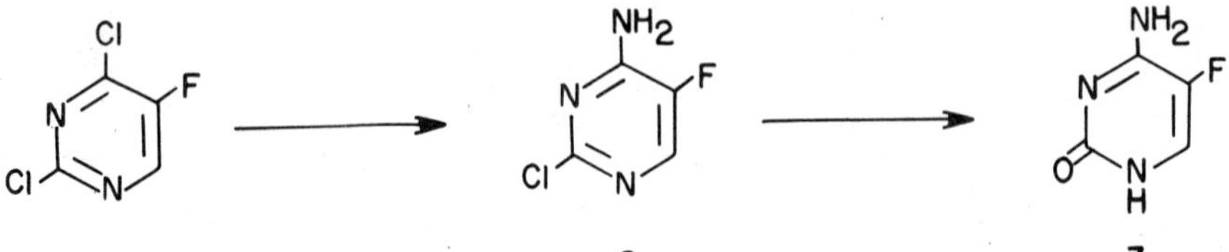

1 (167.0) 2 (147.5) 3 (129.1)

To a solution of 2,4-dichloro-5-fluoropyrimidine[7] (*1*) (10 g, 0.06 mol) in ethanol (100 ml) is slowly added concentrated, aqueous ammonia (25 ml) with stirring and cooling so that the temperature of the slightly opalescent solution is maintained below 30°. After about 3 hr a Volhard titration shows that >90% of the chlorine is present in the ionic form. The mixture is evaporated *in vacuo* at 40° (bath), giving a white sludge that is slurried with water (75 ml). The suspension is filtered, and the solid is washed with water until chloride-free, and dried *in vacuo* at 60°. The product melts at 196.5° to 197.5°; yield 6.44 g. Evaporation of the mother liquor affords a second crop of 0.38 g, raising the total yield to 6.82 g (79%); $\lambda_{max}^{0.1\underline{M}\ HCl}$ 235 nm ($\varepsilon_{m\underline{M}}$ 7.40) and 269 to 270 nm ($\varepsilon_{m\underline{M}}$ 6.10).

5-Fluorocytosine (*3*)

A slurry of 4-amino-2-chloro-5-fluoropyrimidine (*2*) (34 g, 231 mmol) in concentrated hydrochloric acid (231 ml) is heated in a water bath at 93° to 95° for ≈2 hr. The progress of the reaction is monitored by the change in the 300/270 UV extinction ratio determined in 0.1M hydrochloric acid. When the ratio has risen from 0.1 (initial) to 0.85, the reaction is stopped.[c] The clear solution is cooled and evaporated to dryness under vacuum at 40°. The residue is slurried three times with water and the slurry reevaporated. To an ice-cooled solution of the residue in water (100 ml) is added concentrated aqueous ammonia (29 ml) with stirring. The resulting precipitate is removed by filtration, washed successively with water (until free of chloride), ethanol, and ether, and dried *in vacuo* at 65° to give 22.3 g of *3*. An additional 6.35 g is recovered by evaporation of the mother liquor; total yield, 28.65 g (96%). This product is recrystallized from 8 volumes of water (>90% recovery); m.p. ≃295° (dec.); $\lambda_{max}^{0.1M\ HCl}$ 210 nm (ε_{mM} 10.20) and 285 to 286 nm (ε_{mM} 9.10); $\lambda_{max}^{pH\ 7}$ 210 nm (ε_{mM} 10.70) and 276 nm (ε_{mM} 5.80); $\lambda_{max}^{0.1M\ KOH}$ 292 nm (ε_{mM} 7.80); pK_{a1} 2.90, pK_{a2} 10.71.

REFERENCES

1. A. Bendich, H. Getler, and G. B. Brown, *J. Biol. Chem.*, 177, 565 (1949).

2. J. Kream and E. Chargaff, *J. Am. Chem. Soc.*, 74, 5157 (1952).

3. B. A. Koechlin, F. Rubio, S. Palmer, T. Gabriel, and R. Duschinsky, *Biochem. Pharmacol.*, 15, 435 (1966).

---

[c]Prolonged action of hydrochloric acid and, in particular, more dilute acid, produces 5-fluorouracil, which can be detected by TLC on silica gel, with 4:1:1 1-butanol—water; R_F: 5-fluorocytosine, 0.40; 5-fluorouracil, 0.58.

4. J. Malbica, L. Sello, B. Tabenkin, J. Berger, E. Grunberg, J. H. Burchenal, J. J. Fox, I. Wempen, T. Gabriel, and R. Duschinsky, *Fed. Proc.*, 21, 384 (1962); E. Grunberg, E. Titsworth, and M. Bennet, *Antimicrob. Agents Chemother.*, 566 (1963).

5. H. J. Scholer, *Chemotherapy*, (Basel), 22, 103 (1976).

6. R. Duschinsky, U.S. Patent, 3,040,026 (1962); C. A., 56, 8724 (1962).

7. G. J. Durr, in W. W. Zorbach and R. S. Tipson (eds.), *Synthetic Procedures in Nucleic Chemistry*, Vol. 1, Interscience, New York, 1968, pp. 92 to 93.

8. R. Duschinsky, E. Pleven, and C. Heidelberger, *J. Am. Chem. Soc.*, 79, 4559 (1957).

9. G. E. Hilbert and T. B. Johnson, *J. Am. Chem. Soc.*, 52, 1152 (1930).

[11] PREPARATION OF 2,5-DIAMINO-4,6-DICHLOROPYRIMIDINE VIA N-(4,6-DICHLORO-5-NITROPYRIMIDIN-2-YL)ACETAMIDE

The Preparation of 2-Aminopyrimidine Intermediates

Carroll Temple, Jr., Buford H. Smith, and John A. Montgomery

Kettering-Meyer Laboratory, Southern Research Institute, Birmingham, Alabama 35205

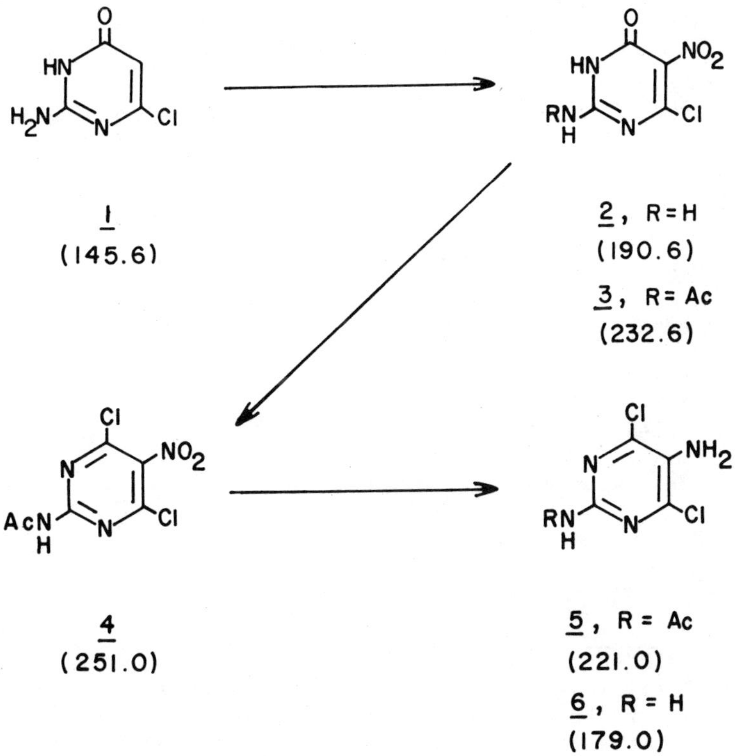

INTRODUCTION

The low yield of an intermediate obtained by the method reported for the preparation of 2,5-diamino-4,6-dichloropyrimidine[1] prompted the synthesis of the latter from *1*.[2] The 5-amino-4-chloro- and 4-chloro-5-nitro-pyrimidines provide a convenient source of starting materials for the synthesis of a variety of condensed heterocycles containing the 2-aminopyrimidine moiety.[3]

PROCEDURE

N-(*6-Chloro-5-nitro-4-pyrimidinone-2-yl*)*acetamide* (*3*)

Solid 2-amino-6-chloropyrimidin-4-one[a] (*1*, 10 g, 69 mmol) is added with stirring to 140 ml of 1:1 conc. sulfuric acid—nitric acid (d 1.49 to 1.50) at 17.5°. The solid dissolves within several minutes, and the temperature rises to 35°. After 50 min the solution is poured with stirring onto crushed ice (500 g). The solid that deposits is collected by filtration, washed with water, and dried *in vacuo* over phosphorus pentoxide. The resulting crude *2*[4] is extracted with acetone (4 × 500 ml), and the extracts are combined and evaporated to dryness *in vacuo*. The resulting solid is suspended in acetic anhydride (100 ml) containing 2 drops of concentrated sulfuric acid, and the suspension is heated for 30 min

---

[a]Aldrich Chemical Co. The melting point of this material is raised by treating the sample with dilute base, removing some insoluble material by filtration, and acidifying the filtrate with acetic acid; m.p. 259° to 261° (95% recovery).

in a preheated oil bath at 90°. The resulting solution is filtered hot, and the filtrate is cooled to deposit *3*; yield, 8.7 g, m.p. 260° (dec.) (Kofler Heizbank). On concentration of the filtrate to ≃15 ml a second crop is deposited; yield, 1.4 g; m.p. 260° (dec.); $\lambda_{max}^{pH\ 7}$ in nm ($\varepsilon_{\underline{mM}}$), pH 7, 242 (10.2); 278 sh (3.72), 356 (2.02); ν_{max} 1680 cm^{-1}.

Analysis calc. for $C_6H_5ClN_4O_4$: C, 30.95; H, 2.15; Cl, 15.27; N, 24.10. Found: C, 30.71; H, 2.29; Cl, 14.83; N, 23.73.

N-*(4,6-Dichloro-5-nitropyrimidin-2-yl)acetamide (4)*

A mixture of *3* (10.0 g, 43 mmol) and phosphoryl chloride (200 ml) is boiled under reflux for 4 hr. The resulting solution is concentrated under diminished pressure to 50 ml, and the concentrate is poured, with vigorous stirring, into ice-water (1 l). The brown solid that precipitates is collected by filtration, washed with water (1 l), and dried *in vacuo* over phosphorous pentoxide; yield, 7.3 g (67%); m.p. 206° to 207° (dec.). Although this material has an analysis agreeing with that calculated for *4*, the melting point is increased by recrystallization of a sample (0.5 g) from hot acetic anhydride; yield, 0.25 g; m.p. 224° (Mel-Temp). This sample is homogeneous in TLC on silica gel with ethyl acetate (*R_F* 0.9), except for a spot at the origin; $\lambda_{max}^{pH\ 7}$ in nm ($\varepsilon_{\underline{mM}}$); 240 (13.0), 247 sh (12.9), 275 sh (7.57), and 340 br sh (0.95); ν_{max} 1705 cm^{-1}; M^+, 250, 252, 254.

Analysis calculated for $C_6H_4Cl_2N_4O_3$: C, 28.71; H, 1.61; N, 22.32. Found: C, 28.88; H, 1.63; N, 22.48.

N-(5-Amino-4,6-dichloropyrimidin-2-yl)acetamide (5)

A solution of *4* (7.2 g, 29 mmol) in a mixture of *N,N*-dimethylacetamide (40 ml) and ethanol (750 ml) is hydrogenated in the presence of Raney nickel (7 g wet, washed with water and ethanol) at room temperature and atmospheric pressure. The theoretical amount of hydrogen is absorbed within 5 hr. The suspension is filtered (Celite) under nitrogen, and the colored filtrate is evaporated *in vacuo* to give a gummy solid. This residue is triturated with ether, and the brown solid is collected by filtration and dried *in vacuo* over phosphorus pentaoxide for 4 hr; yield, 2.8 g (44%), m.p. 135° to 138° (dec.) (Mel-Temp); $\lambda_{max}^{pH\ 7}$ in nm ($\varepsilon_{m\underline{M}}$) 263 (12.1) and 318 (4.79); ν_{max} 1665 cm^{-1}.

Analysis calculated for $C_6H_6Cl_2N_4O$: C, 32.60; H, 2.74; N, 25.34. Found: C, 32.37; H, 2.94; N, 25.09.

The ether wash is evaporated to dryness, giving a yellow solid (3.9 g). Thin-layer chromatography on silica gel with 9:1 chloroform—methanol indicates this residue to be a mixture of *5* and the corresponding deacetylated compound *6*.

2,5-Diamino-4,6-dichloropyrimidine (6)

The mixture of *5* and *6* (3.9 g) obtained in the preparation of *5* is suspended in 6M ethanolic hydrogen chloride (40 ml), and the suspension is stirred for 3 hr at room temperature. The solid is collected by filtration, washed with ether, and dried *in vacuo* over phosphorus pentoxide; yield, 3.0 g, m.p. > 264° (Mel-Temp).

Analysis calculated for $C_4H_4Cl_2N_4 \cdot HCl \cdot 0.25\ C_2H_6O$: C, 23.81; H, 2.36; N, 24.68. Found: C, 23.86; H, 2.63; N, 24.80.

An additional amount of *6* (0.2 g) is obtained from the ethanol filtrate.

A portion (1.0 g) of the first crop is suspended in water (25 ml), and the pH is adjusted to 7 by addition of solid sodium hydrogen carbonate. The mixture is stirred for 10 min, then heated for 15 min at 50°, and cooled; the cream-colored solid is collected by filtration, washed with water, and dried *in vacuo* over phosphorus pentoxide; yield, 0.6 g; m.p. 188° to 191° (dec.) (Mel-Temp) (lit.[1] m.p. 260° to 261°); (for some samples, decomposition occurs before melting unless the melting point is taken rapidly); $\lambda_{max}^{pH\ 7}$ in nm ($\varepsilon_{\underline{mM}}$); 245 (13.0) and 341 (4.84); ν_{max} 1645 cm^{-1}; M^+ 178, 180, and 182.

Analysis calculated for $C_4H_4Cl_2N_4$: C, 26.84; H, 2.25; N, 31.30. Found: C, 26.94; H, 2.49; N, 31.56.

REFERENCES

1. R. K. Robins, K. L. Dille, and B. E. Christensen, *J. Org. Chem.*, 19, 930 (1954).

2. C. Temple, Jr., B. H. Smith, and J. A. Montgomery, *J. Org. Chem.*, 40, 3141 (1975).

3. R. D. Elliott, C. Temple, Jr., J. L. Frye, and J. A. Montgomery, *J. Med. Chem.*, 18, 492 (1975).

4. J. Davoll and D. D. Evans, *J. Chem. Soc.*, 5041 (1960).

[12] SUBSTITUTED (2-AMINOPURINE-6-THIYL)PYRIMIDINES

A General Method for the Preparation of S-Substituted Pyrimidine Derivatives of Thioguanine

Harry C. Koppel, Robert H. Springer, Roland K. Robins, and C. C. Cheng

Midwest Research Institute, Kansas City, Missouri 64110

INTRODUCTION

A convenient preparation has been devised wherein a variety of substituted pyrimidines are attached to the sulfur atom of 6-thioguanine. The condensation is performed in aqueous solution by treatment of 6-thioguanine with an appropriate chloropyrimidine

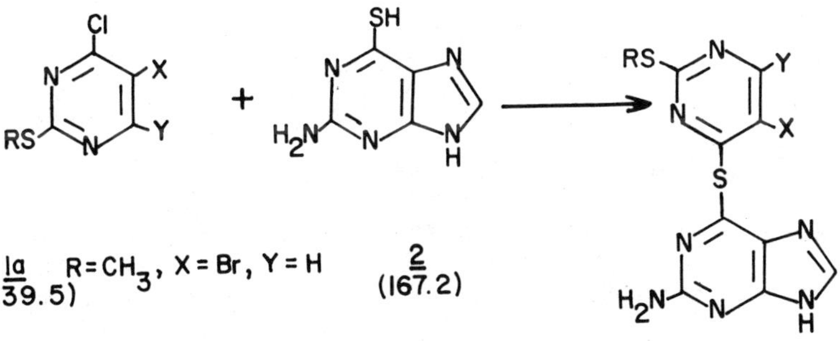

1a $R = CH_3, X = Br, Y = H$ (239.5)

2 (167.2)

1b $R = C_2H_5, X = H, Y = Cl$ (209.1)

3a $R = CH_3, X = Br, Y = H$ (370.3)

3b $R = C_2H_5, X = H, Y = Cl$ (339.8)

under alkaline conditions.[1] Aqueous ammonia, which causes fewer side reactions, is more suitable than sodium (or potassium) hydroxide for reactions of this type.

4-(2-Aminopurine-6-thiyl)-5-bromo-2-methylthiopyrimidine (3a)

To warm, concentrated, aqueous ammonia (200 ml) is added 6-thioguanine (*2*) (16.72 g, 0.1 mol). The resulting solution is heated at 70° while a solution of 5-bromo-4-chloro-2-methylthiopyrimidine[2] (*1a*) (23.95 g, 0.1 mol) in 1,4-dioxane (100 ml) is added, with stirring. The mixture is then stirred for 90 min at 70°, during which time the desired product *3a* precipitates. The product is collected by filtration of the hot mixture and washed well with aqueous ammonia to remove unreacted 6-thioguanine. Recrystallization of the product from water—dimethylformamide gives a 30% yield of analytically pure *3a*, m.p. 220° (dec.); $\lambda_{max}^{pH\ 11}$ in nm ($\varepsilon_{m\underline{M}}$) 227 (21.00), 263 (21.20), and 324 (9.50).

4-(2-Aminopurine-6-thiyl)-6-chloro-2-ethylthiopyrimidine (3b)

Compound *3b* is prepared from *2* (16.72 g, 0.1 mol) and 4,6-dichloro-2-ethylthiopyrimidine (*1b*) (20.91 g, 0.1 mol) as for the preparation of *3a*. The product is recrystallized from a mixture of dimethylformamide—water, to give a 29% yield of *3b*, m.p. 240° (dec.); $\lambda_{max}^{pH\ 11}$ in nm ($\varepsilon_{m\underline{M}}$) 227 (25.10), 258 (21.10), and 326 (10.00).

REFERENCES

1. H. C. Koppel, R. H. Springer, R. K. Robins, and C. C. Cheng, *J. Med. Pharm. Chem.*, 5, 639 (1962).

2. T. B. Johnson and A. W. Joyce, *J. Am. Chem. Soc.*, 38, 1557 (1916).

[13] 1-(TETRAHYDRO-2-FURYL)PYRIMIDINES

A General Method for the Preparation of 1-Substituted Uracils, Cytosines, and Related Compounds by a Modified Hilbert—Johnson Method

C. Wayne Noell and C. C. Cheng

Midwest Research Institute, Kansas City, Missouri 64110

INTRODUCTION

The preparation of 1-(tetrahydro-2-furyl)-substituted derivatives of uracil, cytosine, and related compounds (analogs of uridine, deoxyuridine, cytidine, and deoxycytidine) has been accomplished by condensation of 2-chlorotetrahydrofuran with the appropriate 2,4-dimethoxypyrimidines.[1] Owing to the extreme lability of the tetrahydrofuryl moiety in an acidic environment, condensations are conducted in the presence of anhydrous sodium carbonate.

PROCEDURE

1,2-Dihydro-4-methoxy-2-oxo-1-(tetrahydro-2-furyl)pyrimidine (3a)

A moderate stream of dry hydrogen chloride is passed through a solution (cooled in an ice bath) of 2,3-dihydrofuran[2] (10 g) in dry dichloromethane (80 ml) for 15 min. To the solution is added anhydrous sodium carbonate (20 g), the mixture is stirred for 5 min, and 2,4-dimethoxypyrimidine[3] (*1a*; 20 g) is added in one

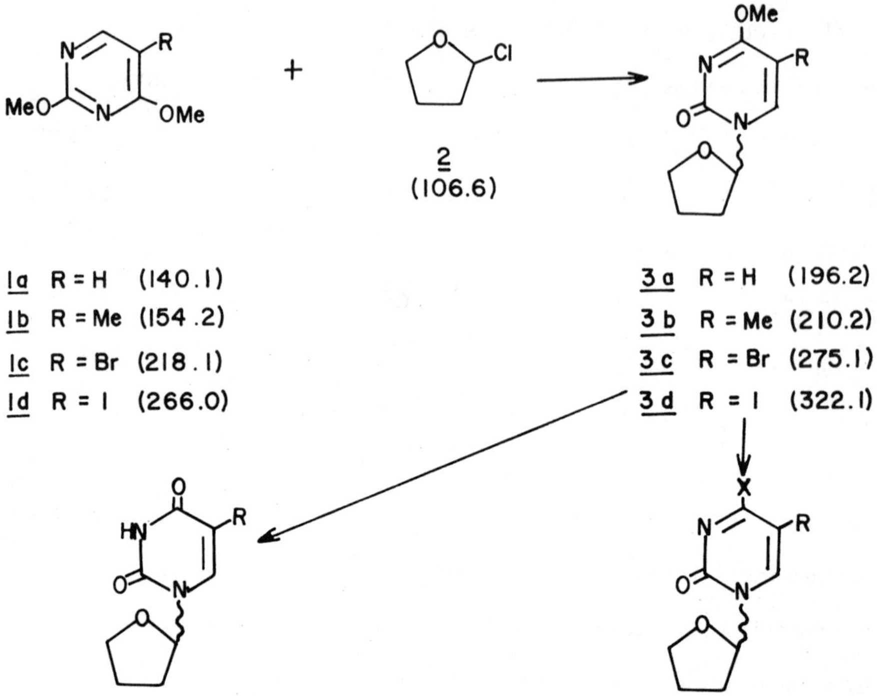

2 (106.6)

1a R = H (140.1)
1b R = Me (154.2)
1c R = Br (218.1)
1d R = I (266.0)

3a R = H (196.2)
3b R = Me (210.2)
3c R = Br (275.1)
3d R = I (322.1)

4a R = H (182.2)
4b R = Me (196.2)
4c R = Br (261.1)
4d R = I (308.1)

5a R = H, X = NH_2 (181.2)
5b R = H, X = NHMe (195.2)
5c R = H, X = NHOH (197.2)
5d R = Me, X = NH_2 (195.2)
5e R = Me, X = NHMe (209.2)
5f R = Me, X = NHOH (211.2)
5g R = Br, X = NH_2 (260.2)
5h R = I, X = NH_2 (307.1)

portion. The mixture is then stirred for an additional 15 min, and the ice bath is removed. Stirring is continued for 20 hr at room temperature, the mixture is filtered, and the filtrate is evaporated to a clear oil in a rotary, flash evaporator. The oil is triturated with warm petroleum ether (b.p. 35° to 60°), the solvent is decanted, and the residue is covered with anhydrous ether (100 ml). The mixture is triturated for 5 min by means of a glass rod, and then kept for 2 hr at 5°. The resulting crystalline solid is collected by filtration, washed with petroleum ether, and dried over phosphorus pentoxide in a vacuum desiccator to give 24 g (86% yield) of *3a*, m.p. 68° to 70°. An analytical sample is obtained by recrystallization from ether, m.p. 74° to 75°; λ_{max} in nm ($\varepsilon_{\underline{mM}}$) at pH 1, 275 (7.90); at pH 11, 273 (7.90).

1-(Tetrahydro-2-furyl)uracil (4a)

Method A. A solution of *3a* (6 g) in water (40 ml) containing concentrated hydrochloric acid (2 ml) is kept for 12 hr at 3°, while the major absorption maxima undergo a hypsochromic shift to 263 nm. The solution is carefully rendered neutral with 2$\underline{M}$ sodium hydroxide solution and evaporated to dryness *in vacuo*. The resulting solid mass is boiled with dry ethyl acetate (160 ml), sodium chloride is removed by filtration, and the filtrate is concentrated to 50 ml. On cooling, *4a* crystallizes; it is collected by filtration and recrystallized from ethyl acetate to give 3 g (54% yield) of analytically pure product, m.p. 103° to 104°; λ_{max} in nm ($\varepsilon_{\underline{mM}}$) at pH 1, 263 (10.80); and at pH 11, 225 (6.50), and 263 (8.00).

Method B. A mixture of *3a* (3 g) and 2M sodium hydroxide (30 ml) is warmed for 10 min at 80° to 90° and allowed to cool slowly to room temperature. The solution is kept for 20 min at room temperature and then carefully neutralized with 1M hydrochloric acid at <15°. The resulting solution is evaporated to dryness *in vacuo* at <15°, and the residue is extracted with ethyl acetate and processed as described for method A to give 2.1 g (70%) of *4a*. The product is identical with that obtained by method A.

1-(Tetrahydro-2-furyl)cytosine (5a)

A mixture of *3a* (10 g) and 200 ml of methanolic ammonia (methanol presaturated with ammonia at 0°) is heated for 12 hr at 100° in a sealed container. The mixture is then evaporated to dryness under diminished pressure, the semicrystalline residue is triturated with warm ethyl acetate (100 ml) and cooled, and the crystalline mass is collected by filtration. One recrystallization from ethyl acetate containing a trace of methanol gives 4 g (43%) of analytically pure *5a*, m.p. 193° to 194°, λ_{max} in nm (ε_{mM}) at pH 1, 280 (14.80); and at pH 11, 231 (7.80) and 271 (10.00).

1,2-Dihydro-4-(methylamino)-2-oxo-1-(tetrahydro-2-furyl)pyrimidine (5b)

A mixture of *3a* (10 g), 35% aqueous methylamine (40 ml) and water (50 ml) is stirred for 2 hr at room temperature and evaporated to dryness *in vacuo*. The residue is triturated with ether, and the resulting crystalline mass is collected by filtration, to give, after drying for 16 hr at 70°, 7.5 g (75%) of a white solid, m.p. 123° to 125°. One recrystallization from benzene gives 5.4 g (54%) of analytically pure *5b*, m.p. 130° to 131°; λ_{max} in nm ($\varepsilon_{\underline{mM}}$) at pH 1, 281 (15.80); at pH 11, 234 (9.40), and 271 (11.90).

1,2-Dihydro-4-(hydroxyamino)-2-oxo-1-(tetrahydro-2-furyl)pyrimidine (5c)

The pH of a solution of hydroxylamine hydrochloride (7.2 g) in water (50 ml) is adjusted to 7 by means of 4$\underline{M}$ sodium hydroxide, *3a* (5 g) is added, and the mixture is boiled under reflux for 1 hr. The hot solution is then evaporated to dryness in a rotary evaporator, and the crystalline residue is triturated with cold water (50 ml). The resulting crystalline solid is recrystallized from water (100 ml), to give 3 g (60%) of analytically pure product, m.p. 173° to 175°; λ_{max} in nm ($\varepsilon_{\underline{mM}}$) at pH 1, 281 (14.40); and at pH 11, 239 (11.60).

1,2-Dihydro-4-methoxy-5-methyl-2-oxo-1-(tetrahydro-2-furyl)pyrimidine (3b)

Compound *3b* is prepared from 2,3-dihydrofuran (20 g), sodium carbonate (40 g), and 2,4-dimethoxy-5-methylpyrimidine[4] (44.2 g) as for the preparation of *3a*. The product is recrystallized from hexane—acetone to give 24 g (40%) of *3b*, m.p. 99° to 100°. Further recrystallization from the same solvent pair raises the m.p. to 100° or 101°; λ_{max} in nm ($\varepsilon_{m\underline{M}}$) at pH 1, 279 (6.90), and at pH 11, 279 (7.10).

1-(Tetrahydro-2-furyl)thymine (4b)

Method A. To a solution of *3b* (5 g) in water (50 ml) is added concentrated hydrochloric acid (5 ml). The solution is stirred for 4 hr and the precipitate collected by filtration and recrystallized from water to give 2.3 g (50%) of analytically pure *4b*, m.p. 183° to 185°; λ_{max} in nm ($\varepsilon_{m\underline{M}}$) at pH 1, 268 (10.40); and at pH 11, 225 (7.80) and 267 (8.20).

Method B. A mixture of *3b* (2 g) and 2$\underline{M}$ sodium hydroxide (20 ml) is boiled gently for 10 min and then allowed to cool to room temperature. The pH of the solution is brought to 6 with glacial acetic acid. The solution is cooled, and the resulting precipitate is collected by filtration, washed successively with water and acetone, and dried at 80° to give 0.9 g (48%) of *4b*, which is identical with that prepared by method *A*.

5-Methyl-1-(tetrahydro-2-furyl)cytosine (5d)

Compound *5d* is prepared from *3b* (6 g) amd methanolic ammonia (180 ml) as for the preparation of *5a*. The product is recrystallized from acetonitrile to give 4 g (72%) of analytically pure product, m.p. 217° to 219°; λ_{max} in nm ($\varepsilon_{m\underline{M}}$) at pH 1, 288 (12.60); and at pH 11, 230 (7.50) and 278 (8.80).

1,2-Dihydro-5-methyl-4-methylamino-2-oxo-1-(tetrahydro-2-furyl)-pyrimidine (5e)

Compound *5e* is prepared from *3b* (10 g) and 35% aqueous methylamine (150 ml) as for the preparation of *5b*. Recrystallization from benzene—methanol gives 9.2 g (93%) of white crystals, m.p. 173° to 174°; λ_{max} in nm ($\varepsilon_{\underline{mM}}$) at pH 1, 286 (11.60); and at pH 11, 233 (6.70) and 276 (8.50).

1,2-Dihydro-4-hydroxyamino-5-methyl-2-oxo-1-(tetrahydro-2-furyl)-pyrimidine (5f)

Compound *5f* is prepared from hydroxylamine hydrochloride (13.2 g) and *3b* (10 g) as for the preparation of *5c*. Recrystallization from water gives 7 g (70%) of *5f*, m.p. 170° to 172°; λ_{max} in nm ($\varepsilon_{\underline{mM}}$) at pH 1, 285 (13.80); and at pH 11, 240 (10.40).

5-Bromo-1,2-dihydro-4-methoxy-2-oxo-1-(tetrahydro-2-furyl)pyrimidine (3c)

Compound *3c* is prepared from 2,3-dihydrofuran (12 g), of sodium carbonate (30 g), and 5-bromo-2,4-dimethoxypyrimidine[5] (*1c*; 25 g) as for the preparation of *3a*. The product is recrystallized from water, to give 9.8 g (31%) of *3c*, m.p. 110° to 111°; λ_{max} in nm ($\varepsilon_{\underline{mM}}$) at pH 1, 289 (5.40); and at pH 11, 289 (5.20).

5-Bromo-1-(tetrahydro-2-furyl)uracil (4c)

Compound *4c* is prepared from *3c* (1.5 g) by essentially the same procedure as that used for the preparation of *4b*. After recrystallization of the crude product from ethanol—water a 70% yield of *4c* is obtained; m.p. 205° to 207°; λ_{max} in nm ($\varepsilon_{\underline{mM}}$) at pH 1, 280 (8.90); and at pH 11, 276 (6.50).

5-Bromo-1-(tetrahydro-2-furyl)cytosine (5g)

Compound *5g* is prepared from *3c* (7 g) and methanolic ammonia (200 ml) as for the preparation of *5a*. Recrystallization from water gives 3.5 g (53%) of *5g*, m.p. 192° to 193°; λ_{max} in nm ($\varepsilon_{\underline{mM}}$) at pH 1, 300 (11.20); and at pH 11, 286 (7.80).

1,2-Dihydro-5-iodo-4-methoxy-2-oxo-1-(tetrahydro-furyl)pyrimidine (3d) and 1,2-Dihydro-5-iodo-4-methoxy-2-oxopyrimidine

From 2,3-dihydrofuran (20 g), sodium carbonate (30 g), and 5-iodo-2,4-dimethoxypyrimidine[6] (*1d*) (70 g) is obtained, by a procedure similar to that used for the preparation of *3a*, 40 g of a solid material that consists of *3d* plus 1,2-dihydro-5-iodo-4-methoxy-2-oxopyrimidine. It is dissolved in cold ethyl acetate (200 ml), and the solution is boiled for a few minutes, whereupon a precipitate begins to separate. After the mixture has been boiled for 15 min, the insoluble material is collected by filtration, washed with hot ethyl acetate, and recrystallized from acetonitrile. The yield of 1,2-dihydro-5-iodo-4-methoxy-2-oxopyrimidine is 4.6 g, m.p. 228° to 229°; λ_{max} in nm ($\varepsilon_{\underline{mM}}$) at pH 1, 227 (18.20) and 290 (3.90); at pH 11, 234 (17.60) and 291 (5.50).

When treated with an excess of methanolic ammonia at 100° in a sealed container, this product gives 5-iodocytosine[7] in almost quantitative yield.

The filtrate and washings are combined and evaporated to dryness under diminished pressure, and the residue is crystallized from water (1 l) to give 29.2 g of *3d*, m.p. 141° to 143°. One recrystallization from the same solvent yields an analytically pure sample, m.p. 145° to 146°; λ_{max} in nm ($\varepsilon_{\underline{mM}}$) at pH 1, 227 (16.40) and 295 (5.80); at pH 11, 230 (14.80) and 295 (5.80).

5-Iodo-1-(tetrahydro-2-furyl)uracil (4d)

Compound *4d* is prepared from *3d* (1.0 g) as for the preparation of *4b*. The crude product is recrystallized from ethanol—water; yield 94%; m.p. 209° to 211°; λ_{max} in nm ($\varepsilon_{\underline{mM}}$) at pH 1, 289 (7.40); at pH 11, 279 (5.60).

5-Iodo-1-(tetrahydro-2-furyl)cytosine (5h)

Compound *5h* is prepared from *3d* (8 g) and methanolic ammonia (180 ml) as for the preparation of *5a*. Recrystallization from water gives 3.2 g (42%) of *5h*, m.p. 187° to 189°; λ_{max} in nm ($\varepsilon_{\underline{mM}}$) at pH 1, 308 (8.90); at pH 11, 277 (10.40) and 294 (5.20).

REFERENCES

1. C. W. Noell and C. C. Cheng, *J. Heterocycl. Chem.*, 5, 25 (1968).

2. R. Paul, M. Fluchaire, and G. Collardeau, *Bull. Soc. Chim. Fr.*, 668 (1950).

3. G. E. Hilbert and T. B. Johnson, *J. Am. Chem. Soc.*, 52, 2001 (1930).

4. Y.-F. Chi and S.-F. Chen, *Hua Hsueh Hsueh Pao*, 22, 194 (1956); *Sci. Sin.*, 6, 111 (1957).

5. G. E. Hilbert and E. F. Jansen, *J. Am. Chem. Soc.*, 56, 132 (1934).

6. M. Prystaš and F. Šorm, *Collect. Czech. Chem. Commun.*, 29, 121 (1964).

7. T. B. Johnson and C. O. Johns, *J. Biol. Chem.*, 1, 305 (1905/1906).

[14] A GENERAL SYNTHESIS OF 6-ALKYL(OR ARYL)-2-(CARBAMOYLMETHYL)-4-PYRIMIDINONES

Primary Synthesis from Amidine and a β-Keto-ester

Desmond J. Brown, John C. Gamble, and Paul Waring

John Curtin School of Medical Research, Canberra, Australia 2600

INTRODUCTION

The only 6-alkyl-2-(carbamoylmethyl)-4-pyrimidinone previously described is the 6-methyl derivative (5, R' = Me); this was obtained in mediocre yield by condensation of the amidine *3* with

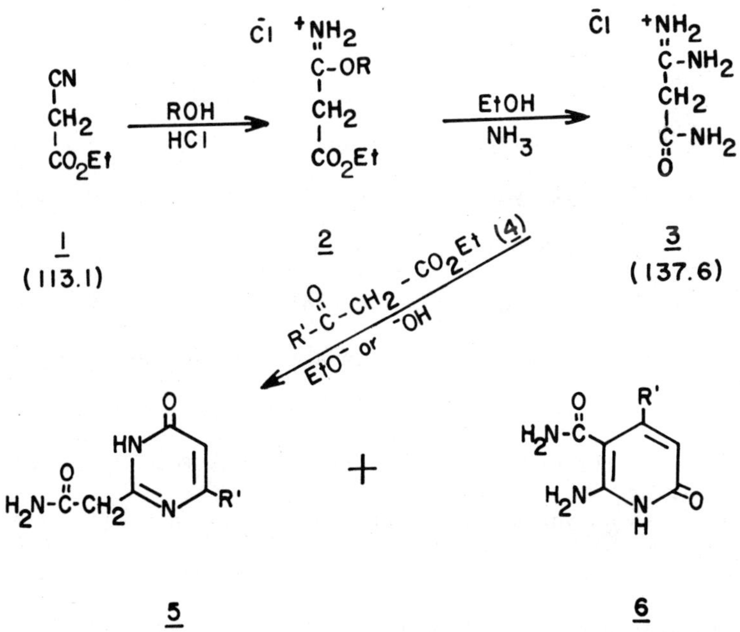

ethyl acetoacetate (*4*, R' = Me) in water[1,2] or by reductive ring-fission of 5-methylpyrazolo[1,5-*a*]pyrimidine-2,7(1*H*,4*H*)-dione.[3] The arylpyrimidinone (*5*, R' = Ph) has been described,[1] but its freedom from the isomeric pyridone (*6*, R' = Ph) as an impurity is doubtful. The intermediate *3* has been prepared by ammonolysis of the imidic ester hydrochloride (*2*, R = Et),[1,4] itself best prepared by treatment of ethyl cyanoacetate (*1*) with ethanolic hydrogen chloride for 16 days.[5] The present, integrated procedure (*1* → *5*) is general, simple, and relatively rapid. The proportion of isomeric pyridone (*6*), formed as a separable by-product, varies according to the identity of the alkyl or aryl group (R') and the conditions of the condensation.

PROCEDURE

Ethyl(methoxyformimidoyl)acetate hydrochloride (*2*, R = Me)

A mixture of ethyl cyanoacetate (*1*) (300 g, 2.65 mol), anhydrous methanol (122 ml, 3.0 mol), and anhydrous ether (3 l) in a 5-l flask fitted with a guard tube is cooled in ice to <5°. Dry hydrogen chloride is introduced in a steady stream during 3 hr, and the mixture is then kept at <5° for 18 hr. The white solid is removed by filtration using a sintered-glass funnel, washed well thereon with anhydrous ether, and then dried in a vacuum desiccator (calcium chloride) with the pump running for at least 2 hr. The yield of pure (hygroscopic) imidic ester hydrochloride is

304 g (63%), m.p. 88° to 90°.

Analysis calculated for $C_6H_{12}ClNO_3$: C, 39.7; H, 6.7. Found: C, 39.7; H, 6.7.

Amidinoacetamide hydrochloride (*3*)

Anhydrous ethanol (3 l) is cooled externally with ice while dry ammonia gas is introduced in a rapid stream to saturation (≃3 hr). Compound *2* (R = Me) (300 g, 1.65 mol) is added, and the mixture is gently agitated with a magnetic stirrer for 24 hr at 20° to 25°. The white solid is removed by filtration and washed with ethanol. The filtrate and washings are combined and concentrated to a small bulk, and a second crop is removed by filtration. The crystals are combined and recrystallized in four batches from the same ethanol (1 l), to give compound *3* (165 g, 73%), m.p. 173° to 176° (lit.[4] 174°).

*2-(Carbamoylmethyl)-6-methyl-4(3*H*)-pyrimidinone* (*5*, *R'* = *Me*)

Amidinoacetamide hydrochloride (10 g, 73 mmol), ethyl acetoacetate (9.5 g, 75 mmol), and ethanolic 0.5<u>M</u> sodium hydroxide (150 ml, 75 mmol) are mixed and agitated with a magnetic stirrer in a 400-ml lipless beaker covered with an upturned petri dish for 18 hr at room temperature. Every few hours the crusty solid is broken up. The slurry is evaporated to dryness (rotary evaporator), ice-cold water (50 ml) is added, and the pH is adjusted to between 3 and 4 with hydrochloric acid. The solid is removed by filtration and washed on the filter with ice-cold ethanol (10 ml). The dried residue[a] (≃10 g) is recrystallized from boiling methanol (300 ml) to give the methylpyrimidinone (6.0 g). Evaporation of the filtrate and dissolution of the residue in boiling methanol (100 ml) with subsequent cooling to only 20° to 25° gives a further crop of pyrimidinone (1.2 g).[b] Recrystallization of the combined crops (6.2 g, 51%) from water gives pure *5*, m.p. 245° to 247°; NMR data [$(CD_3)_2SO$]: δ 2.35 (s, Me), 3.60 (s, CH_2; eliminated by D_2O), 6.15 (s, 5-H), 7.2 (br, NH), and 7.7 (br, NH).

Analysis calculated for $C_7H_9N_3O_2$: N, 25.1. Found N, 24.9.

2-(Carbamoylmethyl)-6-phenyl-4(3H)-pyrimidinone (5, R' = Ph)

The foregoing procedure for the preparation of *5* (R' = Me) is repeated, using ethyl benzoylacetate (*4*, R' = Ph) (14.05 g, 73 mmol) instead of ethyl acetoacetate, and a stirring time of 14 hr. The dry, crude solid[c] (≃8 g) is dissolved in boiling

---

[a]This residue contains the pyrimidinone (*5*, R' = Me) and the pyridone (*6*, R' = Me) in the ratio 8:1 by NMR spectroscopy; when hot ethanolic sodium ethoxide is used for the condensation, as for *5*, (R' = CMe_3), the crude product is almost pure pyridone (*6*, R' = Me).

[b]Evaporation of the mother liquors, and recrystallization of the residue from water (30 parts) gives 6-amino-5-carbamoyl-4-methyl-2(1*H*)-pyridone (*6*, R' = Me) as pale-yellow crystals, m.p. 263° to 265° (dec.); NMR data [$(CD_3)_2SO$]: δ 2.10 (s, Me) and 5.30 (s, 3-H).

Analysis calculated for $C_7H_9N_3O_2$: C, 50.3; H, 5.4; N, 25.1. Found: C, 50.4; H, 5.4; N, 25.0.

[c]This solid contains the pyrimidinone (*5*, R' = Ph) and the pyridone (*6*, R' = Ph) in the ratio 1:2 by NMR spectroscopy; the ratio is slightly changed in favor of the pyridone by using hot ethanolic sodium ethoxide as the condensing medium.

methanol (150 ml), and the solution is cooled to room temperature and allowed to stand for 1 hr to yield the phenylpyrimidinone (2.0 g). The filtrate is then cooled at -5° for 3 hr to yield the isomeric pyridone (*6*, R' = Ph) (4.5 g). The mother liquors are evaporated to dryness and again fractionally recrystallized from methanol (40 ml) to give the pyrimidinone (0.5 g) and the pyridone[d] (1.0 g). The total yield of phenylpyrimidinone is 2.5 g (15%), m.p. 251° to 253°; NMR data [$(CD_3)_2SO$]: δ 3.70 (s, CH_2), 6.9 (s, 5-H), 7.55 (m, 3H of Ph), and 8.10 (m, 2H of Ph).

Analysis calculated for $C_{12}H_{11}N_3O_2$: C, 62.9; H, 4.8; N, 18.3. Found: C, 62.9; H, 4.9; N, 18.4.

*6-*tert*-Butyl-2-(carbamoylmethyl)-4(3H)-pyrimidinone (5, R' = CMe$_3$)*

Sodium (1.1 g, 0.05 mol) is allowed to react with absolute ethanol (30 ml) in a 200-ml, triple-necked flask equipped with a stirrer and a guard tube. To the resulting solution is added amidinoacetamide hydrochloride (6.7 g, 0.05 mol) and ethyl pivaloylacetate (*4*, R' = CMe_3) (8.6 g, 0.05 mol). The mixture is heated with stirring for 24 hr in a bath at 70° and then evaporated. The dry residue is slurried with water (15 ml), and the pH is adjusted to 3 by dropwise addition of concentrated hydrochloric acid with

---

[d]The 6-amino-5-carbamoyl-4-phenyl-2(1*H*)-pyridone (*6*, R' = Ph) (5.5 g, 33%) has m.p. 244° to 245° (from methanol); NMR data [$(CD_3)_2SO$]: δ 5.60 (s, 3-H) and 7.60 (s, 5 H of Ph).

Analysis calculated for $C_{12}H_{11}N_3O_2$: C, 62.9; H, 4.8; N, 18.3. Found: C, 62.9; H, 4.9; N, 18.7.

stirring. The mixture is refrigerated and the crude greenish product (3.7 g, 36%) is removed by filtration and dried in a vacuum desiccator (calcium chloride). Recrystallization from 4-methyl-2-pentanone (240 ml) with added carbon gives the *tert*-butylpyrimidinone 5 (2.45 g, 24%) as colorless plates, m.p. 196° to 199°. A specimen (from ethanol) has m.p. 200° to 202°; NMR data [$(CD_3)_2SO$]: δ 1.75 (s, CMe_3), 3.50 (s, CH_2), 6.10 (s, 5-H), 7.1 (br, NH), and 7.5 (br, NH).

Analysis calculated for $C_{10}H_{15}N_3O_2$: C, 57.4; H, 7.2; N, 20.1. Found: C, 57.5; H, 7.0; N, 20.2.

REFERENCES

1. A. Pinner, *Ber.*, 28, 473 (1895).

2. A. Dornow and E. Neuse, *Arch. Pharm. Ber. Deut. Pharm. Ges.*, 287, 361 (1954).

3. J.-L. Imbach, R. Jacquier, and J.-L. Vidal, *Bull. Soc. Chim. Fr.*, 1929 (1970).

4. S. M. McElvain and B. E. Tate, *J. Am. Chem. Soc.*, 73, 2760 (1951).

5. S. M. McElvain and J. E. Schroeder, *J. Am. Chem. Soc.*, 71, 40 (1949).

[15] SUBSTITUTED 4-(BENZYLOXY)-6(1*H*)-PYRIMIDINONES

Selective, Acid-Catalyzed Dealkylation of 4,6-Di(benzyloxy)-pyrimidines

Miroslav Prystasz

Institute of Organic Chemistry and Biochemistry, Czechoslovak Academy of Sciences, 166 10 Prague 6, Czechoslovakia

INTRODUCTION

Substituted 4-(benzyloxy)-6(1*H*)-pyrimidinones (*6—10*) have been shown[1,2] to be very useful aglycons in studies on the course of glycosylation by the mercuri procedure and subsequent O→N trans-glycosylation. Their synthesis consists[1,3] in a facile

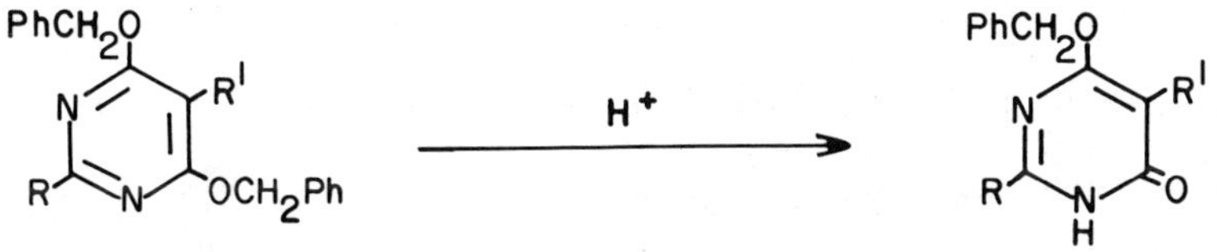

1, R = R' = H (292.3)

2, R = Me, R' = H (306.4)

3, R = H, R' = Me (306.4)

4, R = Ph, R' = H (368.4)

5, R = H, R' = Ph (368.4)

6, R = R' = H (202.2)

7, R = Me, R' = H (216.2)

8, R = H, R' = Me (216.2)

9, R = Ph, R' = H (278.3)

10, R = H, R' = Ph (278.3)

dealkylation of only one benzyloxy group of 4,6-di(benzyloxy)-pyrimidines (*1*—*5*) by use of one equivalent of hydrogen chloride in anhydrous acetonitrile at room or elevated temperature. Processing of the reaction mixture is very simple. On prolonged heating with an excess of hydrogen chloride the substituted 4-hydroxy-6(1*H*)pyrimidinones are formed.[3]

PROCEDURE

4-(Benzyloxy)-6(1H)-pyrimidinone (6)

A solution of 4,6-di(benzyloxy)pyrimidine[4] (*1*) (14.62 g, 0.05 mol) in 0.25<u>M</u> hydrogen chloride in acetonitrile (200 ml, 0.05 mol) is kept for 60 hr at room temperature to deposit crystals, which are collected by suction. The mother liquor is concentrated to one third to give an additional crop of the pyrimidinone *6*; total yield, 8.30 g (82%). This material is sufficiently pure for synthetic purposes. Recrystallization from toluene gives analytically pure pyrimidinone *6*, m.p. 196° to 197°; λ_{max}^{EtOH} 212 nm ($\varepsilon_{m\underline{M}}$ 20.90), 265 (3.65), and 230 (infl.) (5.00); NMR data ($CDCl_3$): δ 5.34 (s, CH_2), 5.81 (s, H-5), 7.30—7.48 (m, C_6H_5), and 8.05 (s, H-2).

4-(Benzyloxy)-2-methyl-6(1H)-pyrimidinone (7)

A solution of 4,6-di(benzyloxy)-2-methylpyrimidine[1] (*2*) (9.19 g, 30 mmol) in anhydrous (molecular sieves) acetonitrile (30 ml) is treated with 2<u>M</u> hydrogen chloride in acetonitrile (15 ml, 30 mmol) and the mixture is kept for 30 hr at room temperature to afford 3.76 g (58%) of the pyrimidinone *7*, m.p. 177° to 179°, unchanged on recrystallization from toluene. Concentration of the mother liquors gives an additional crop; overall yield, 4.47 g (69%); λ_{max}^{EtOH} 216 nm ($\varepsilon_{\underline{mM}}$ 13.80), 265 (5.00); NMR data ($CDCl_3$): δ 2.47 (s, CH_3), 5.28 (s, CH_2), 5.65 (s, H-5), and 7.30—7.47 (m, C_6H_5).

*4-(Benzyloxy)-5-methyl-6(1*H*)-pyrimidinone (8)*

A mixture of 4,6-di(benzyloxy)-5-methylpyrimidine[4] (*3*) (1.53 g, 5.0 mmol) and 0.25M hydrogen chloride in acetonitrile (20 ml, 5.0 mmol) is heated in a sealed tube for 1 hr at 120° and cooled to deposit crystals of *8*. Recrystallization from boiling toluene affords 0.94 g (87%) of the pyrimidinone *8*, m.p. 222°; NMR data ($CDCl_3$): δ 2.02 (s, CH_3), 5.44 (s, CH_2), 7.30 to 7.48 (m, C_6H_5), and 7.99 (s, H-2).

Heating a mixture of the pyrimidine *3* (306 mg, 1.0 mmol) and 0.25M hydrogen chloride in acetonitrile (40 ml, 10.0 mmol) for 6 hr at 120° affords 4-hydroxy-5-methyl-6(1*H*)-pyrimidinone in 82% yield.[3]

*4-(Benzyloxy)-2-phenyl-6(1*H*)-pyrimidinone (9)*

A solution of 4,6-di(benzyloxy)-2-phenylpyrimidine[1] (*4*) (3.68 g, 10.0 mmol) in 0.25M hydrogen chloride in acetonitrile (40 ml, 10.0 mmol) is kept for 80 hr at room temperature to deposit 2.20 g (79%) of *9*, m.p. 231° to 233°. Recrystallization from toluene gives the pyrimidinone *9*, m.p. 233° to 234°; λ_{max}^{EtOH} 214 nm (ε_{mM} 25.70), 242 (13.50), and 294 (7.80); NMR data ($CDCl_3$): δ 5.48 (s, CH_2), 5.85 (s, H-5), and 7.31 to 7.60 (m, $2 \times C_6H_5$).

4-(Benzyloxy)-5-phenyl-6(1H)-pyrimidinone (10)

By the procedure described for the synthesis of compound *6*, 3.68 g (10.0 mmol) of the pyrimidine[4] *5* affords 2.34 g (84%) of the pyrimidinone *10*, m.p. 192° to 193° (from acetonitrile); NMR data ($CDCl_3$): δ 5.43 (s, CH_2), 7.27 to 7.60 (m, $2 \times C_6H_5$), and 7.93 (s, H-2).

REFERENCES

1. M. Prystasz and F. Sorm, *Collect. Czech. Chem. Commun.*, 33, 210 (1968).

2. M. Prystasz, *Collect. Czech. Chem. Commun.*, 33, 1813 (1968).

3. M. Prystasz, *Collect. Czech. Chem. Commun.*, 32, 4260 (1967).

4. M. Prystasz, *Collect, Czech. Chem. Commun.*, 32, 4241 (1967).

[16] C-METHYLATED 4-METHOXY-2(1*H*)-PYRIMIDINONES

Application of the Hilbert—Johnson Reaction; Preparation of 4-Methoxy-2(1H)-pyrimidinones via Displacement of the 2-Methylsulfonyl Group

Miroslav Prystasz

Institute of Organic Chemistry and Biochemistry, Czechoslovak Academy of Sciences, 166 10 Prague 6, Czechoslovakia

INTRODUCTION

Glycosylation of 4-alkoxy-2(1*H*)-pyrimidinones constitutes an alternative synthesis of the Hilbert—Johnson products.[1,2] Synthesis of 4-alkoxy-2(1*H*)-pyrimidinones has been accomplished by several methods involving the cleavage of 2,4-dialkoxypyrimidines under alkaline[3] or acidic[4] conditions and rearrangement of 4-alkoxypyrimidine 1-oxides.[5,6] By using method 1, 2,4-dimethoxypyrimidines *1* and *2* are treated with acetyl chloride to give the acetyl derivatives *3* and *4*, which are methanolyzed[2,7] without isolation to the intermediates *5* and *6*. Method 2 is of more general character and value,[6] utilizing a simultaneous oxidation of substances such as 4-methoxy-2-(methylthio)pyrimidines *7* and *8* with aqueous hydrogen peroxide under alkaline conditions and displacement of the methylsulfonyl group of compounds *9* and *10* by a

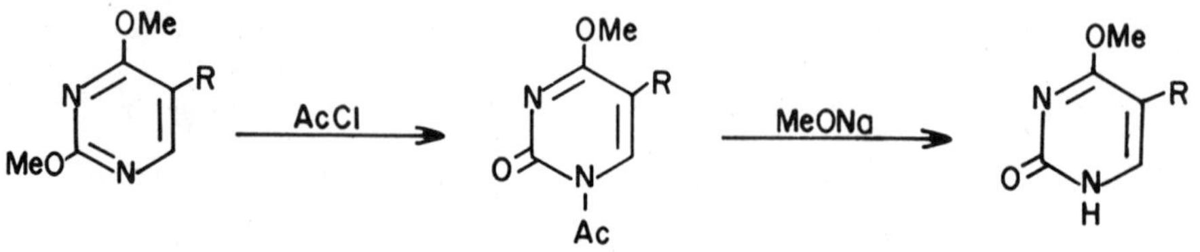

1, R = H (140.1); 2, R = Me (154.1)

3, R = H (168.2); 4, R = Me (182.2)

5, R = H (126.1); 6, R = Me (140.1)

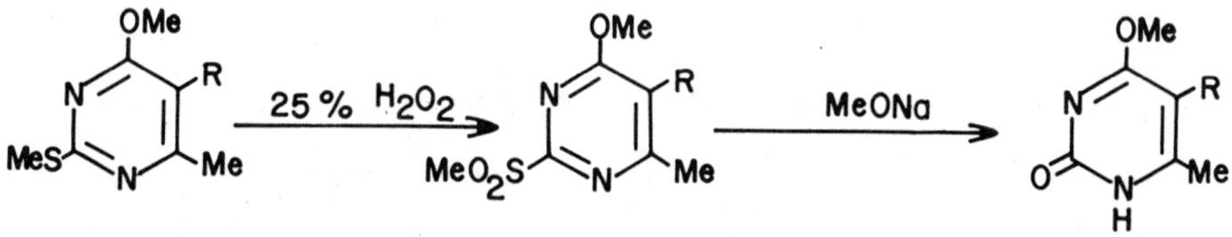

7, R = H (170.3); 8, R = Me (184.3)

9, R = H (202.3); 10, R = Me (216.3)

11, R = H (140.1); 12, R = Me (154.1)

hydroxyl group to give the pyrimidinones *11* and *12*. This method avoids isolation of the methylsulfonyl intermediates, in contrast to earlier procedures.[8,9]

PROCEDURE

4-Methoxy-2(1H)-pyrimidinone (5)

A solution of 2,4-dimethoxypyrimidine (*1*) (2.80 g, 20.0 mmol) in acetyl chloride (15 ml) is kept for 2 days at room temperature and evaporated *in vacuo*. The residue is coevaporated with toluene (2 × 20 ml), and then dissolved in M methanolic sodium methoxide (25 ml). The mixture is kept for 1 hr at room temperature, neutralized with Dowex-50 (H^+) ion-exchange resin, and evaporated under diminished pressure. The residue is taken up in hot chloroform (30 ml) and the suspension is filtered. The filtrate is evaporated *in vacuo*, and the crystalline residue is recrystallized from methanol to afford pyrimidinone *5*, m.p. 215° to 216° (lit.[4] m.p. 206° to 208°). Crystallization from methanol—ether of the residue obtained by evaporating the mother liquors gives a second crop; total yield 1.56 g (62%); NMR data ($CDCl_3$): δ 3.98 (s, OCH_3), 5.95 (d, H-5, $J_{5,6}$ 7.0 Hz), and 7.55 (d, H-6).

4-Methoxy-5-methyl-2(1H)-pyrimidinone (6)

By method 1, 3.08 g (20.0 mmol) of 2,4-dimethoxy-5-methylpyrimidine (*2*) affords 2.30 g (82%) of pyrimidinone *6*, m.p. 225° to 227° (from methanol); further recrystallization from methanol does not change the m.p. (lit.[7] m.p. 220° to 221°); NMR data ($CDCl_3$): δ 1.96 (s, 5-CH_3), 4.01 (s, OCH_3), and 7.36 (s, H-6).

4-Methoxy-6-methyl-2(1H)-pyrimidinone (11)

A mixture of the pyrimidine[8] *7* (17.0 g, 0.1 mol) and 1.3M methanolic sodium methoxide (100 ml, 0.13 mol) is treated dropwise during 20 min with 25% aqueous hydrogen peroxide (55 ml, 0.4 mol). The resulting crystalline suspension passes into solution on refluxing for 1 hr. The solution is rendered neutral by Dowex-50 (H^+) ion-exchange resin and evaporated *in vacuo*. The residue is dissolved in chloroform (200 ml), the solution treated with charcoal, filtered, and the filtrate evaporated. The residue is crystallized from ethanol to give the pyrimidinone *11*, m.p. 208° to 211°. Concentration of the mother liquors affords a second crop; total yield 5.15 g (37%). On recrystallization from ethanol, m.p. 211° to 213° (lit.[5,8] m.p. 209° to 211°, and 210° to 212°); NMR data ($CDCl_3$): δ 2.32 (s, 6-CH_3), 3.95 (s, OCH_3), and 5.76 (s, H-5).

The final mother liquors are evaporated *in vacuo*, and the residue is coevaporated with toluene (100 ml) and then chromatographed on a column of neutral alumina (100 g, Brockmann activity II to III). Elution with benzene (400 ml), evaporation of the eluate, and distillation of the residue (130°/10 torr) affords 4.2 g (25%) of the starting pyrimidine 7. The procedure consisting in isolation of the methylsulfonyl intermediate *9* and its subsequent reaction with sodium hydroxide has afforded the pyrimidinone *11* in 31% overall yield.[8]

4-Methoxy-5,6-dimethyl-2-(methylthio)pyrimidine (8)

A mixture of 4-chloro-5,6-dimethyl-2-(methylthio)pyrimidine[10] (18.9 g, 0.10 mol) and 1.2M methanolic sodium methoxide (100 ml) is boiled under reflux for 1 hr, cooled, evaporated under diminished pressure, and the residue shaken with a mixture of benzene (100 ml) and water (100 ml). The benzene layer is dried (anhydrous sodium sulfate) and evaporated under diminished pressure, and the residue is sublimed at 70°/15 torr to afford 17.5 g (95%) of the pyrimidine *8*, m.p. 40° to 41°.

4-Methoxy-5,6-dimethyl-2(1H)-pyrimidinone (12)

By the second method described for the synthesis of compound *11*, 18.4 g (0.1 mol) of the pyrimidine *8* gives 7.3 g (48%) of the pyrimidinone *12*, m.p. 249° to 250° (from methanol). On recrystallization from methanol, the m.p. remains unchanged; $\lambda_{max}^{pH\ 7}$ 205 nm ($\varepsilon_{\underline{mM}}$ 13.80), 278 (7.40); NMR data ($CDCl_3$): δ 1.93 (s, 5-CH_3), 2.32 (s, 6-CH_3), and 3.99 (s, OCH_3). Analogous processing of the final mother liquors yields 4.05 g (22%) of the starting pyrimidine *8*. As with the pyrimidinones *5* and *6*, 10% yield of compound *12* is obtained by using method 1.

REFERENCES

1. E. Wittenburg, *Chem. Ber.*, 101, 1095 (1968).
2. M. Prystasz, *Collect. Czech. Chem. Commun.*, 40, 1786 (1975).
3. G. E. Hilbert and E. P. Jansen, *J. Am. Chem. Soc.*, 57, 522 (1935).
4. C. W. Noell and C. C. Cheng, *J. Heterocycl. Chem.*, 5, 25 (1968).
5. E. Ochiai and M. Yamanaka, *Chem. Pharm. Bull.*, 3, 175 (1955).

6. M. Prystasz and F. Sorm, *Collect. Czech. Chem. Commun.*, 34, 2316 (1969).

7. J. L. Wong and D. S. Fuchs, *J. Org. Chem.*, 35, 3786 (1970).

8. J. F. W. McOmie, E. R. Sayer, and J. Chesterfield, *J. Chem. Soc.*, 1830 (1957).

9. E. Wittenburg, *J. Prakt. Chem.*, 33, 165 (1966).

10. R. Hull, J. B. Lovell, H. T. Openshaw, and A. R. Todd, *J. Chem. Soc.*, 41 (1947).

[17] BIS(THYMIN-5-YL) SULFONE[a]

Thiation and Oxidation of Thymine

Alfredo Giner-Sorolla and Lillian Medrek

Drug Development Laboratory, Memorial Sloan-Kettering Cancer Center, New York, New York 10021

INTRODUCTION

5-(Mercaptomethyl)uracil (*4*) exerts antitumor effect in several transplanted rodent-tumors[1] and its 2-deoxy-D-*erythro*-pentosyl derivative inhibits replication of certain viruses.[2] Bis(thymin-5-yl) sulfone (*6*) was synthesized as a potential antibacterial agent. Its synthesis involves the reaction of 5-(hydroxymethyl)-uracil (*1*) with hydrogen chloride to yield 5-(chloromethyl)uracil (*2*). Treatment of *2* with thioacetamide in *N,N*-dimethylformamide affords 5-(acetiminothiomethyl)uracil hydrochloride (*3*), from which 5-(mercaptomethyl)uracil (*4*) is obtained by alcoholysis. Condensation of *2* with *4* gives bis(thymin-5-yl) sulfide (*5*), which is oxidized to its sulfone (*6*) with hydrogen peroxide.[1] Previous procedures for the synthesis of *2* by chloromethylation[3,4] of uracil or from *1* and hot hydrochloric acid[3] and *1* and thionyl chlorides[5] gave a less satisfactory product.

---

[a]This work was supported in part by the National Cancer Institute (Grant CA-08748).

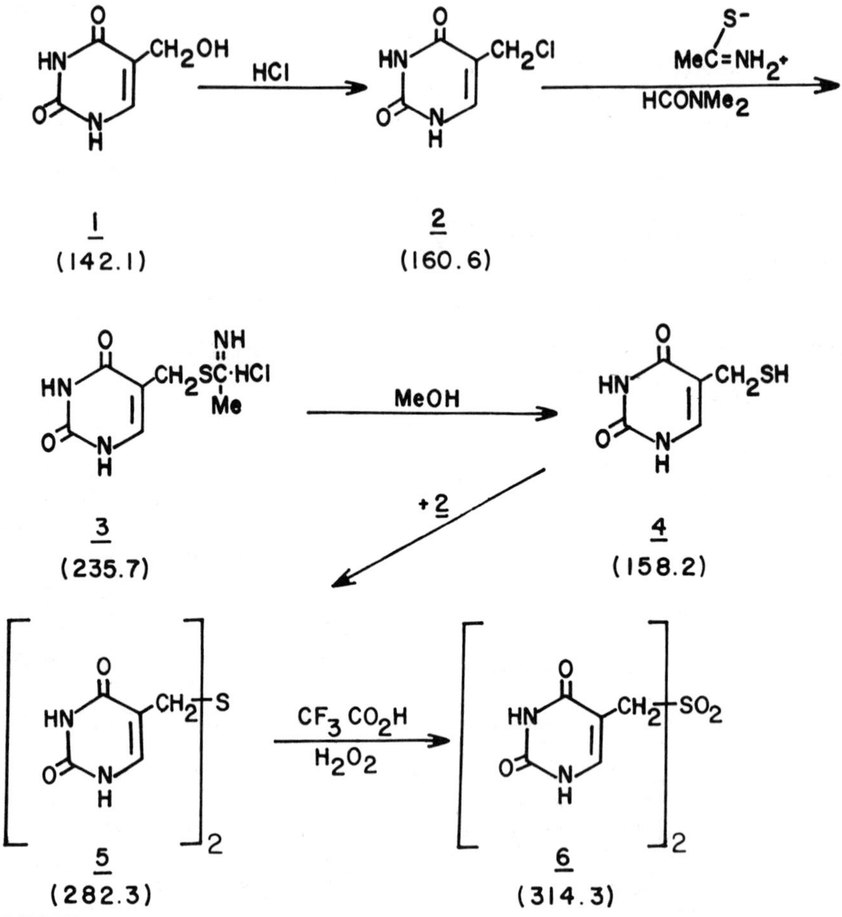

PROCEDURE

5-Chloromethyluracil (*2*)

5-(Hydroxymethyl)uracil[b] (*1*, 5.68 g, 0.04 mol) is dissolved in concentrated hydrochloric acid (50 ml) at 25°. A copious crystalline precipitate appears after a few minutes; after stirring

---

[b]Supplied by Calbiochem, Los Angeles, Calif.

for 30 min at 25°, it is collected on a sintered-glass filter, washed with a little concentrated hydrochloric acid, and dried *in vacuo* over phosphorus pentaoxide; yield 5.76 g (90%) of colorless crystals, m.p. 350° to 352° (effervescence).[c]

5-(Acetiminothiomethyl)uracil Hydrochloride (3)

Solutions of 5-(chloromethyl)uracil (*2*, 6.40 g, 40 mmol) and thioacetamide (6.0 g, 80 mmol) in *N*,*N*-dimethylformamide (30 ml and 20 ml, respectively)[d] and combined, and the mixture is heated slowly to 40°. A copious crystalline precipitate appears,[e] and the mixture is slowly heated to 70°, kept for 30 min at 70°, cooled, and filtered, and the crystals are thoroughly washed with ether to yield 5.81 g (62%)[f] of colorless needles, m.p. 290° to 292° (eff.).[g]

---

[c]Compound *2* remains unaltered if kept in a desiccator.

[d]In contrast to the reported[5] insolubility of *2* in *N*,*N*-dimethylformamide, we were able to prepare 30% solutions of *2* in that solvent at 40° to 50°. *N*,*N*-Dimethylformamide solutions of *2* made by the method here described were uniformly clear, while those of *2* obtained from uracil, trioxymethylene, and hydrochloric acid contain variable proportions of amorphous insoluble material (in some instances up to 25% of the specimen of compound *2* employed).

[e]The reagents must be anhydrous, as *3* does not precipitate otherwise, and the mixture must be concentrated *in vacuo*, with a decrease in the yield and quality of *3*.

[f]The temperature should never exceed 70°, as a lower yield is obtained otherwise.

[g]Catalytic effects of *N*,*N*-dimethylformamide (on reactions) used as the solvent have been observed by several authors.[6,7]

5-(Mercaptomethyl)uracil (4)

A suspension of compound *3* (3.70 g, 15.7 mmol) in methanol (75 ml) is boiled under reflux for 2 hr, giving a precipitate. The suspension is cooled and filtered, and the solid is washed successively with cold methanol and ether to yield 2.1 g (84%) of white crystals, m.p. 270° (after melting, it resolidifies at 280° and decomposes at 312° to 314°); $\lambda_{max}^{pH\ 1.0}$ 266 ($\varepsilon_{\underline{mM}}$ 8.29); $\lambda_{max}^{pH\ 6.9}$ 267 (7.93) and 287 nm (5.65); infrared (IR) spectrum showed a characteristic SH bank, ν 2620 cm^{-1}.

Bis(thymin-5-yl) Sulfide (5)

Solutions of *2* (3.21 g, 20 mmol), and *4* (3.16 g, 20 mmol) in *N,N*-dimethylformamide (25 ml each) are combined, and the mixture is heated for 3 hr in a water bath at 90° to 100°. The mixture is cooled, washed with ether (4 × 30 ml), and dried *in vacuo* over phosphorus pentaoxide to yield 5.58 g (quantitative) of colorless microneedles, m.p. 332° to 334° (dec.); λ_{max} in nm ($\varepsilon_{\underline{mM}}$) at pH 1.0, 264 (7.20); at pH 10.5, 283 (15.00).

Bis(thymin-5-yl) Sulfone (6)

A 30% solution of hydrogen peroxide (6 ml) is added dropwise with stirring to a solution of *5* (2.0 g, 7 mmol) in trifluoroacetic acid (20 ml). The resulting suspension is slowly heated to 60° during 30 min and kept overnight at 25°. The precipitate is removed by filtration and washed successively with cold water, ethanol, and ether to yield 1.05 g (46%) of colorless microcrystalline *6*, m.p. 340°; $\lambda_{max}^{pH\ 10.5}$ 282 nm ($\varepsilon_{m\underline{M}}$ 14.80).[h]

REFERENCES

1. A. Giner-Sorolla and L. Medrek, *J. Med. Chem.*, 9, 97 (1966).

2. V. S. Gupta, G. L. Bubbar, J. B. Meldrum. and J. R. Saunders, *J. Med. Chem.*, 18, 973 (1975).

3. W. A. Skinner, M. G. M. Schelstraete, and B. R. Baker, *J. Org. Chem.*, 25, 149 (1960).

4. J. H. Burckhalter, R. J. Seiwald, and H. C. Scarborough, *J. Am. Chem. Soc.*, 82, 991 (1960).

5. J. A. Carbon, *J. Org. Chem.*, 25, 1731 (1960).

6. T. P. Johnston, L. B. Holum, and J. A. Montgomery, *J. Am. Chem. Soc.*, 80, 6265 (1958).

7. R. W. Balsiger, A. L. Fikes, T. P. Johnston, and J. A. Montgomery, *J. Org. Chem.*, 26, 3386 (1961).

---

[h]The extremely low solubility at nonalkaline pH values does not permit further UV determinations.

[18] 5-FORMYLURACIL DIMETHYL ACETAL

Oxidation of 5-(Hydroxymethyl)uracil

Eric C. Ressner, Arieh Kampf, and Mathias P. Mertes

Department of Medicinal Chemistry, University of Kansas, Lawrence, Kansas 66045

INTRODUCTION

The synthesis of 5-formyluracil (*2*) by the Reimer—Tiemann reaction results in a low yield and the product[1] is difficult to separate.

The preparation of *2* from 5-(hydroxymethyl)uracil (*1*) has been reported by Brossmer and Ziegler.[2] An adaptation of their method using ammonium ceric nitrate according to a procedure of Trahanovsky and co-workers[3] is described here. 5-Formyluracil dimethyl acetal (*3*) is the precursor for the synthesis of a nucleoside.

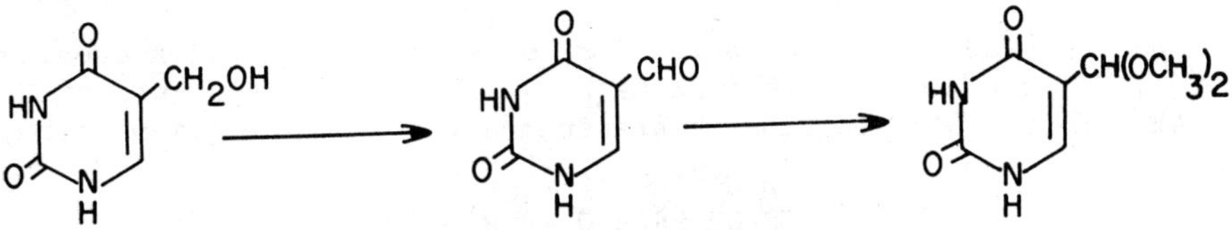

1 (142.1) 2 (140.1) 3 (174.2)

PROCEDURE

5-Formyluracil (2)

A slurry of 5-(hydroxymethyl)uracil[4] (*1*) (3.55 g, 25 mmol) in water (55 ml) is stirred vigorously, and M ammonium ceric nitrate[a] (55 ml) (a 10% excess) is added in one portion. Stirring is continued while the suspension is heated on a hot plate to 60°. Heating is then discontinued, and the mixture is allowed to cool[b] and is chilled in a refrigerator, giving up to 1.5 g of analytically pure *2* as a white powder that is removed by filtration, m.p. 304° to 306° (dec.; darkens at 280°) [lit.[2] m.p. 305° (dec.)].

5-Formyluracil Dimethyl Acetal (3)

A solution of *2* (12.0 g, 85 mmol) in dry methanol (500 ml) containing *p*-toluenesulfonic acid monohydrate (400 mg) is boiled overnight under reflux and cooled to room temperature, affording 6.5 g (42%) of compound *3*, m.p. 194° to 195°. Concentration of the mother liquor to 90 ml yields an additional 3.8 g of *3* (25%), m.p. 194° to 195°. After three recrystallizations from methanol,

---

[a]Reagent-grade ammonium ceric nitrate (G. Frederick Smith Chemical Co., Columbus, Ohio), formula $Ce(NH_4)_2(NO_3)_6$, molecular weight 548.2. The half reactions representing the redox reaction are:

$$2\ Ce^{4+} + 2e \rightarrow 2\ Ce^{3+}$$
$$RCH_2OH \rightarrow RCH{=}O + 2\ H^+ + 2e$$

Thus a M solution contains 548 $mg \cdot ml^{-1}$ if the ceric salt is 100% pure.

[b]The oxidation is rapid, and heating until the dissolution of 5-(hydroxymethyl)uracil is complete is normally sufficient for attaining good yields.

m.p. 199° to 200°.

REFERENCES

1. R. H. Wiley and Y. Yamamoto, *J. Org. Chem.*, 25, 1906 (1960).

2. R. Brossmer and D. Ziegler, *Tetrahedron Lett.*, 5253 (1966).

3. W. S. Trahanovsky, L. B. Young, and G. L. Brown, *J. Org. Chem.*, 32, 3865 (1967).

4. R. E. Cline, R. M. Fink, and K. Fink, *J. Am. Chem. Soc.*, 81, 2521 (1959).

[19] POTASSIUM 1,2,3,4-TETRAHYDRO-2,4-DIOXO-*s*-TRIAZINE-6-CARBOXYLATE (POTASSIUM 5-AZA-OROTATE)

Cyclocondensation of Biuret with Ethyl Potassium Oxalate

Alois Pískala and Jirí Gut

Institute of Organic Chemistry and Biochemistry, Czechoslovak Academy of Sciences, 166 10 Prague 6, Czechoslovakia

INTRODUCTION

The title compound *3* is a strong inhibitor of the *de novo* synthesis of pyrimidines.[1] The present procedure[2] exemplifies the preparation of *s*-triazinediones by cyclocondensation of biuret with esters of aliphatic acids. This method has also been successfully used in the preparation of [2,4-^{14}C]5-azaorotate, the decarboxylation of which yields [2,4-^{14}C]5-azauracil.[1,2] Potassium 5-azaorotate (*3*), designated in the older literature as potassium oxonate, is also accessible by alkaline oxidation of uric acid with hydrogen peroxide.[3]

PROCEDURE

Potassium 5-Aza-orotate (*3*)

Ethanolic potassium ethoxide [prepared from potassium (1.56 g, 0.04 mol) and absolute ethanol (40 ml)] is added to a hot solution of biuret (2.06 g, 0.02 mol) and potassium ethyl oxalate (3.12 g,

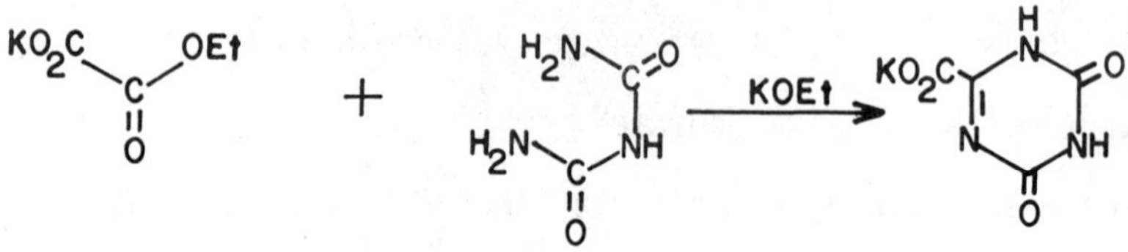

1 (156.2) 2 (103.1) 3 (195.2)

0.02 mol) in absolute[a] ethanol (200 ml). The mixture is magnetically stirred, boiled under reflux for 6 hr and allowed to cool.[b] The potassium salts are collected with suction and taken up in water (50 ml). The suspension is filtered, and the filtrate is cooled in an ice bath and acidified by adding glacial acetic acid (10 ml) with stirring. The precipitate is collected with suction, washed with ice-cold water, and dried under diminished pressure; yield of compound 3, 2.5 g (64%), m.p. 343° to 345° (dec.)[c]; $\lambda_{max}^{pH\ 9.0}$ 255 nm (log ε 3.83), λ_{min} 224 (log ε 3.34).[4]

REFERENCES

1. For leading references, see A. Cihák and F. Sorm, *Biochem. Pharmacol.*, 21, 607 (1972).

2. A. Piskala and J. Gut, *Collect. Czech. Chem. Commun.*, 27, 1562 (1962).

---

[a] The ethanol is dried with sodium ethoxide and diethyl phthalate.

[b] The reaction is performed with exclusion of atmospheric moisture and carbon dioxide.

[c] This melting-point value has not been reported in the literature.

3. F. J. Moore and R. M. Thomas, *J. Am. Chem. Soc.*, 40, 1120 (1918).

4. H. Brandenberger and R. H. Brandenberger, *Helv. Chim. Acta*, 37, 2207 (1954).

[20] 6-AMINO-8-METHYL-8-AZAPURINE (7-AMINO-2-METHYL-1,2,3-TRIAZOLO[4,5-*d*]PYRIMIDINE)

Example of a General Reaction for Preparing 6-Amino-8-azapurines from 5-Cyano-4-(dimethylaminomethylene)amino-1,2,3-triazoles

Adrien Albert

Department of Pharmacological Sciences, Health Sciences Center, State University of New York at Stony Brook, New York 11794

INTRODUCTION

6-Amino-8-methyl-8-azapurine (*2*) was prepared[1] by heating 8-methyl-6-(methylthio)-8-azapurine with ethanolic ammonia in a sealed tube at 150°. However, it is much more conveniently prepared as described here,[2] by heating 5-cyano-4-(dimethylaminomethylene)-amino-2-methyl-1,2,3-triazole (*1*) with aqueous ammonium acetate. The preparation[2] of the amidine (*1*) from *N,N*-dimethylformamide and 4-amino-2-methyl-1,2,3-triazole-5-carboxamide is also described here. The latter has been prepared[1] by the formylation of 4-amino-1,2,3-triazole-5-carboxamide[3] (obtained by the condensation of benzyl azide with cyanoacetamide to give 4-amino-3-benzyl-1,2,3-triazole-5-carboxamide followed by debenzylation[3]) followed by methylation and deformylation.

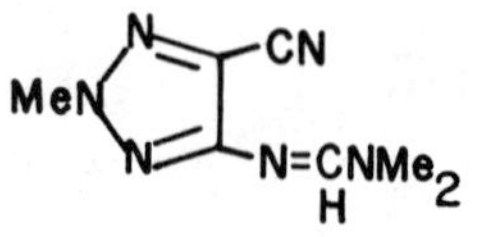

1
(178.2)

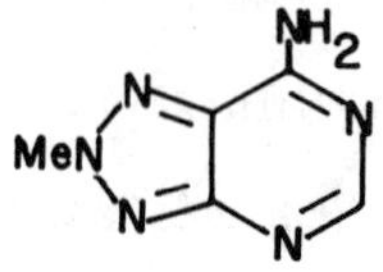

2
(150.2)

Examples of the general application of the reaction are the conversion of: (a) the amidine (1) into 8-methyl-6-(methylamino)-8-azapurine with aqueous methylamine acetate[2] and into the 6-butylamino analog with butylamine acetate[4] and (b) the 1-methyl isomer of *1*, i.e., 5-cyano-4-(dimethylaminomethylene)amino-1-methyl-1,2,3-triazole with ammonium (and methylamine) acetate to give 6-amino-(and 6-methylamino)-7-methyl-8-azapurine, respectively. The reaction does not proceed with 3-alkyltriazoles sufficiently for preparative use. Hence, 6-amino-9-methyl(and 9-benzyl)-8-azapurines are better prepared by the action of formamidine acetate on 4-amino-5-cyano-3-methyl(and 3-benzyl)-1,2,3-triazole[5]; this exemplifies another general reaction.

The 6-amino-8-azapurines are of interest in connection with the adenine-reversible inhibition of the growth of bacteria, fungi, 8-azaguanine-resistant malignant tumors, and viruses.[5]

PROCEDURE

5-Cyano-4-(dimethylaminomethylene)amino-2-methyl-1,2,3-triazole (1)

4-Amino-2-methyl-1,2,3-triazole-5-carboxamide[1] (1.4 g, 0.01 mol) is stirred at 25° with *N,N*-dimethylformamide (5 ml) while phosphoryl chloride (1.7 ml, 0.02 mol) is added dropwise. The viscous solution is then stirred at 25° for 2 hr. Ice (8 g) is added, and when it has melted and the mixture is homogeneous the container is placed in a bath maintained at 10°. The mixture is stirred while aqueous 14M ammonia (≃6 ml) is added to pH 5.5 and is filtered after 15 min. The precipitate is dried at 20° and is then heated with boiling benzene (10 parts) under reflux for 15 min. The suspension is cooled, kept for 1 hr at ≃25°, and filtered. The filtrate is evaporated to dryness to give the amidine (*1*) in 85% yield, m.p. 92° (from 40 parts of water, or 80 of cyclohexane). At 25°, 1 part is soluble in 5 parts of benzene, or 22 parts of ethanol.

6-Amino-8-methyl-8-azapurine (2)

A mixture of amidine (*1*) (1.8 g, 0.01 mol), ammonium acetate (3 g, 0.04 mol), and water (30 ml) is boiled under reflux for 30 min, refrigerated, and filtered, giving an 80% yield of 6-amino-8-methyl-8-azapurine, m.p. 266° (from 50 parts of boiling water).

REFERENCES

1. A. Albert, *J. Chem. Soc.*, C, 2076 (1968).

2. A. Albert, *J. Chem. Soc. Perkin Trans.*, I, 461 (1972).

3. J. R. Hoover and A. R. Day, *J. Am. Chem. Soc.*, 78, 5832 (1956).

4. A. Albert, *J. Chem. Soc. Perkin Trans.*, I, 2030 (1974).

5. A. Albert, *J. Chem. Soc. Perkin Trans.*, I, 345 (1975).

[21] 9-BENZYL-1,6-DIHYDRO-8-AZAPURINE (3-BENZYL-6,7-DIHYDRO-1,2,3-TRIAZOLO[4,5-*d*]PYRIMIDINE)

General Method for Preparing 1,6-Dihydro-8-azapurines, from 4-Amino-5-(aminomethyl)-1,2,3-triazoles and an Amidine

Adrien Albert

Department of Pharmacological Sciences, Health Sciences Center, State University of New York at Stony Brook, New York 11794

INTRODUCTION

The well-known instability of 1,6-dihydropurines toward hydrolysis and oxidation[1] lends interest to the study of their analogs, the 1,6-dihydro-8-azapurines (e.g., *2*), which are much more stable. Previously, these had been obtained only by reduction of 8-azapurines, and the position of the hydrogen atoms was assigned by NMR spectroscopy. In this way the parent substance (1,6-dihydro-8-azapurine)[2] and its 2-methyl,[2] 2-mercapto,[2] and 2-hydroxy[3] derivatives were made, and they comprised the sole known examples. In a more convenient procedure,[4] 4-amino-5-(aminomethyl)-1,2,3-triazoles are condensed with aliphatic amidines. The preparation of 9-benzyl-1,6-dihydro-8-azapurine (*2*) from 4-amino-5-(aminomethyl)-3-benzyl-1,2,3-triazole (*1*) and formamidine acetate is described, together with that of the intermediate triazole[5] (*1*).

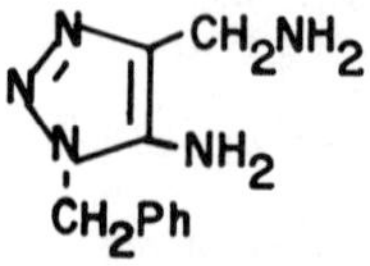

1
(203.2)

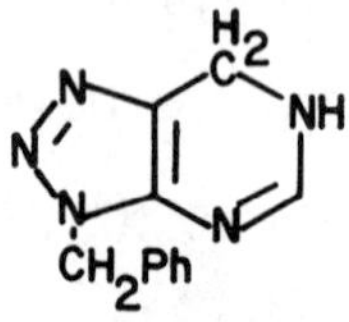

2
(213.2)

The new reaction has also been used to prepare 1,6-dihydro-7-methyl-, -8-methyl-, -2,7- and -2,8-dimethyl-, 9-benzyl-1,6-dihydro-2-methyl-, and -2-(trichloromethyl)-8-azapurine, all in excellent yields.[4]

PROCEDURE

4-Amino-5-(aminomethyl)-3-benzyl-1,2,3-triazole (1)

4-Amino-3-benzyl-5-cyano-1,2,3-triazole[a] (4.8 g, 24 mmol) and Raney nickel (9.6 g, wet weight) are hydrogenated in ethanolic, 3M ammonia (180 ml) for 5 hr at 70° and 4 atm. The suspension is filtered, the solid is boiled with ethanol (50 ml), and the filtrates are combined, evaporated to dryness, and dried at 60°/25 torr. The resultant glass is homogenized in hot water (48 ml), and the suspension is refrigerated and then filtered from the corresponding secondary amine[5] (15% discarded). The filtrate and washings are combined and mixed at 100° with 0.5M phosphoric acid sufficient to give pH 7.5, and the suspension is chilled and

---

[a]A. Albert, J. *Chem. Soc.*, (C), 230 (1970).

filtered. The precipitated phosphate $[(C_{10}H_{13}N_5)_2 \cdot H_3PO_4]$ is washed with water, suspended in boiling water (20 ml), 10M sodium hydroxide (≃2 ml) is added until pH 12 is reached and maintained, and the solution is cooled and extracted with chloroform (3 × 50 ml). The chloroform extracts are combined, dried (potassium carbonate) and evaporated, giving compound *1* in 66% yield. It is usually pure (m.p. 105°),[b] but may be recrystallized from 30 parts of benzene to yield two crops.

9-Benzyl-1,6-dihydro-8-azapurine (2)

Compound *1* (2.03 g, 0.01 mol), formamidine acetate[c] (3.1 g, 0.03 mol), and sieve-dried butanol (20 ml) are heated under reflux for 4 hr and the solvent is removed at 90°/25 torr. The residue is stirred with warm water (20 ml) and chilled to deposit *2* in 85% yield. It may be recrystallized from 60 parts of benzene or 15 parts of water, but with the latter the operation must be performed quickly to prevent hydrolysis to the starting material (*1*); compound *2* has m.p. 142° when pure.

REFERENCES

1. J. H. Lister, *The Purines*, Wiley-Interscience, London, 1971, p. 433.

---

[b]A low-melting product signifies incomplete hydrogenation. A product that does not melt completely has probably absorbed carbon dioxide from the air.

[c]Fluka's product was used; this intermediate is also available from Aldrich. It was dried in air at 110°.

2. A. Albert, *J. Chem. Soc.*, B, 427 (1966).

3. A. Albert and W. Pendergast, *J. Chem. Soc. Perkin Trans.*, I, 457 (1972).

4. A. Albert, *J. Chem. Soc. Perkin Trans.*, I, 345 (1975).

5. A. Albert, *J. Chem. Soc. Perkin Trans.*, I, 1634 (1973).

[22] 1-SUBSTITUTED 5-AZAURACILS AND 1-SUBSTITUTED 5-AZA-2,4-DITHIOURACILS

A General Method for the Preparation of 1-Substituted s-*Triazine-2,4(1*H*,3*H*)-diones and the Corresponding* s-*Triazine 2,4(1*H*,3*H*)-dithiones*

Darrell E. O'Brien, Fred Baiocchi, and C. C. Cheng

Midwest Research Institute, Kansas City, Missouri 64110

INTRODUCTION

Although 5-azauracil [*s*-triazine-2,4(1*H*,3*H*)-dione] can be obtained by mild oxidation of uric acid,[1] *N*-substituted uric acids under similar reaction conditions fail to yield *N*-substituted 5-azauracils. 1-Methyl-5-azauracil was prepared by the base-catalyzed reaction of ethyl formate with 1-methylbiuret.[2] Because of the instability of 1-substituted 5-azauracils in aqueous base, it is important that the reaction be performed under anhdyrous conditions. It was found that with the use of propyl alcohol as the solvent, a number of 1-alkyl- and 1-(aralkyl)-5-azauracils can be readily prepared.[3] The corresponding 1-substituted 5-aza-2,4-dithiouracils are prepared by the condensation of two equivalents of thiocyanic acid with one equivalent of an appropriate isocyanide at low temperature.[4]

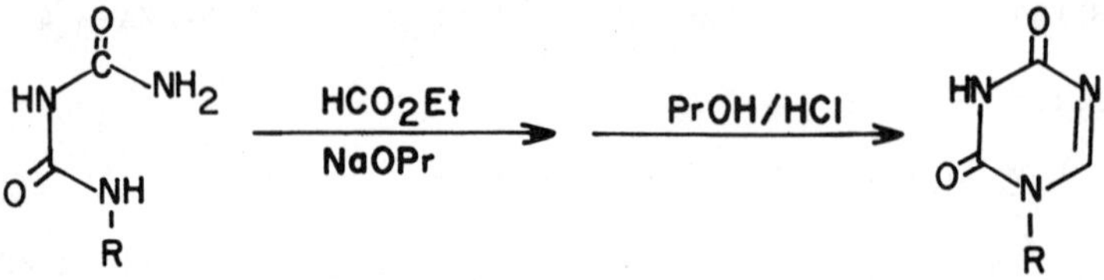

1a (131.1) R=Et

1b (185.2) R=cyclohexyl

1c (183.2) R=furfuryl

2a (141.1) R=Et

2b (195.2) R=cyclohexyl

2c (193.2) R=furfuryl

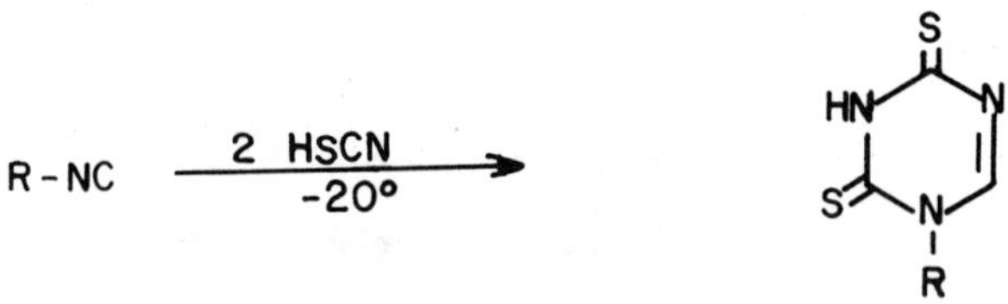

3a (85.1) R=cyclopentyl

3b (99.2) R=cyclohexyl

3c (123.1) R=p-methoxyphenyl

4a (213.3) R=cyclopentyl

4b (227.4) R=cyclohexyl

4c (251.3) R=p-methoxyphenyl

PROCEDURE

1-Furfurylbiuret (1c)

A mixture of furfurylamine (19.4 g, 0.2 mol) and nitrobiuret[5]

(29.6 g, 0.2 mol) in water (500 ml) is heated for 2 hr at 85°. The resulting solution is boiled under reflux for 30 min, treated with charcoal, and filtered. On cooling, the filtrate deposits 22.6 g (62% yield) of *1c*, m.p. 151° to 153°. Recrystallization from water raises its melting point to 153° to 154°.

1-Ethyl-5-azauracil (2a)

To a solution prepared from sodium (2.8 g, 0.1 g-atom) and 200 ml of 1-propanol is added a solution of 1-ethylbiuret[5,6] (*1a*) (16 g, 0.12 mol) in hot (55°) 1-propanol (150 ml). The clear yellow solution is heated until refluxing, and ethyl formate (13.6 g, 0.18 mol) is added dropwise. The resulting mixture is boiled under reflux with gentle stirring for 5 hr and is then allowed to cool for 14 hr. To the mixture is added, with vigorous stirring, a freshly prepared solution of hydrogen chloride in 1-propanol. The mixture is boiled under reflux for 15 min, and the precipitated sodium chloride removed by filtration. Evaporation of the filtrate gives crystals that, after crystallization from ethanol, afford 16.9 g (72% yield) of *2a*, m.p. 133° to 135°; λ_{max}^{EtOH} in nm ($\varepsilon_{\underline{mM}}$) 243 (3.38).

1-Cyclohexyl-5-azauracil (2b)

Compound *2b* is prepared from 1-cyclohexylbiuret[7] (*1b*) (22.2 g) and ethyl formate (13.6 g) in 1-propanol as for the preparation of *2a*. The product is recrystallized from ethanol, to give a 48% yield of

2b, m.p. 265° to 266°, λ_{max}^{EtOH} in nm ($\varepsilon_{\underline{mM}}$) 247 (3.42).

1-Furfuryl-5-azauracil (2c)

Compound *2c* is prepared from 1-furfurylbiuret (*1c*) (22.0 g) and ethyl formate (13.6 g) in 1-propanol as for the preparation of *2a*. The product is recrystallized from ethanol to give a 42% yield of *2c*, m.p. 186° to 187°, $\lambda_{max}^{pH\ 1}$ in nm ($\varepsilon_{\underline{mM}}$) 240 (3.36).

Cyclopentyl Isocyanide (3a)

A solution of *N*-cyclopentylformamide[8] (76.3 g, 0.68 mol) in pyridine (340 ml) and heptane (210 ml) is treated with phosphoryl chloride (62.1 g, 0.41 mol) by the method of Ugi and Meyr.[9] The pure isocyanide (*3a*), which weighs 36 g (56% yield), is obtained by distillation at 48° to 49°/8 torr.

1-Cyclopentyl-5-aza-2,4-dithiouracil (4a)

To a solution of thiocyanic acid (11.8 g, 0.2 mol) in anhydrous ether (200 ml) cooled to -20° is added dropwise, with vigorous stirring, cyclopentyl isocyanide (*3a*) (8.5 g, 0.1 mol). The resulting solution is stirred for 2 hr at -20° ± 2°; during this time, the product (*4a*) gradually precipitates from the mixture. The product is removed by filtration, washed with anhydrous ether (3 × 50 ml), and recrystallized from aqueous ethanol to give 13.2 g (62% yield) of *4a*, m.p. 190° to 191°; λ_{max}^{EtOH} in nm ($\varepsilon_{\underline{mM}}$) 248 (4.14) and 290 (4.15), $\lambda_{max}^{pH\ 1}$ in nm ($\varepsilon_{\underline{mM}}$) 276 (3.32).

1-Cyclohexyl-5-aza-2,4-dithiouracil (*4b*)

Compound *4b* is prepared from cyclohexyl isocyanide (*3b*) (10 g) and thiocyanic acid (11.8 g) as for the preparation of *4a*. Recrystallization from aqueous ethanol gives a 64% yield of *4b*, m.p. 202° to 204°; λ_{max}^{EtOH} in nm ($\varepsilon_{\underline{mM}}$) 248 (4.05) and 288 (4.17), $\lambda_{max}^{pH\ 1}$ in nm ($\varepsilon_{\underline{mM}}$) 251 (4.07) and 280 (4.19).

1-(p-Methoxyphenyl)-5-aza-2,4-dithiouracil (*4c*)

Compound *4c* is prepared from *p*-(methoxyphenyl) isocyanide (12.3 g) and thiocyanic acid (11.8 g) as for the preparation of *4a*. Recrystallization from ethanol gives a 54% yield of *4c*, m.p. 202° to 203°; λ_{max}^{EtOH} in nm ($\varepsilon_{\underline{mM}}$) 290 (4.32), $\lambda_{max}^{pH\ 1}$ in nm ($\varepsilon_{\underline{mM}}$) 282 (4.39).

REFERENCES

1. H. Brandenberger and R. H. Brandenberger, *Helv. Chim. Acta*, 37, 2207 (1954).

2. A. Piskala and J. Gut, *Collect. Czech. Chem. Commun.*, 26, 2519 (1961).

3. D. E. O'Brien, F. Baiocchi, and C. C. Cheng, *Biochemistry*, 2, 1203 (1963).

4. I. Ugi and K. Rosendahl, *Angew. Chem.*, 73, 656 (1961).

5. J. Thiele and E. Uhlfelder, *Justus Liebigs Ann. Chem.*, 303, 93 (1898).

6. F. B. Dains and E. Wertheim, *J. Am. Chem. Soc.*, 42, 2303 (1920).

7. J. Bougault and J. Leboucq, *Bull. Soc. Chim. Fr.*, 47, 594 (1930).

8. A. N. Kost and G. A. Shvekhgeimer, *Vestn. Mosk. Univ.*, 5, No. 9, *Ser. Fiz.-Mat. Estestven Nauk*, No. 6, 51 (1950).

9. I. Ugi and R. Meyr, *Chem. Ber.*, 93, 239 (1960).

[23] 4-ACETAMIDOPYRAZOLO[3,4-*d*]PYRIMIDINE AND 4-DIMETHYLAMINOMETHYLENEAMINOPYRAZOLO[3,4-*d*]PYRIMIDINE[a]

The Synthesis of Certain Pyrazolopyrimidine Heterocycles Related to 4-APP

Raymond P. Panzica and Leroy B. Townsend

Department of Chemistry and Department of Biopharmaceutical Sciences, University of Utah, Salt Lake City, Utah 84112

INTRODUCTION

In a quest for pyrazolo[3,4-*d*]pyrimidines that might be candidates in glycosylation reactions, three substances were synthesized: (a) 4-acetamido-1(2)-acetylpyrazolo[3,4-*d*]pyrimidine (*2*), (b) 4-acetamidopyrazolo[3,4-*d*]pyrimidine (*3*), and (c) 4-[(dimethylamino)methylene]aminopyrazolo[3,4-*d*]pyrimidine (*4*). Like many 4-substituted pyrazolo[3,4-*d*]pyrimidines,[1,2] the heterocycles *2* and *4* exhibited antitumor activity.[a] Both *2* and *4* were

---

[a]This research was supported by research contract NO1-CM-43806 from the Division of Cancer Treatment, National Cancer Institute, National Institutes of Health, Department of Health, Education, and Welfare.

[b]Preliminary results against lymphoid leukemia L-1210 are as follows: *2*; host BDF_1; dose (mg/kg), 200 mg; T/C = 125; *4*; host, BDF_1; dose (mg/kg), 25 mg; T/C = 147. The evaluation of the activity is in accordance with the criteria of the Division of Cancer Treatment, National Cancer Institute, National Institutes of Health, Department of Health, Education, and Welfare.

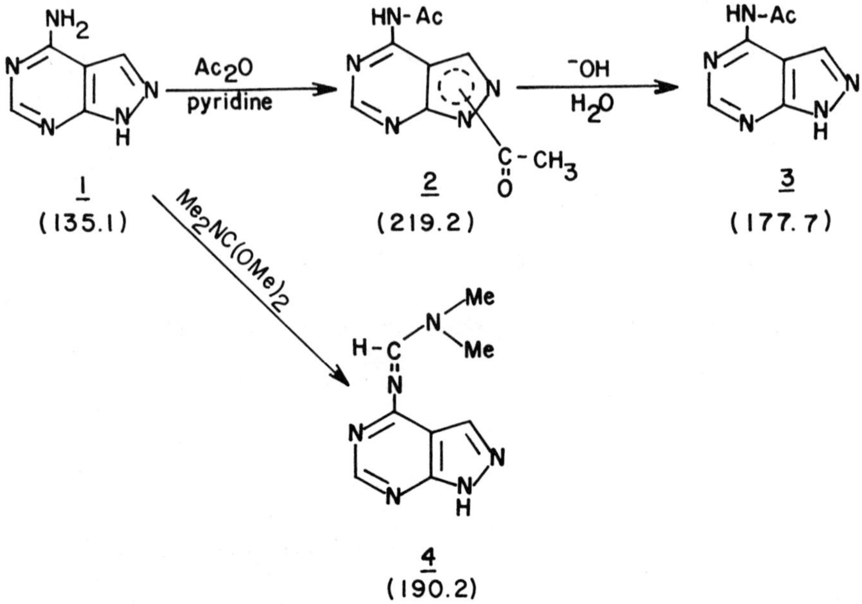

synthesized directly from 4-aminopyrazolo[3,4-*d*]pyrimidine (*1*).[3] Mild, basic hydrolysis of *2* furnished *3* in almost quantitative yield.

PROCEDURE

4-Acetamido-1(2)-acetylpyrazolo[3,4-d]pyrimidine (2)

To a suspension of 4-aminopyrazolo[3,4-*d*]pyrimidine (*1*) (5.0 g, 37 mmol)[c] in dry pyridine (50 ml) is added acetic anhydride (25 ml) in one portion, and the mixture is stirred and boiled under

---

[c]Dried for 6 hr in a vacuum oven at 120°.

reflux[d] under anhydrous conditions for 1 hr. The mixture is allowed to cool, the excess solvents are removed *in vacuo* (water bath at 50°), the greyish residue is coevaporated with xylene (3 × 30 ml) to remove the last traces of pyridine and acetic anhydride, and the solid is dissolved[e] in hot dimethyl sulfoxide (Me_2SO, 100 ml), treated with charcoal (Norit), and the suspension filtered through a dry-packed pad of Celite. The filter pad is washed with hot Me_2SO (2 × 10 ml). The solution (≃120 ml) is allowed to cool to room temperature, distilled water (40 ml) is added, crystallization is induced by scratching the inside of the beaker with a glass rod, and the suspension is kept for 7 hr. Crystalline *2* is removed by filtration, washed with distilled water (4 × 25 ml), and air-dried; yield 5.9 g (73%)[f] of white needles, m.p. 224° to 226°; $\lambda_{max}^{pH\ 1}$ nm ($\varepsilon_{\underline{mM}}$) 274 (14.25), 228.5 (22.27), $\lambda_{min}^{pH\ 1}$ nm, 245.5 (5.17), λ_{max}^{MeOH} nm, 228 sh (7.45), 278 sh (9.03), 267 (10.28), 227.5 (10.24), λ_{min}^{MeOH} nm, 242 (3.97), $\lambda_{max}^{pH\ 11}$

[d]When the reflux temperature is reached (≃10 min), the suspension turns yellow, and becomes very thick, almost stopping the stirring (magnetic). Continued heating of the thick suspension effects dissolution. This transition requires approximately 25 to 30 min, and the reddish solution is refluxed for an additional 30 min (total time, 1 hr).

[e]This step is optional. The solid is sufficiently pure (m.p. 221° to 223°) for use in other reactions. When the reaction is stopped at this point, the solid is removed by filtration, thoroughly washed with water, and air dried; yield 7.3 g (90%). Proton magnetic resonance (PMR) and UV spectra are identical to those for the recrystallized product.

[f]The average yield from this procedure is 75%.

nm, 313.5 sh (3.51), 275 (8.48), 233.5 (16.11), $\lambda_{min}^{pH\ 11}$ nm, 251 (4.16); PMR data (Me_2SO-d_6): δ 2.32 (s, 3, CH_3), 285 (s, 3, CH_3), 8.72 and 8.92 (2 s, 1, H-3,6), and 11.56 (s, 1, NH).

Analysis calculated for $C_9H_9N_5O_2$: C, 49.31; H, 4.14; N, 31.95. Found: C, 49.12; H, 4.43; N, 31.99.

4-Acetamidopyrazolo[3,4-d]*pyrimidine* (*3*)

A suspension of 4-(acetamido)-1(2)-acetylpyrazolo[3,4-*d*]pyrimidine (*2*) (2.0 g, 9.14 mmol) in M sodium hydroxide (20 ml)[g] is stirred at room temperature; dissolution is effected in ≈5 min, and the beaker containing the pink-to-purple solution is then placed in an ice-bath and the solution stirred for an additional 5 min. Next, the pH of the cold solution is carefully adjusted to 6 by dropwise addition of glacial acetic acid.[h] During this process *3* is precipitated. Water (5 ml) is added to the suspension, which is stirred with cooling for 10 min. The solid is removed by filtration, washed thoroughly with cold distilled water, and dried at 100° under vacuum; yield 1.55 g (96%), m.p. 292° to 294° (dec.) $\lambda_{max}^{pH\ 1}$ nm, (ε_{mM}) 274 sh (7.39), 262 (9.27), 233 sh (9.39), $\lambda_{min}^{pH\ 1}$ nm, 241.5 (5.49), λ_{max}^{MeOH} nm, 301 sh (1.74), 282.5 sh (7.25), 270.5 (10.15), 258 (9.58), λ_{min}^{MeOH} nm, 237.5 (4.30), $\lambda_{max}^{pH\ 11}$ nm, 301 sh (2.55), 281 sh (6.55), 266 (8.33), 258

---

[g] Two milliequivalents per millimole are required.

[h] Short-range pH paper is used for monitoring the neutralization.

sh (8.03), 227.5 (10.13), $\lambda_{min}^{pH\ 11}$ nm, 246 (5.88); PMR (Me_2SO-d_6): δ 2.34 (s, 3, CH_3), 8.58 and 8.73 (2s, 1, H-3,6), 11.28 and 13.99 (2 s, 1, NH).

Analysis calculated for $C_7H_7N_5O$: C, 47.46; H, 3.98; N, 39.53. Found: C, 47.17; H, 3.87; N, 39.43.

4-[(Dimethylamino)methylene]aminopyrazolo[3,4-d]*pyrimidine* (*4*)

To a suspension of 4-aminopyrazolo[3,4-*d*]pyrimidine (*1*) (2.7 g, 20 mmol) in dry *N,N*-dimethylformamide (50 ml) is added *N,N*-dimethylformamide dimethyl acetal (10 ml, 100 mmol) in one portion. The suspension is stirred at room temperature under anhydrous conditions. After 1.5 hr dissolution occurs, stirring is continued for 16.5 hr (total 18 hr), and the solution is evaporated *in vacuo* (water bath, 40°) to a light-yellow, microcrystalline solid. This solid is triturated with cold ethanol (30 ml), the suspension is filtered, and the solid is washed with cold ethanol (10 ml), and air-dried; yield, 2.85 g. The filtrate and washings are combined, evaporated, and refrigerated (4°) to afford an additional[i] 0.29 g of *4*; yield 3.85 g (83%), m.p. 208° to 210°; $\lambda_{max}^{pH\ 1}$ nm (ε_{mM}) 324 sh (22.25), 315.5 (23.28), and 232.5 sh (11.53); $\lambda_{min}^{pH\ 1}$ nm, 252 (6.62); λ_{max}^{MeOH} nm, 312 (27.73), and 231.5

---

[i]Examination of the stability of *4* in ethanol and aqueous solutions at room temperature, monitored by thin-layer chromatography on SilicAR 7GF with 7:3 (v/v) chloroform—methanol, shows no breakdown of *4* to occur in ethanol after 16 hr; however, during the same period breakdown to *1* is observed in aqueous solution.

(9.36); λ_{min}^{MeOH} nm, 256 (4.95) and 222 (8.45); $\lambda_{max}^{pH\ 11}$ nm, 301 (21.34), 258.5 (8.22), and 225 (14.46); $\lambda_{min}^{pH\ 11}$ nm, 269 (7.04) and 245 (6.77); PMR data ($Me_2SO\text{-}d_6$): δ 3.23 (s, 3, CH_3), 3.28 (s, 3, CH_3), 8.22 and 8.55 (2 s, 1, H-3,6), 9.01 (s, 1, =CH), and 13.64 (s, 1, NH).

Analysis calculated for $C_8H_{10}N_6$: C, 50.52; H, 5.30; N, 44.18. Found: C, 50.88; H, 5.36; N, 44.02.

REFERENCES

1. J. M. Venditti, E. Frei, and A. Goldin, *Cancer*, 13, 959 (1960).

2. J. F. Henderson and K. G. Junga, *Cancer Res.*, 21, 173 (1961).

3. R. K. Robins, *J. Am. Chem. Soc.*, 75, 6407 (1957).

[24] SUBSTITUTED 4-(ALKYLAMINO)PYRAZOLO[3,4-*d*]PYRIMIDINES

Synthesis of 2-Methyl-4-(3-methylbutylamino)pyrazolo[3,4-d]pyrimidine-3-carbonitrile

R. Bruce Frye and Dieter Werner

Department of Chemistry, Massachusetts Institute of Technology, Cambridge, Massachusetts 02139

INTRODUCTION

The 4-(alkylamino)pyrazolo[3,4-*d*]pyrimidines have demonstrated a wide range of biological activities.[1-5] These compounds can be prepared by several methods,[6-8] and a convenient and concise, synthetic route from the readily available 3-amino-1-methylpyrazole-4,5-dicarbonitrile[9-11] (*1*) is described here. The preparation of 1-methyl-4-(3-methylbutylamino)pyrazolo[3,4-*d*]pyrimidine-3-carbonitrile[a] (*3*) from 3-amino-1-methylpyrazole-4,5-dicarbonitrile and

---

[a]Conversion of the carbonitrile into the corresponding 3-carboxamide is effected by suspending the material in concentrated ammonium hydroxide and heating for 2 hr at 70°. Evaporation affords a residue that is crystallized from methanol.

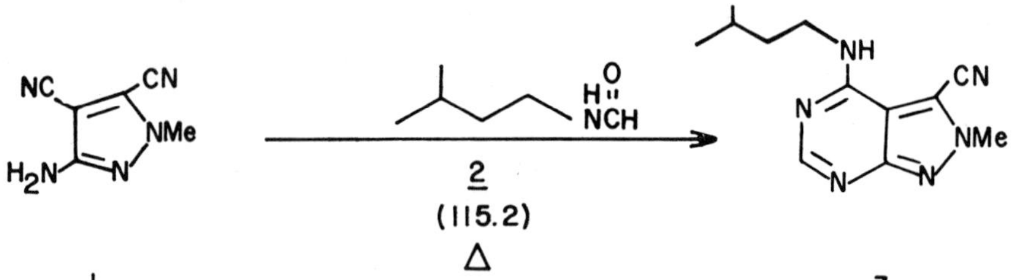

1 (147.1)

3 (244.3)

N-(3-methylbutyl)formamide[b,c] illustrates the general[d] procedure.[12]

PROCEDURE

N-*(3-Methylbutyl)formamide (2)*

To 3-methylbutylamine (2.25 g, 3.0 ml, 34.4 mmol), cooled to 0°, is added formic acid (4.5 ml) at a rate that maintains the temperature of the solution below 60°. The solution is then heated for 5 hr at 50°, and the excess of formic acid is removed by distillation at atmospheric pressure. The residue is distilled under diminished pressure to afford *N*-(3-methylbutyl)formamide as a liquid (2.24 g, 76%), b.p. 72°/0.15 torr; *m/e* 115, 100, 86, 72, 58, and 44.

---

[b]The condensation is also successful with *N*-methylformamide, *N*-ethylformamide, *N*-methylacetamide, *N*-(3-methyl-2-butenylformamide, and *N*-hexylformamide.

[c]Other *N*-substituted formamides are commercially available or may be prepared from formic acid and the appropriate amine.

[d]The alkylformamides also condense with aminopyrrole carbonitriles, to afford 4-(alkylamino)pyrrolo[2,3-*d*]pyrimidines, although these reactions require somewhat longer reaction times, and the yields tend to be lower.

2-Methyl-4-(3-methylbutylamino)pyrazolo[3,4-d]pyrimidine-3-carbonitrile (3)

A solution of 5-amino-1-methylpyrazole-3,4-dicarbonitrile[10] (*1*) (175 mg, 1.19 mmol) in *2* (2.5 ml) is boiled under reflux for 1 hr, the excess of amide *2* is removed under diminished pressure, and the residue is purified by chromatography on neutral silica gel (elution with ether) to afford *3* as white crystals; yield 192 mg (66%), m.p. 100.5° to 103°; λ_{max}^{EtOH} (pH 1) 322 nm ($\epsilon_{m\underline{M}}$ 16.00), 311 (17.90), 281 (8.70), 272 (sh), 259 (6.60), and 233 (sh), λ_{min} 318 (14.50), 285 (8.30), 260 (6.60), and 241 (5.70); λ_{max}^{EtOH} (pH 7) 320 (10.70), 255 (8.30) and 230 (12.80); λ_{min} 285 (4.00), 243 (5.70), and 212 (10.00); λ_{max}^{EtOH} (pH 12) 314 (11.10) and 263 (10.40), λ_{min} 290 (7.30) and 244 (7.90); λ_{max}^{Nujol} 3400, 2960 to 2850, 2240, 1615, and 1560 cm^{-1}; *m/e* 244, 229, 201, 188, 174, 149, and 147.

Analysis calculated for $C_{12}H_{16}N_6 \cdot 0.5\ H_2O$: C, 56.89; H, 6.76. Found: C, 56.60; H, 7.15.

REFERENCES

1. R. K. Robins, *J. Am. Chem. Soc.*, 78, 784 (1956).

2. F. R. White, *Cancer Chemother. Rep.*, 3, 26 (1959).

3. J. L. Scott, Jr., and L. V. Foye, *Cancer Chemother. Rep.*, 20, 73 (1962).

4. R. K. Robins, *J. Med. Chem.*, 7, 186 (1964) and references cited therein.

5. S. M. Hecht, R. B. Frye, D. Werner, S. D. Hawrelak, F. Skoog, and R. Y. Schmitz, *J. Biol. Chem.*, 250, 7343 (1975).

6. R. K. Robins, in R. C. Elderfield (ed.), *Heterocyclic Compounds*, Wiley, New York, Vol. 8, 1967, pp. 406 to 421 and references cited therein.

7. E. C. Taylor and P. K. Loeffler, *J. Am. Chem. Soc.*, 82, 3147 (1960).

8. P. Schmidt, K. Eichenberger, M. Wilhelm, and J. Druey, *Helv. Chim. Acta*, 42, 763 (1959).

9. C. L. Dickinson, J. K. Williams, and B. C. McKusick, *J. Org. Chem.*, 29, 1919 (1964) and references cited therein.

10. S. M. Hecht, D. Werner, D. D. Traficante, M. Sundaralingam, P. Prusiner, T. Ito, and T. Sakurai, *J. Org. Chem.*, 40, 1815 (1975).

11. R. A. Earl, R. J. Pugmire, G. R. Revankar, and L. B. Townsend, *J. Org. Chem.*, 40, 1822 (1975).

12. S. M. Hecht and D. Werner, *J. Chem. Soc. Perkin Trans.*, I, 1903 (1973).

[25] 3,6-DIMETHYLPYRAZOLO[4,3-*d*]PYRIMIDIN-7-ONE[a]

Synthesis of Pyrazolo[4,3-d]pyrimidines via the Versatile Intermediate Ethyl 3-Methyl-4-nitropyrazole-5-carboxylate

Arthur F. Lewis and Leroy B. Townsend

Division of Medicinal Chemistry, Department of Biopharmaceutical Sciences and Department of Chemistry, University of Utah, Salt Lake City, Utah 84112

INTRODUCTION

Ethyl 3-methyl-4-nitropyrazole-5-carboxylate (*6*) has proved to be a versatile intermediate for the synthesis of various 3-methylpyrazolo[4,3-*d*]pyrimidines.[1-3] These 3-methylpyrazolo[4,3-*d*]-pyrimidines have been used extensively as model compounds for the characterization of compounds derived from the nucleoside antibiotics formycin and formycin B. The procedure given here is for the large-scale preparation of *6* and includes several improvements over procedures previously published.[1,2] The utility of this intermediate (*6*) is illustrated by the synthesis of 3,6-dimethylpyrazolo[4,3-*d*]pyrimidin-7-one (*9*).

---

[a]This work was supported by Research Contract NO1-CM-43806 from the Division of Cancer Treatment, National Cancer Institute, National Institutes of Health, Department of Health, Education, and Welfare.

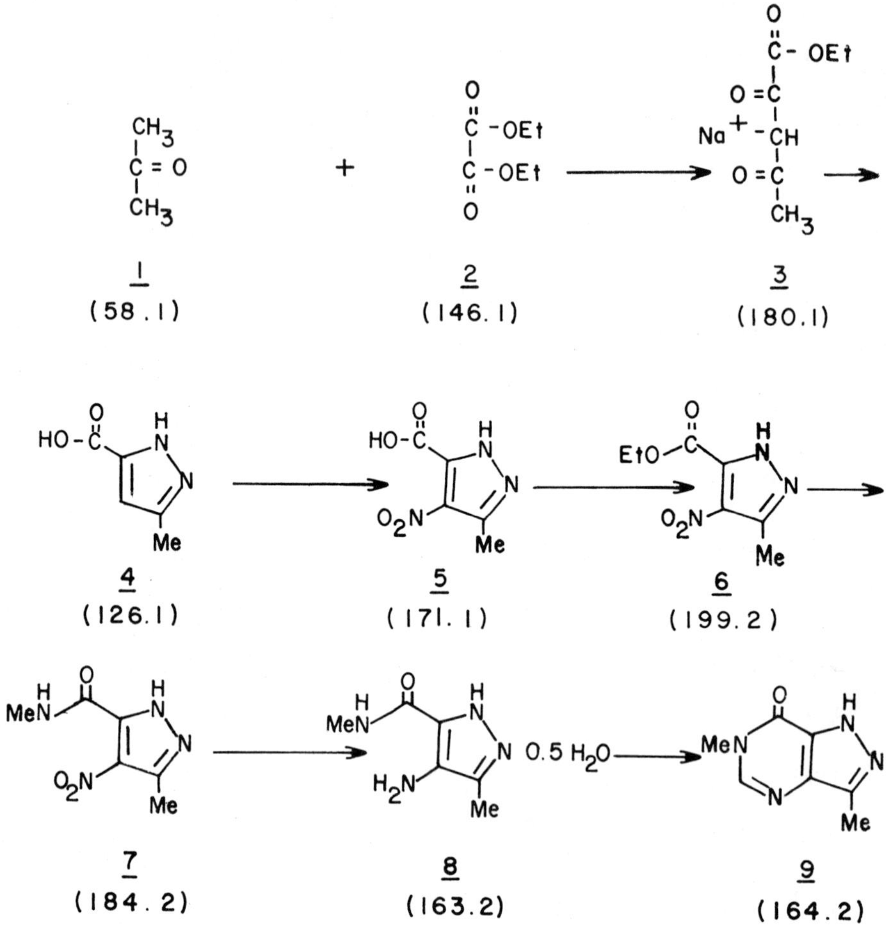

PROCEDURE

Sodium Ethyl Acetopyruvate[4] *(3)*

Sodium metal (500 g, 21.8 mol) is added to absolute ethanol (11.2 1) at a rate that maintains brisk refluxing,[b] and the resulting solution is allowed to cool to 40° to 45°. A mixture

---

[b]The reaction is conducted in a 22-1, round-bottomed flask fitted

of diethyl oxalate (*2*) (2.69 l, 2.84 kg, 19.4 mol) and anhydrous *1* (AR) (1.45 l, 1.16 kg, 19.9 mmol) is added, with vigorous stirring, at a rate that maintains a reaction temperature of 40° to 45°. The thick, yellow paste is stirred for an additional 20 min and then poured into 16 l of vigorously stirred, anhydrous diethyl ether.[c] The yellow mixture is stirred for 10 min, and compound *3* is collected by filtration.[d] The filter cake is washed with ether (≃8 l), pressed as dry as possible, and air dried. The yield averages 2.5 to 3 kg (72% to 86%).

3-Methylpyrazole-5-carboxylic Acid (4)

A solution of compound *3* (2.745 kg, 15.2 mol) in water (7.4 l) is cooled[e] to 27° and, with good stirring, 85% hydrazine hydrate (872 ml) is added slowly while the reaction temperature is kept[e] at 25° to 30°. The solution is then stirred at 25° to 30° for 30 min, the cooling bath is removed, a solution of potassium hydroxide (207 g) in water (370 ml) is slowly added, and the solution

---

with a removable, four-necked head, and equipped with a heavy-duty mechanical stirrer driven by an air-powered stirring motor (the evolution of copious amounts of hydrogen *precludes* the use of an electrical stirring motor). Two large condensers are used during the addition of the sodium. One condenser is then replaced by a large pressure-equalizing dropping funnel for the addition of the diethyl oxalate-acetone. The reaction is protected from moisture by drying tubes (calcium chloride).

[c]A large stainless-steel or polyethylene bucket is used, and an air-driven stirring motor is again required.

[d]Filtered on two 32-cm, Büchner funnels.

[e]A dry ice—acetone bath maintained at 0° is used.

is stirred for 3 hr at room temperature and acidified to pH 2 with cold, dilute sulfuric acid while the temperature is kept[e] below 20°. The solid is collected by filtration, washed with ice-water (5 × 2 l), and air-dried to yield 1.216 kg (9.64 mol, 63%) of *4*, m.p. 233° to 235°.

3-Methyl-4-nitropyrazole-5-carboxylic Acid (5)

A mixture of 90% nitric acid (1.1 l) and fuming sulfuric acid (15% SO_3) (940 ml) is stirred and cooled to 20° to 25°, and compound *4* (1.126 kg, 8.94 mol) is slowly added, with continuous stirring and cooling, while the temperature[f,g] is maintained at 20° to 25°. The mixture is stirred for 1 hr at room temperature, heated on a steam bath for 3 hr with vigorous stirring,[h] and poured, with efficient stirring, onto an excess of crushed ice. The solid product is collected by filtration, washed with cold water until neutral, and air dried to give 656 g (3.84 mol, 43%). The m.p. ranges from 190° to 210° for various preparations and appears to be dependent on the amount of acid removed by the water.

---

[f]It is critical that this temperature be maintained by cooling during the addition of the solid *4*.

[g]This mixture is divided into halves, and the two portions are treated simultaneously, each in a 5-l, wide-necked, round-bottomed reaction-vessel. The halves are combined when they are poured onto crushed ice.

[h]Copious evolution of nitrogen oxides is to be expected. The solution should be only pale yellow when the reaction is complete.

Ethyl 3-Methyl-4-nitropyrazole-5-carboxylate (6)

A solution of compound 5 (374 g, 2.19 mol) in a mixture of absolute ethanol (3.74 l) and triethyl orthoformate (374 ml) is heated on a steam bath for 3 hr, while anhydrous hydrogen chloride is bubbled into the solution. The resulting solution is poured into a large evaporating dish, the solvent is allowed to evaporate spontaneously at atmospheric pressure, and the solid residue is washed with water[i] until the washings are neutral and is air-dried to furnish 339 g (1.70 mol, 78%) of *6*, m.p. 83° to 86°. The product may be recrystallized from benzene—petroleum ether.

3-Methyl-4-nitropyrazole-5-carboxylic Acid Methylamide (7)

Compound *6* (21 g, 105 mmol) is added to 40% aqueous methylamine (300 ml), and the mixture is heated on a steam bath under reflux condenser for 1.5 hr. The condenser is then removed, heating is continued for 0.5 hr, and the resulting solution is evaporated to dryness under diminished pressure. The residue is dissolved in water (70 ml), and the pH is adjusted to ≃6 by the addition, with stirring, of dilute hydrochloric acid. The mixture is heated to boiling, and sufficient water is slowly added to dissolve all of the solid (final volume, ≃500 ml). The boiling solution is treated with charcoal, the suspension is filtered, and the filter cake is washed with boiling water (75 ml). The filtrate and

---

[i]Additional material can be obtained by extraction of the water washings with ether.

washings are combined and refrigerated for 16 hr. The crystalline solid is collected by filtration, washed with water, and air-dried to yield 14.22 g (77 mmol, 73%) of 7, m.p. 216° to 220°. A second crop (1.6 g, 8.6 mmol; m.p. 210° to 218°) is obtained by concentrating the filtrate to ≈100 ml and refrigerating the concentrate. A pure sample (m.p. 218° to 220°) is obtained by recrystallizing from water (80% recovery) and drying *in vacuo* at 110°; λ_{max} in nm ($\varepsilon_{\underline{mM}}$), at pH 1, 277 (7.92); at pH 11, 317 (10.50); in methanol, 271 (7.00).

Analysis calculated for $C_6H_8N_4O_3$ (184.2): C, 39.14; H, 4.38; N, 30.42. Found: C, 39.21; H, 4.10; N, 30.61.

4-Amino-3-methylpyrazole-5-carboxylic Acid Methylamide, Hemihydrate (8)

Compound 7 (11 g, 59.7 mmol) is dissolved, by heating, in a mixture of ethanol (400 ml) and water (50 ml), and the solution is allowed to cool to room temperature. A slurry consisting of 10% Pd/C (1 g) in water (50 ml) is added, and the mixture is shaken in a hydrogen atmosphere (40 lb·in^{-2}, initial pressure) for 2 hr. The catalyst is removed by filtration (Celite bed), and washed with boiling, 3:2 (v/v) ethanol—water (200 ml). The filtrate and washing are combined and evaporated to dryness *in vacuo*. The residue is coevaporated with ethanol (2 × 50 ml), to afford 9 g (55.1 mmol, 92%) of a light-purple solid, m.p. 179° to 181°.

A pure sample (m.p. 181° to 182°) is obtained by three recrystallizations from water (60% recovery), followed by drying *in vacuo* at 80° λ_{max} in nm ($\varepsilon_{\underline{mM}}$), at pH 11, 232 (6.69), and 280 (4.98), in methanol, 280 (4.57). Thin-layer chromatography on SilicAR-7GF: with water R_7 0.73; in ethyl acetate—1-propanol—water (4:1:2, v/v, upper phase), R_7 0.60.

Analysis calculated for $C_6H_{10}N_4O \cdot 0.5\ H_2O$ (163.2): C, 44.16; H, 6.79; N, 34.34. Found: C, 44.21; H, 7.26; N, 34.17.

3,6-Dimethylpyrazolo[*4,3*-d]*pyrimidin-7-one* (*9*)

Method A. A mixture of compound *8* (1 g, 6.13 mmol) with 99% formamide) (6 ml) is placed in a preheated oil bath at 180°. The temperature of the oil bath is slowly raised to 200° during 1 hr and kept at 200° to 205° for 1.75 hr. The mixture is cooled to <100°, water (20 ml) is added to the dark-colored solution, and the mixture is refrigerated for 16 hr. The dark-brown solid is collected by filtration, and recrystallized twice (with charcoal treatment) from water to yield a solid (410 mg, 2.5 mmol, 41%), m.p. 229° to 231°, which is dried *in vacuo* at 110°; λ_{max} in nm ($\varepsilon_{\underline{mM}}$), at pH 11, 281.5 (6.40), and 233 (18.60); at pH 1, 281.5 (7.14); in ethanol, 271.4 (6.81). Thin-layer chromatography on SilicAR-7GF: with water, R_8 0.79; with ethanol, R_8 0.87, with ethyl acetate—1-propanol—water (4:1:2, v/v, upper phase), R_8 1.11.

Method B. A mixture of compound *8* (2 g, 12.2 mmol) with diethoxymethyl acetate[5] (20 ml) is stirred and heated under reflux for 5 hr. The resulting solution is evaporated under reduced pressure on a hot-water bath, ethanol (25 ml) is added, and the solution is evaporated *in vacuo*. Recrystallization of the residue from water, with charcoal treatment, yields 1.31 g (7.98 mmol, 65%) of a solid that is identical (UV, m.p.) with the product obtained by method *A*.

REFERENCES

1. R. K. Robins, L. B. Holum, and F. W. Furcht, *J. Org. Chem.*, 21, 833 (1958).

2. C. Musante, *Gazz, Chem. Ital.*, 75, 121 (1945).

3. R. A. Long, J. F. Gerster, and L. B. Townsend, *J. Heterocycl. Chem.*, 7, 683 (1970).

4. C. S. Marvel and E. E. Dreger, *Org. Synth., Coll. Vol.*, 1, 238 (1941).

5. J: A. Montgomery and N. E. Fitzgibbon, Jr., contribution #169 of this volume.

[26] DIETHYL 3-AMINO-4-CHLOROPYRIDINE-2,6-DICARBAMATE[a]

A New Intermediate for the Preparation of 2,6-Disubstituted 1-Deazapurines and 2,6-Disubstituted-1-deaza-8-azapurines

Barry L. Cline, Raymond P. Panzica, and Leroy B. Townsend

Division of Medicinal Chemistry, Department of Biopharmaceutical Sciences and Department of Chemistry, University of Utah, Salt Lake City, Utah 84112

INTRODUCTION

The recent interest in 1-deazaguanine like heterocycles and nucleoside derivatives [1-4] makes this improved preparation of ethyl 7-chloro-*v*-triazolo[4,5-*b*]pyridine-5-carbamate[5] (*6*) and ethyl 7-chloroimidazo[4,5-*b*]pyridine-5-carbamate[6] (*3*) of considerable significance. The literature procedures[7] leading to the preparation of *3* and *6* involved hydrolysis of the 2-carbamate group of *1*[b]; reduction of the nitro group afforded an *ortho* diamine that was quite susceptible to auto oxidation. The present procedure eliminates this hydrolysis step by using a stable,

---

[a]This research was supported by research contract NO1-CM-43806 from the Division of Cancer Treatment, National Cancer Institute, National Institutes of Health, Department of Health, Education, and Welfare.

[b]The preparation of diethyl 4-chloropyridine-2,6-dicarbamate, the precursor to compound *1*, from chelidamic acid (4-pyridone-2,6-dicarboxylic acid) is described by Markees and Kidder.[8]

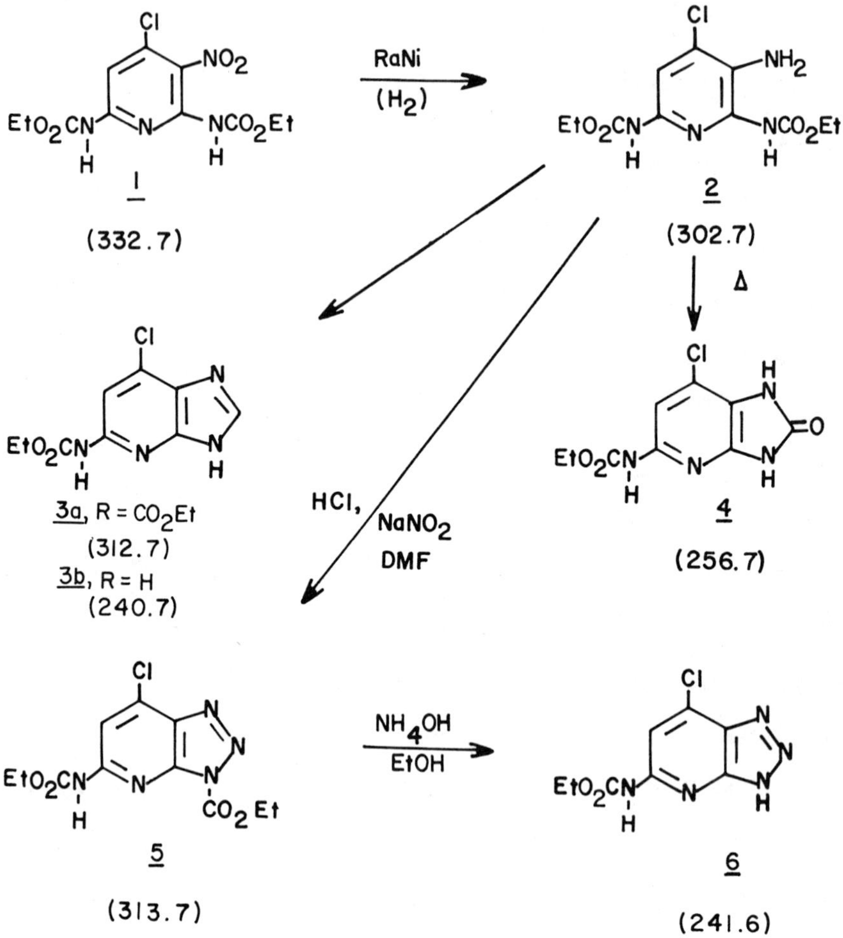

crystalline intermediate (*2*) in the ring-closure reactions to provide an improved overall yield.

PROCEDURE

Diethyl 3-Amino-4-chloropyridine-2,6-dicarbamate (*2*)

Raney nickel[c] (50 g, wet weight) was added to a solution of diethyl 4-chloro-3-nitropyridine-2,6-dicarbamate[7] (50 g, 0.15 mol) (*1*) in 1 l of absolute ethanol in a 2-l pressure bottle and the mixture was placed on a Paar shaker under hydrogen (42 psi) for 1.5 hr. The mixture was filtered through a Celite bed and the bed was washed with hot ethanol (400 ml). The combined filtrates were concentrated *in vacuo* at room temperature to 500 ml and refrigerated at 5° for 12 hr. The crystalline mass was collected by filtration, washed with ethanol-water [1:1, v/v, 200 ml, room temperature (RT)] and dried over P_2O_5 *in vacuo* at room temperature for 8 hr to afford 34 g of *2*. Concentration of the original filtrate to 100 ml and refrigeration at 5° for 12 hr furnished an additional 9.7 g of product; yield 43.7 g (96%); m.p. 140° to 142°; PMR (DMSO-d_6), δ 9.67 (1, s, 6-NH), 9.20 (1, s, 2-NH), 7.64 (1, s, H5), 4.89 (2, br s, 3-NH_2), 4.31 (4, q, CH_2), 1.22 (6, t, CH_3); UV $\lambda_{max}^{pH\ 1}$ nm, 360 (ε, 10,000), 267 (ε, 14,500), $\lambda_{max}^{pH\ 11}$ nm, 315 (ε, 7,700), 252 (ε, 17,400).

Analysis calculated for $C_{11}H_{15}N_4O_4Cl$: C, 43.64; H, 4.99; N, 18.51. Found: C, 43.49; H, 4.93; N, 18.56.

Ethyl 3-Carbethoxy-7-chloroimidazo[4,5-b]*pyridine-5-carbamate* (*3a*)

A slurry of diethyl 3-amino-4-chloropyridine-2,6-dicarbamate (*2*)

---

[c]Raney nickel #28 was purchased from W. R. Grace. The Raney nickel was weighed wet and washed with distilled water until neutral as indicated by pH paper.

(20 g, 66.1 mmol) in triethyl orthoformate (500 ml) was mechanically stirred and concentrated HCl (10 ml) was added. The mixture was stirred at room temperature for one hr. The white, crystalline precipitate was collected by filtration, washed with ethyl ether and dried over KOH *in vacuo* at room temperature for 12 hr to afford 16 g of product. The filtrate was concentrated *in vacuo* to 100 ml and kept at 5° for 12 hr to afford an additional 5 g of 5. Total yield: 21 g (100%); m.p. 158° to 159°; PMR (DMSO-d_6) δ 10.28 (1, s, 5-NH), 8.55 (1, s, H2), 7.87 (1, s, H6), 4.41 (2, q, 3-CH_2), 4.13 (2, q, 5-CH_2), 1.43 (3, t, 3-CH_3), 1.27 (3, t, 5-CH_3); UV, $\lambda_{max}^{pH\ 1}$ nm, 298 (ε 13,400), 258 (ε 10,700), λ_{max}^{EtOH} nm, 299 (ε 13,800), 267 sh (ε 8,500), 259 (ε 10,300), 254 sh (ε 9,400).

Analysis calculated for $C_{12}H_{13}N_4O_4Cl$: C, 46.09; H, 4.19; N, 17.92. Found: C, 46.03; H, 4.24; N, 17.92.

Ethyl 7-Chloroimidazo[4,5-b]pyridine-5-carbamate (3b)

A slurry of ethyl 3-carbethoxy-7-chloroimidazo[4,5-*b*]pyridine-5-carbamate (*3a*) (20 g, 64.0 mmol) in methanol (500 ml) was stirred at room temperature and concentrated NH_4OH (10 ml) was added. The mixture was stirred for 45 min and then concentrated *in vacuo* to 150 ml. Water (50 ml) was added and the mixture was neutralized (paper) with 1<u>N</u> HCl. The product was collected by filtration, washed with water and dried *in vacuo* over P_2O_5 for 12 hr. Yield: 15 g (97%, hemihydrate); m.p. loses water at 226°, then >300°; PMR (DMSO-d_6), δ 10.27 (1, br s, NH), 8.41 (1, s, H2), 7.95 (s, 1, H6),

4.26 (2, q, CH_2), 1.33 (3, t, CH_3); UV, $\lambda_{max}^{pH\ 1}$ nm, 299 (ε 15,800), 252 sh (ε 5,700), 245 sh (ε 8,900), 240 (ε 9,500), $\lambda_{max}^{pH\ 11}$ nm, 303 (ε 14,400), 260 (ε 3,600).

Analysis calculated for $C_9H_9N_4O_2Cl \cdot \frac{1}{2}H_2O$: C, 43.29; H, 4.03; N, 22.44. Found: C, 43.25; H, 4.43; N, 22.56.

Ethyl 7-Chloroimidazo[4,5-b]*pyridin-2-one-5-carbamate (4)*

One gram[d] (3.9 mmol) of finely ground *2* in a 25-ml pear-shaped flask was heated for 5 min in a 150° oil bath (the solid melts with bubbling then solidifies). The product was dissolved in 10 ml 1N KOH; the solution was diluted to 50 ml with water and neutralized with 1N HCl. The precipitate was collected by filtration, washed with water (100 ml, RT), and dried *in vacuo* over P_2O_5 at room temperature to afford 0.81 g (95%) of *4*; m.p. >300°; PMR (DMSO-d_6), δ 11.18 (2, br s, 1, 3-NH), 9.82 (1, s, 5-NH), 7.37 (1, s, 6H), 4.11 (2, q, CH_2), 1.27 (3, t, CH_3); UV, $\lambda_{max}^{pH\ 1}$ nm, 312 (16,800), 249 (9,700), $\lambda_{max}^{pH\ 11}$ nm, 306 (16,200), 260 (6,400).

Analysis calculated for $C_9H_9N_4O_3Cl$: C, 42.11; H, 3.53; N, 21.83. Found: 41.99; H, 3.91, N, 21.60.

Ethyl 3-Carbethoxy-7-chloro-v-triazolo[4,5-b]*pyridine-5-carbamate (5)*

Diethyl 3-amino-4-chloropyridine-2,6-dicarbamate (*2*) (20 g,

---

[d] When this reaction is run on a larger scale, the starting material should be placed in a wide container in a 150° oven for best results.

66 mmol) was dissolved in a mixture of 600 ml of DMF and 160 ml of concentrated HCl. The mechanically stirred solution was cooled to 0° using an ice—salt-water bath. A solution of 9.1 g (0.13 mol) of $NaNO_2$ in 40 ml of water was added dropwise over a 15-min period and the thick mixture was allowed to stir at room temperature for 1 hr. The white precipitate was collected by filtration, washed with water (200 ml, RT) and dried over P_2O_5 *in vacuo* at room temperature for 8 hr; yield, 18.6 g. The filtrate was concentrated *in vacuo* at 30° to 150 ml, 200 ml of water was added, and the mixture was refrigerated at 5° for 12 hr. The white precipitate was treated as above to afford an additional 1.8 g of *5*; yield 20.4 g (99%); m.p. 141° to 142°; PMR (DMSO-d_6), δ 11.02 (1, s, 5-NH), 8.31 (1, s, H6), 4.64 (2, q, 3-CH_2), 4.26 (2, q, 5-CH_2), 1.50 (3, t, 3-CH_3), 1.31 (3, t, 5-CH_3); UV, $\lambda_{max}^{pH\ 1}$ nm, 306 (17,800) 269 (9,300), λ_{max}^{EtOH} nm, 310 sh (15,400), 304 (16,600) 267 sh (7,900), 264 (8,500).

Analysis calculated for $C_{11}H_{12}N_5O_4Cl$: C, 42.11; H, 3.86; N, 22.33. Found: C, 41.92; H, 3.91; N, 22.67.

Ethyl 7-Chloro-v-triazolo[4,5-b]pyridine-5-carbamate (6)

To ethyl 3-carbethoxy-7-chloro-*v*-triazolo[4,5-*b*]pyridine-5-carbamate (*4*) (20.4 g, 65 mmol) suspended in 500 ml of methanol was added 10 ml of concentrated ammonium hydroxide (28%). The mixture was stirred at room temperature for 30 min and the clear solution was evaporated to dryness under diminished pressure at room temperature. The residue was slurried with 200 ml of ethanol and the solution again evaporated to dryness at room temperature. The product was dissolved in 300 ml of hot methanol, 100 ml of water was added, and the solution allowed to stand at room temperature for 12 hr. The white crystalline mass was collected by filtration, washed with water (100 ml, RT), and dried over P_2O_5 *in vacuo* for 8 hr; yield, 14.2 g. Concentration of the filtrate *in vacuo* at room temperature afforded an additional 1.3 g of product to give a quantitative yield of *6*; m.p. 211°; PMR (DMSO-d_6), δ 10.73 (1, s, 5-NH), 8.15 (1, s, H6), 4.28 (2, q, CH_2), 1.33 (3, t, CH_3); UV, $\lambda_{max}^{pH\ 1}$ nm, 307 sh (13,900), 301 (14,900), 270 sh (6,100), 262 (6,600) $\lambda_{max}^{pH\ 11}$ nm, 303 (14,200).

REFERENCES

1. J. W. Schelling and C. A. Salemink, *Recl. Trav. Chem. Pays-Bas*, 93, 160 (1974).

2. B. L. Cline, R. P. Panzica and L. B. Townsend, *J. Heterocycl. Chem.*, 12, 603 (1975).

3. B. L. Cline, R. P. Panzica and L. B. Townsend, *Proc. Utah Acad. Sci.*, 52, 79 (1975).

4. J. E. Schelling and C. A. Salemink, *Recl. Trav. Chem. Pays-Bas*, 94, 153 (1975).

5. C. Temple, B. H. Smith, and J. A. Montgomery, *J. Org. Chem.*, 38, 1095 (1973).

6. C. Temple, B. H. Smith, and J. A. Montgomery, *J. Org. Chem.*, 38, 613 (1973).

7. J. A. Montgomery and N. F. Wood, *J. Org. Chem.*, 29, 734 (1964).

8. D. G. Markees and G. W. Kidder, *J. Am. Chem. Soc.*, 75, 4130 (1956).

[27] PYRROLO[3,2-*c*]PYRIDIN-4,6(5*H*,7*H*)-DIONE (3,7-DIDEAZAXANTHINE)

Replacement of N-3 and N-7 of Xanthine by Methine Groups

Stewart W. Schneller and Ramachandra S. Hosmane

Department of Chemistry, University of South Florida, Tampa, Florida 33620

INTRODUCTION

The synthesis of pyrrolo[3,2-*c*]pyridin-4,6(5*H*,7*H*)-dione (3,7-dideazaxanthine) (*3*) has been achieved by intramolecular cyclization of 3-(ethoxycarbonyl)pyrrole-2-acetamide (*2*) with sodium ethoxide. In view of the moderate yields by this procedure and the convenient accessibility of the requisite amide *2* from ethyl 3-(ethoxycarbonyl)pyrrole-2-acetate[1] (*1*) and ammonium hydroxide, a significant xanthine analog is available: (a) for biological evaluation as an inhibitor of xanthine oxidase and (b) as a valuable starting compound in the synthesis of other 3,7-dideazapurines.

PROCEDURE

3-(Ethoxycarbonyl)pyrrole-2-acetamide (2)

Ethyl 3-(ethoxycarbonyl)pyrrole-2-acetate[1] (3 g, 13.3 mmol) is

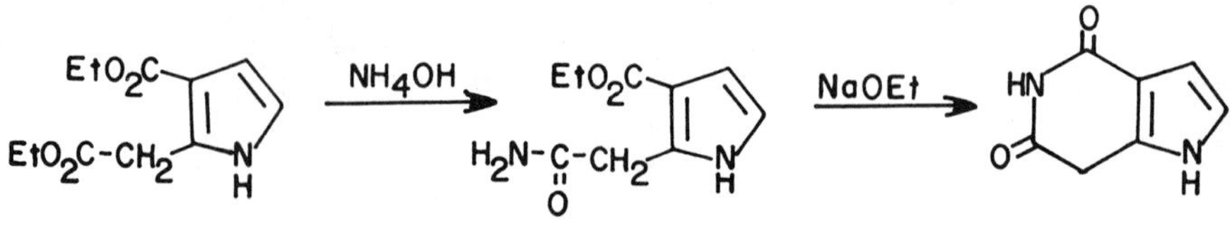

1 (225.3) 2 (196.2) 3 (150.1)

added in small portions to refluxing[a] 28% ammonium hydroxide solution (50 ml). After the addition is complete the brownish solution is boiled under reflux for an additional 10 min and filtered, and the filtrate is cooled in an ice bath, made acidic with 20% sulfuric acid, and refrigerated overnight. The white solid that has separated is removed by filtration and recrystallized from benzene—petroleum ether as silky white needles (1.6 g, 8.2 mmol, 62%), m.p. 143° to 144°; PMR[b] data (chloroform-*d*): δ 1.33 (t, 3, *J* 7.0 Hz, CH_3), 4.31 (q, 2, *J* 7.0 Hz, CH_2), 3.95 (s, 2, CH_2), 6.55 (t, 1, *J* 3.0 and 2.7 Hz,[c] H-4 *or* H-5), 6.85 (t, 1, *J* 3.0 and 2.7 Hz,[c] H-4 or H-5); IR spectrum[d] ν_{max}^{KBr} 3325 to 3125 (broad, NH) and 1700 to 1625 (broad, 2 C=O) cm^{-1}; mass spectrum

---

[a]The use of an oil bath allows for a more even reflux.

[b]Spectrum recorded with a Varian A-60, 60-MHz spectrometer.

[c]Triplet due to coupling with adjacent aromatic proton and the pyrrole ring NH. This accounts for the two coupling-constants reported.

[d]Spectrum recorded with a Perkin-Elmer model 337 spectrophotometer.

(70 eV)[e]: *m/e* 196 (M^+, 22%), 179 (M^+-NH_3, 46), 153 (M^+-CONH, 39), 152 (M^+-$CONH_2$, 13), 151 (M^+-$CONH_3$, 28), and 124 [M^+-$(CO)_2NH_2$, 100].

Pyrrolo[3,2-c]pyridin-4,6(5H,7H)-dione (3,7-Dideazaxanthine) (3)

3-(Ethoxycarbonyl)pyrrole-2-acetamide (6 g, 0.03 mol) is added to a solution prepared from sodium (4.2 g, 0.18 g-atom) and absolute ethanol (170 ml). The blue-green solution is boiled under reflux[f] on a steam bath for 30 min, cooled, and evaporated on a rotary evaporator. The residual solid is dissolved in water (100 ml) and the solution is boiled with decolorizing charcoal, the suspension filtered, and the filtrate acidified with glacial acetic acid, and cooled overnight in a refrigerator. The blue solid that separates is recrystallized from glacial acetic acid to give tiny buff-colored needles[g] (2.5 g, 17 mmol, 57%), m.p. >300°; PMR[b] data (Me_2SO-d_6): δ 4.1 (s, 2, CH_2), 6.75 (t, 1, *J* 3.0 and 2.7 Hz,[c] H-2 or H-3), 7.25 (t, 1, *J* 3.0 and 2.7 Hz,[c] H-2 or H-3); ν_{max}^{KBr} 3300 to 3170 (broad, NH) and 1725 to 1685 (broad, 2 C=O) cm^{-1}; mass spectrum (70 eV)[e]: *m/e* 150 (M^+, 100%), 107 (M^+-CONH, 54), and 79 (M^+-$(CO)_2NH$, 73).

---

[e]Mass spectrum recorded with a Varian MAT CH-7 spectrometer (at Indiana University, Bloomington).

[f]A large-scale reaction should be conducted in a fume hood, as evolution of ammonia is detectable at these higher concentrations.

[g]These crystals appear to be hygroscopic, as they become blue in the hydrated form.

REFERENCE

1. G. A. Swan and A. Waggott, *J. Chem. Soc.*, C, 285 (1970).

[28] 2,4-DIAMINO-6,7-DIHYDRO-5*H*-PYRROLO[3,4-*d*]PYRIMIDINE

Synthesis of the Pyrrolo[3,4-d]pyrimidine Ring Structure

Philip L. Southwick, R. Madhav, Venkataraman Amarnath, and Tuvia Sheradsky

Department of Chemistry, Carnegie-Mellon University, Pittsburgh, Pennsylvania 15213

INTRODUCTION

2,4-Diamino-6,7-dihydro-5*H*-pyrrolo[3,4-*d*]pyrimidine (*6*) has produced a statistically significant (≃60%) increase in the survival time of mice in a standardized, antimalarial assay utilizing *Plasmodium berghei*.[1] The 6-(3,4-dichlorophenylthiocarbamoyl) derivative of this compound is an active antimalarial in the same assay, and other 6-substituted derivatives are active against experimental avian malaria (*Plasmodium gallinaceum*).[1-3] This synthesis of 2,4-diamino-6,7-dihydro-5*H*-pyrrolo[3,4-*d*]pyrimidine (*6*) from ethyl *N*-(cyanoethyl)glycinate (*1*) (the cyanoethylation product of ethyl glycinate)[4] is a modification[1] of the method of Sheradsky and Southwick.[5]

PROCEDURE

1-Acetyl-3-amino-4-cyano-3-pyrroline (4)

Acetic anhydride (100 ml) is added with cooling to ethyl *N*-(cyano-

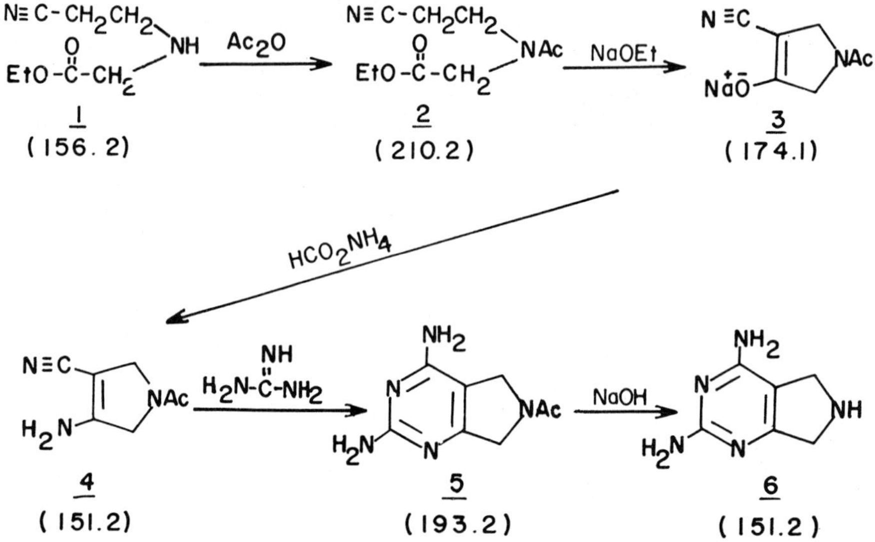

ethyl)glycinate[8] (*1*) (78 g, 0.5 mol), and the mixture is kept for 24 hr at room temperature. Acetic acid and the excess of acetic anhydride are removed by evaporation under diminished pressure. (If the odor of the residue indicates remaining acetic acid or anhydride, ethanol is added and the evaporation is repeated.) The residual oil is dissolved in absolute ethanol (250 ml), and the solution is added to a solution of sodium ethoxide prepared from sodium (11.5 g, 0.5 mol) and absolute ethanol (500 ml). The mixture is boiled under reflux for 2 hr, and then ammonium formate (79 g, 1.25 mol) is added, together with an additional 500 ml of absolute ethanol, and refluxing is continued for 48 hr on a steam bath, with care to ensure adequate mixing by vigorous refluxing. The mixture is cooled in an ice bath, and the crystalline product (*4*) is collected by filtration; yield 68 g (90%), m.p.[a] ≃236°. Recrystallization from 70% aqueous ethanol raises the melting point[b] to 237° to 239° (lit.[5] m.p. 238° to 239°).

6-Acetyl-2,4-diamino-6,7-dihydro-5H-pyrrolo[3,4-d]pyrimidine (*5*)

An ethanolic solution of guanidine is prepared from guanidine hydrochloride (95.5 g, 1.0 mol), sodium (23 g, 1.0 mol), and absolute ethanol (600 ml). Prior to use, the solution is filtered with the aid of Celite (to remove precipitated sodium chloride), and additional absolute ethanol (≃200 ml) is added in washing the filter cake. A solution of compound *4* (75.5 g, 0.5 mol) in absolute ethanol (500 ml) is added, and the mixture is boiled under reflux on a steam bath for 16 hr, cooled in an ice bath, and filtered to collect the crystalline product; yield 67.5 g (70%), m.p.[c] ≃335°. (An analytical sample, m.p. 335° to 337°, was obtained by recrystallization from *N*,*N*-dimethylformamide.[5])

---

[a] The melting of compounds *4*, *5* and *6* occurs with preliminary softening, darkening and decomposition. Melting points are dependent on the rate of heating and are difficult to duplicate closely.

[b] For *3*, λ_{max}^{Nujol}: 2.92 μm, 2.98 μm, 3.08 μm, 4.52 μm, 5.99 μm, 6.06 μm, 6.5 μm, 6.21 μm, 7.01 μm, 7.35 μm, 9.62 μm and 10.29 μm.

[c] For *5*, λ_{max}^{Nujol}: 2.85 μm, 2.91 μm, 3.14 μm, 5.99 μm, 6.05 μm, 6.15 μm, 6.25 μm, 6.36 μm, 6.76 μm, 7.10 μm, 7.39 μm, 8.00 μm, 8.42 μm, 9.10 μm, 9.61 μm, 10.32 μm, 12.48 μm and 12.72 μm.

2,4-Diamino-6,7-dihydro-5H-pyrrolo[3,4-d]pyrimidine (*6*)

To a solution of *5* (19.3 g, 0.1 mol) in ethanol (1,350 ml) is added a solution of sodium hydroxide (75 g) in water (150 ml). The mixture is boiled under reflux for 12 to 14 hr and concentrated under diminished pressure to ≈500 ml. The concentrate is cooled for several hours in an ice bath and filtered to afford the crystalline product; yield 8.8 g (58%) of crude *6*, m.p. 235° to 240°. Recrystallization from water raises the melting point[d,e] to 258° to 259°.

REFERENCES

1. P. L. Southwick, V. Amarnath, and R. Madhav, *J. Heterocycl. Chem.*, 11, 723 (1974).

2. V. Amarnath, Ph.D. Thesis, Carnegie—Mellon University, July 1973.

3. E. F. Elslager, A. Curry, and L. M. Werbel, *J. Heterocycl. Chem.*, 9, 1123 (1972).

4. A. P. Terent'ev and P. F. Butskus, *Zh. Obshch. Khim.*, 23, 1230 (1953); *Chem. Abstr.*, 47, 12,237h (1953).

5. T. Sheradsky and P. L. Southwick, *J. Org. Chem.*, 30, 194 (1965).

---

[d]For *6*, λ_{max}^{Nujol}: 2.90 µm, 3.00 µm, 3.12 µm, 6.00 µm, 6.10 µm, 6.25 µm, 9.35 µm, 9.80 µm, 11.27 µm and 12.75 µm.

[e]Compound *6* reacts with atmospheric carbon dioxide and water to afford a carbonate that melts at temperatures higher than *6* and is less soluble in neutral solvents. Compound *6* is recovered in the free form by suspending the carbonate in water (10 to 20 ml per g of carbonate), adding hydrochloric acid until a clear solution is obtained, and precipitating *6* by adding an excess of solid potassium hydroxide.

[29] 2,5-ANHYDRO-3,4,6-TRI-*O*-BENZOYL-D-ALLONOTHIOAMIDE AND 3-DEOXY DERIVATIVES

An Improved Synthesis of Glycosylthiocarboxamides

Michael V. Pickering, Prem C. Srivastava, J. T. Witkowski, and Roland K. Robins

ICN Pharmaceuticals, Inc., Nucleic Acid Research Institute, 2727 Campus Drive, Irvine, California 92715

INTRODUCTION

Carbohydrates derived from *O*-acyl-β-D-ribofuranosyl cyanides have been shown to be versatile intermediates for the synthesis of *C*-nucleosides.[1-5] Recently in this laboratory[6] and another,[7] 2,5-anhydro-3,4,6-tri-*O*-benzoyl-D-allonothioamide (*2*) has been utilized in the synthesis of novel thiazole *C*-nucleosides. The synthesis of *2* described earlier[7] gives a significant proportion (14%) of benzoyl-elimination product and, consequently, low yields (20%) of *2*. The general method described here uses liquid hydrogen sulfide and 4-(dimethylamino)pyridine and provides the protected glycosylthiocarboxamides in yields greater than 80%.

PROCEDURE

2,5-Anhydro-3,4,6-tri-O-benzoyl-D-allonothioamide (2)

A mixture of 2,3,5-tri-*O*-benzoyl-β-D-ribofuranosyl cyanide[1] (*1*,

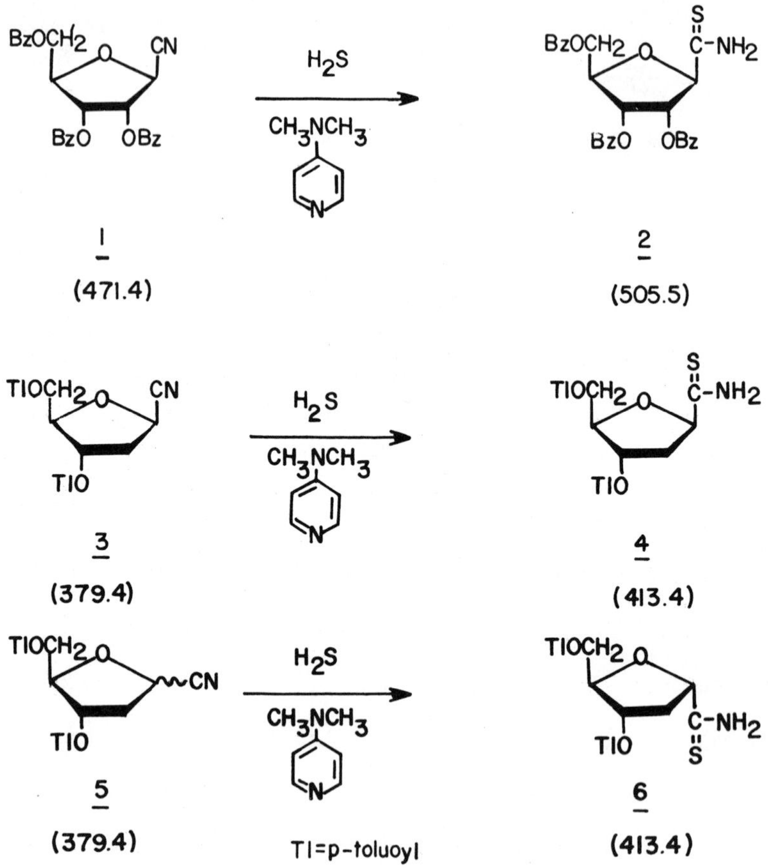

25.0 g, 53 mmol) and 4-dimethylaminopyridine[a] (0.5 g), together with a magnetic stirring bar, is placed in a 300-ml stainless-steel bomb equipped with a pressure-release valve. The mixture is cooled in a dry-ice—isopropyl alcohol bath (≃ -70°). Liquid

---

[a]4-Dimethylaminopyridine is an effective catalyst for this reaction; the cyano sugar *1* is recovered unchanged when treated with liquid hydrogen sulfide in the absence of the catalyst.

hydrogen sulfide[b] (≃75 ml) is carefully poured into the bomb, which is then sealed, and the mixture is stirred for 17 hr at room temperature, and recooled to ≃ −70°. The pressure is gradually released while the mixture is allowed to warm to room temperature. The residue (a foam) is dissolved in dichloromethane (250 ml), and the solution is washed with saline water (2 × 100 ml), dried (magnesium sulfate), and evaporated under reduced pressure to yield 26 g of *2* as a light-yellow foam of sufficient purity for further reactions. It may be purified by chromatography on a column (5 × 70 cm) of silica gel packed in chloroform. Elution with 1:19 (v/v) ethyl acetate—chloroform provides 22 g (83%) of analytically pure product,[c] a light-yellow foam, $[\alpha]_{D}^{25}$ +2.84° (*c* 1, chloroform).

Analysis calculated for $C_{27}H_{23}NO_7S$: C, 64.14; H, 4.58; N, 2.77; S, 6.34. Found: C, 63.88; H, 4.87; N, 2.46; S, 6.33.

*2,5-Anhydro-3-deoxy-4,6-di-*O-p-*toluoyl-*D-ribo-*hexonothioamide* (4)

A mixture of 2-deoxy-3,5-di-*O*-*p*-toluoyl-β-D-*erythro*-pentofuranosyl cyanide[9] (*3*; 7.58 g, 20 mmol), 4-dimethylaminopyridine (150 mg), and liquid hydrogen sulfide (≃50 ml) is stirred in a

---

[b] All operations with hydrogen sulfide must be conducted in a well-ventilated hood.

[c] The product (*2*) is detected on thin-layer chromatograms of silica gel (developed with 10:1 chloroform—ethyl acetate) with a spray of a dilute, ethanolic solution of 2,3-dichloro-1,4-naphthoquinone, followed by exposure to ammonia. A red-to-purple color indicates a positive test for the thiocarboxamide.[8]

sealed bomb for 20 hr at room temperature. The hydrogen sulfide is then allowed to evaporate, the residue (a light-yellow foam) is dissolved in ether (200 ml), and the solution is washed with saline water (2 ml × 100 ml), dried (magnesium sulfate), and evaporated under reduced pressure to a residue which, with the aid of nucleating crystals,[d] is crystallized from ethanol as shining, white needles; yield: 7.0 g (85%), m.p. 154° to 154.5°, $[\alpha]_{D}^{25}$ -12.9° (*c* 1, chloroform).

Analysis calculated for $C_{22}H_{23}NO_5S$: C, 63.90; H, 5.61; N, 3.39; S, 7.75. Found: C, 64.15; H, 5.45; N, 3.19; S, 7.54.

*2,5-Anhydro-3-deoxy-4,6-di-*O-p-*toluoyl-*D-arabino-*hexonothioamide (6)*

A mixture of 2-deoxy-3,5-di-*O*-*p*-toluoyl-D-*erythro*-pentofuranosyl-cyanide[9] (5,[e] 5.0 g, 13 mmol), 4-dimethylaminopyridine (100 mg), and freshly condensed hydrogen sulfide (35 ml) is stirred in a sealed bomb for 20 hr at room temperature. The hydrogen sulfide is then allowed to evaporate the light-yellow foam is dissolved in ethyl ether (200 ml), and the solution is washed with saline water, dried (magnesium sulfate), and evaporated under diminished

---

[d]Nucleating crystals are obtained by chromatography of the crude product on a column of silica gel with 1:19 (v/v) ethyl acetate—chloroform as the eluant. Homogeneous fractions are evaporated, and the product is crystallized from ethanol.

[e]An anomeric mixture (5) is used for the synthesis of *6*, as it is much more convenient to separate the epimers of the thiocarboxamide than the anomers of 5 by chromatography.

pressure to give a light-yellow foam. Chromatography on a column (2.5 cm × 6 cm) of silica gel (packed in chloroform) with 1:19 (v/v) ethyl acetate—chloroform as the eluant separates the two products (*4* and *6*), to provide 2.0 g of *4* as the faster-migrating component, and 1.0 g of *6*, which is crystallized from ethanol, m.p. 152°, $[\alpha]_{D}^{25}$ -39.0° (*c* 1, chloroform).

Analysis calculated for $C_{22}H_{23}NO_5S$: C, 63.90; H, 5.61; N, 3.39; S, 7.75. Found: C, 63.81; H, 5.67; H, 3.29; S, 7.65.

REFERENCES

1. M. Bobek and J. Farkas, *Collect. Czech. Chem. Commun.*, 34, 247 (1969).

2. G. Trummlitz and J. G. Moffatt, *J. Org. Chem.*, 38, 1841 (1973).

3. D. B. Repke, H. P. Albrecht, and J. G. Moffatt, *J. Org. Chem.*, 40, 2481 (1975).

4. J. Igolen, T. Huynh Dinh, A. Kolb, and C. Perreur, *Chim. Therap.*, 3, 207 (1972).

5. A. Kolb, C. Goyette, T. Huynh Dinh, and J. Igolen, *Tetrahedron Lett.*, 2791 (1974).

6. P. C. Srivastava, M. V. Pickering, L. B. Allen, D. G. Streeter, M. T. Campbell, J. T. Witkowski, R. W. Sidwell, and R. K. Robins, *J. Med. Chem.*, 20, 256 (1977).

7. M. Fuertes, M. T. García-López, G. García Muñoz, and R. Madroñero, *J. Carbohydr. Nucleos. Nucleot.*, 2, 277 (1975); M. Fuertes, T. García-López, G. García-Muñoz, and M. Stud, *J. Org. Chem.*, 41, 4074 (1976).

8. M. B. Devani, C. J. Shishoo, H. J. Modi, and P. K. Raja, *J. Pharm. Sci.*, 63, 1471 (1974).

9. A. Kolb, T. Huynh Dinh, and T. Igolen, *Bull. Soc. Chim. Fr.*, 3447 (1973).

[30] ANOMERS OF 1-*O*-ACETYL-2,3,5-TRI-*O*-BENZOYL-D-ARABINOSE

Intermediates in the Synthesis of Modified Nucleosides

I. A. Korbukh, L. N. Abramova, and Maria N. Preobrazhenskaya

Cancer Research Center of the U.S.S.R. Academy of Medical Sciences, Moscow 115478, U.S.S.R

INTRODUCTION

A mixture of the anomers of 1-*O*-acetyl-2,3,5-tri-*O*-benzoyl-D-arabinose has been obtained by acetolysis of methyl 2,3,5-tri-*O*-benzoyl-α-D-arabinoside.[1] The anomers are separated chromatographically.[2]

PROCEDURE

1-O-Acetyl-2,3,5-tri-O-benzoyl-α,β-D-arabinose (2,3)

A solution of methyl 2,3,5-tri-*O*-benzoyl-α-D-arabinoside (7.0 g, 1.47 mmol) in glacial acetic acid (74 ml) and acetic anhydride (14.8 ml) is stirred and kept at 5° to 10°, while concentrated sulfuric acid (4.3 ml) is added dropwise. The solution is kept for 5 hr in a stoppered flask at room temperature and overnight in a refrigerator and then poured into ice water (300 ml), and the mixture is extracted with chloroform (3 ml × 30 ml). The extracts are combined, washed successively with saturated aqueous sodium hydrogen carbonate (5 ml × 40 ml) and water (40 ml), dried

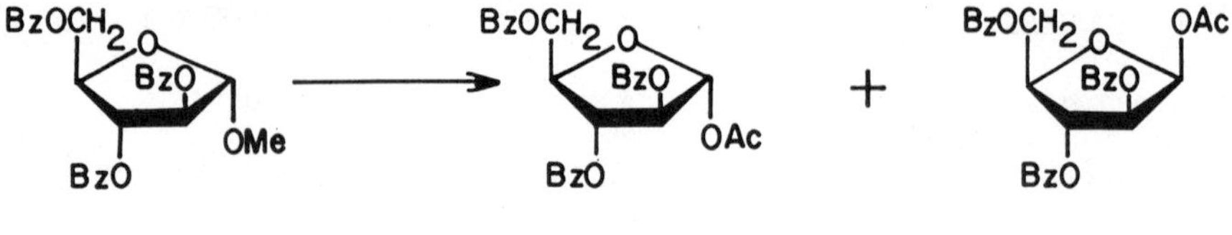

1 (476.5) 2 (504.5) 3 (504.5)

(magnesium sulfate), and evaporated to dryness under reduced pressure. A solution of the resultant syrup in chloroform is added to a column (3 × 70 cm) of silica gel[a] prepacked in chloroform, and the column is eluted with chloroform, 10-ml fractions being collected. The course of the elution is monitored by means of a Uvicord UV spectrometer and by TLC.[b] Fractions 1—6 provide *2* as a syrup (1.68 g), $[\alpha]_D^{20}$ -11.8° (*c* 1.3, chloroform); R_F 0.29 (in chloroform); PMR data (Me_2SO-d_6): δ 8.20 to 7.12 (m, 15, COC_6H_5), 6.32 (s, H-1), 5.50 (m, H 2 + 4), 4.80 to 4.56 (m, H 3 + 5) 2.08 (s, 3, Ac).

Fractions 7—34 are evaporated to dryness, and a solution of the residue in the minimal volume of chloroform is applied to plates (20 × 20 cm) of silica gel[c] that are developed with chloroform (3—4 consecutive developments). The bands containing *2* and *3* are removed, and compound *2* and *3*, respectively, are extracted with methanol. The solvent is evaporated *in vacuo* to give an

---

[a]Silica gel L 40/100 (Lachema, Czechoslovakia).
[b]Silufol UV_{254} (Kavalier Glassworks, Czechoslovakia).
[c]Silica gel LL_{254} 5/40 (Lachema, Czechoslovakia).

additional 2.66 g of *2*, and 0.86 g of *3*. Compound *3* has m.p. 124° to 126° $[\alpha]_{D}^{20}$ -70.0° (*c* 1.36, chloroform); R_F 0.23 (TLC in chloroform); PMR data ($CDCl_3$): δ 8.04 to 7.12 (m, 15, COC_6H_5), 6.56 (d, 1, $J_{1,2}$ 4.8 Hz, H-1), 5.90 (m, H3), 5.72 (q, H2), 4.84 to 4.38 (m, H 4,5,5'), and 1.84 (s, 3, Ac).

REFERENCE

1. H. G. Fletcher, Jr., *Methods Carbohydr. Chem.*, 2, 228 (1963).

2. I. A. Korbukh, L. B. Abramova, and M. N. Preobrazhenskaya, *Zh. Org. Khim.*, 13, 73 (1977).

[31] 2-*O*-*p*-TOLYLSULFONYL-L-ARABINOSE

An Intermediate for the Synthesis of β-L-Ribonucleosides

Antonín Holý

Institute of Organic Chemistry and Biochemistry, Czechoslovak Academy of Sciences, 166 10 Praha 6, Czechoslovakia

INTRODUCTION

2-*O*-*p*-Tolylsulfonyl-L-arabinose (*5*) is conveniently used as the starting material for the stereospecific synthesis of β-L-ribonucleosides.[1] Compound *5* may be obtained by selective *p*-toluenesulfonylation of specifically protected L-arabinose, followed by deprotection. For this purpose L-arabinose is first converted into the benzyl β-L-glycoside (*1*); the 3,4-isopropylidene acetal (*2*) is then obtained therefrom. The free hydroxyl group in *2* is then *p*-toluenesulfonylated, and the isopropylidene group is removed by acid hydrolysis. Finally, in the presence of a palladium catalyst the benzyl group is split off by hydrogenolysis under atmospheric or increased pressure.

PROCEDURE

Benzyl β-L-Arabinopyranoside (1)

Benzyl alcohol (2 l) is saturated with dry hydrogen chloride (100 g) in a 4-l reaction-flask equipped with an efficient

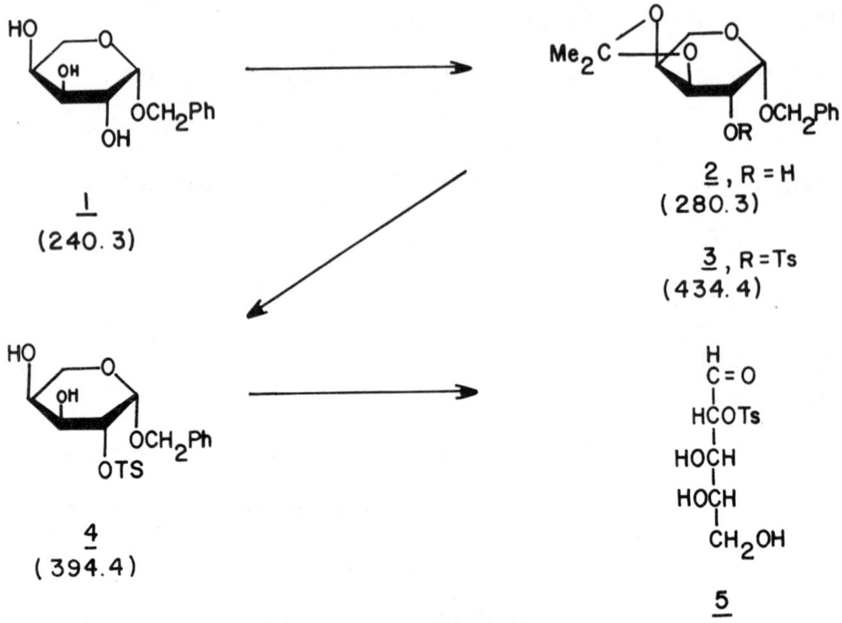

TS = $SO_2C_6H_4Me$-*p*

stirrer, a gas-inlet tube, and a drying tube (calcium chloride). L-Arabinose (500 g, 3.3 mol) is added in one portion, and the mixture is stirred for 2 hr at 50° to 60°. The resulting solution is then kept overnight at 4° to deposit crystalline material, which is removed by filtration. After an additional 2 days at 4°, the mother liquor affords an additional crop of the same material. The crops are combined, washed successively with ether and 1:2 ethanol—ether (500 ml), the same mixture containing 10% of triethylamine, and dried over phosphorus pentoxide *in vacuo*. The overall yield of glycoside *1* amounts to 480 g (50%), m.p. 168° to 169°, $[\alpha]_D^{23}$ +199° (*c* 0.5, *N,N*-dimethylformamide).

Benzyl 2-O-p-Tolylsulfonyl-β-L-arabinopyranoside (4)

In a 4-l reaction-flask equipped with an efficient stirrer and a drying tube (calcium chloride) a mixture of the glycoside *1* (400 g, 1.4 mol), acetone (1.6 l), and ethyl orthoformate (400 ml) is stirred at room temperature. Concentrated sulfuric acid (1.5 ml) is added and the suspension stirred until dissolution occurs (≃15 min) and then for an additional 30 min at room temperature. The mixture is then made alkaline by the addition of triethylamine (25 ml) and is evaporated to dryness at 40°/15 torr. The residue is dissolved in chloroform (1 l), and the solution is washed with water (3 × 200 ml). The washings are combined, extracted with chloroform (2 × 100 ml), and the extracts are combined, dried (magnesium sulfate), and evaporated to dryness. The crude compound *2* thus obtained is dissolved in pyridine (1 l) and treated with *p*-toluenesulfonyl chloride (415 g, 2.2 mol) in a 2-l Erlenmeyer flask. The suspension is stirred magnetically, with exclusion of moisture, until all of the material has dissolved (≃1 hr) and the solution kept overnight at room temperature. The mixture is then slowly poured, with stirring, into water (4 l) in a glass beaker, and the product is extracted with chloroform (3 × 500 ml). The extracts are combined, washed with 1% aqueous hydrochloric acid (3 × 200 ml), and evaporated to dryness at 40°/15 torr. A solution of the residual syrup in ethanol (750 ml), water (600 ml), and concentrated hydrochloric acid (150 ml) is boiled under reflux for 1 hr, cooled, and set aside for crystallization. The product is collected by filtration and dissolved in boiling, 50% aqueous ethanol. To this solution is added ethanol (250 ml), and the mixture is kept at 4° for crystallization. The crystals are removed by filtration, washed with 50% aqueous ethanol (200 ml), and dried *in vacuo* over phosphorus pentoxide. The yield of analytically pure *4* amounts to 500 g (81%), m.p. 122° to 123°, $[\alpha]_{D}^{23}$ +133.6° (*c* 0.5, *N*,*N*-dimethylformamide).

2-O-p-*Tolylsulfonyl*-L-*arabinose* (5)

A suspension of compound *4* (200 g, 0.45 mol) and 10% Pd/C catalyst (2 g) in a mixture of acetic acid (1 l) and concentrated hydrochloric acid (2 ml) is hydrogenolyzed in a 2-l hydrogenation flask at an atmospheric pressure and room temperature until uptake of hydrogen ceases (12 l).[a] The suspension is then filtered through a layer of Celite, the filter is washed with acetic acid (100 ml), and the filtrate and washings are combined and evaporated to dryness at 40°/15 torr. A mixture of the residue

---

[a]Alternatively, compound *4* (500 g) and 10% palladium-on-Celite catalyst (20 g) in 1:1 acetic acid—1,4-dioxane (2.5 l) are stirred for 8 hr at 50° and hydrogen (80 atmospheres) in a stainless steel autoclave. The processing of the mixture is then performed as described.

with 1,4-dioxane (1 l) and 2M hydrochloric acid (1 l)[b] is boiled under reflux for 2 hr, cooled, and evaporated, the residue is dissolved in ethyl acetate (1 l), and the solution is washed with water (100 ml), dried (magnesium sulfate), and evaporated to dryness under reduced pressure. The residue is applied to a column of silica gel (30 to 60 mesh; 800 g) packed in 3:1 benzene—ethyl acetate. Elution is performed with the same mixture, 500-ml fractions being collected. The chromatography is monitored by TLC in 1:1 benzene—ethyl acetate. The fractions containing pure compound *5* are combined, and evaporated at 40°/15 torr, and the residue is crystallized from 1,4-dioxane for 3 to 4 days at room temperature.[c] The product is removed by filtration using suction, washed with a small volume of 1,4-dioxane, and dried *in vacuo* over phosphorus pentaoxide to afford[d] 109 g (80%) of pure crystalline 1,4-dioxane solvate of compound *5*, m.p. 85° to 86°.

REFERENCE

1. A. Holý and F. Šorm, *Collect. Czech. Chem. Commun.*, 34, 3383 (1969).

---

[b]The treatment with acid causes hydrolysis of some 1-*O*-acetyl-2-*O*-*p*-tolylsulfonyl-L-arabinopyranose, formed as a by-product.

[c]The crystallization is greatly accelerated by nucleation with a sample prepared separately.

[d]For application in the synthesis of L-ribonucleosides (see Section III,[92]) the chromatography on silica is unnecessary. The purity of the crude product exceeds 95%, the contaminants being the starting compound *4* and the 1-acetate (cf. footnote *a*).

[32] (±)-(1,4/2)-4-AMINO-2-HYDROXYCYCLOPENTANEMETHANOL AND (±)-(1,4/3)-4-AMINO-3-HYDROXYCYCLOPENTANEMETHANOL

*Carbocyclic Analogs of 2'- and 3'-Deoxy-D-*erythro*-pentofuranosylamine for the Synthesis of Analogs of Nucleosides*

C. Allen O'Dell and Y. Fulmer Shealy

Kettering-Meyer Laboratory, Southern Research Institute, Birmingham, Alabama 35205

INTRODUCTION

(±)-(1,4/2)-4-Amino-2-hydroxycyclopentanemethanol (*7a*) and (±)-(1,4/3)-4-amino-3-hydroxycyclopentanemethanol (*7b*) are key intermediates for the synthesis of racemic, carbocyclic analogs of 2'- and 3'-deoxy-D-*erythro*-pentofuranosides. Routes for the synthesis of *7a* and *7b* from *exo*-bicyclo[2,2.1]hept-5-en-2-ol acetate have been outlined.[1] Separation of the positional isomers formed on opening of the anhydride ring of *2* has been achieved[1] at the amide-ester step (*5a* + *5b*). A simplified route is represented by *5ab* to *7ab*, the entire sequence of steps from *2* to *7ab* being performed without purification of intermediates or separation of positional isomers. Separation of the aminocyclopentanes *7a* and *7b* is then effected by ion-exchange chromatography. The overall yield (44% to 45%) of *7a* (24.3%) plus *7b* (20.1%) from *2* is

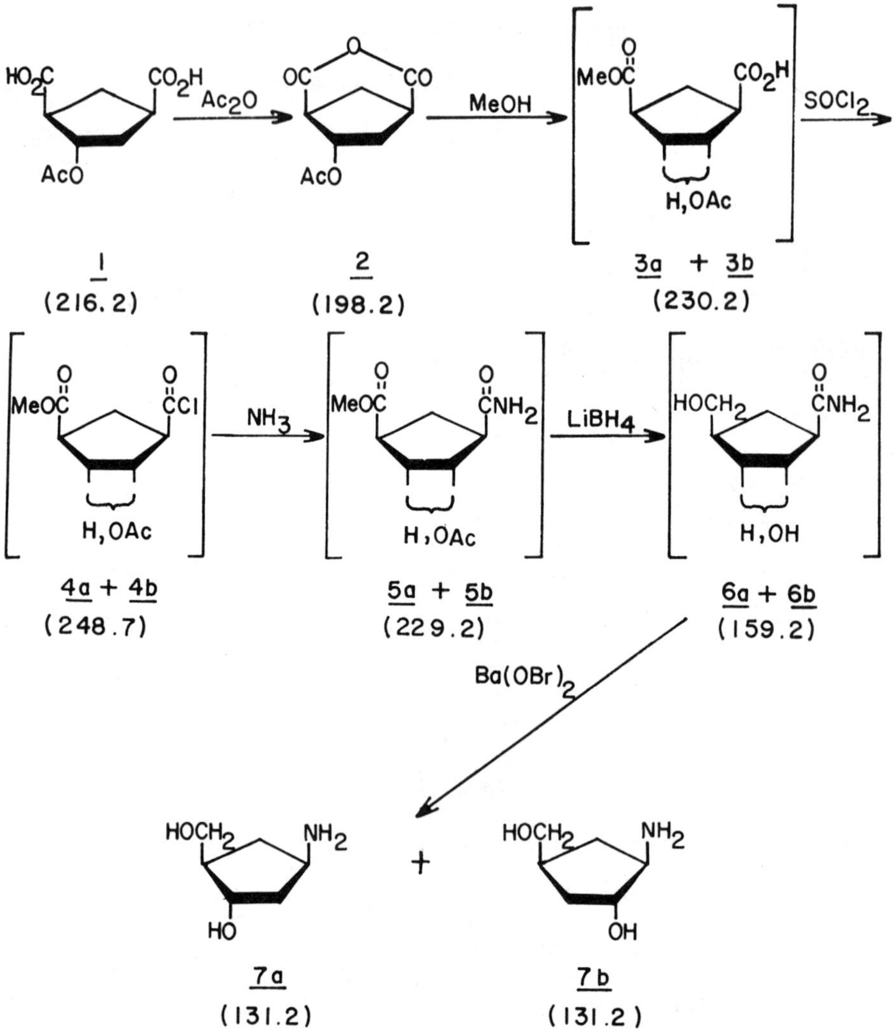

Each structure shown represents one enaniomer of a racemic mixture.

equivalent to an average yield of 85% per step.

PROCEDURE

(1,3/4)-4-(Acetoxy)-1,3-cyclopentanedicarboxylic Anhydride (2)

A mixture of (1,3/4)-4-(acetoxy)-1,3-cyclopentanedicarboxylic acid[1] (*1*, 79.5 g) and acetic anhydride (710 ml) is boiled under reflux (dry nitrogen atmosphere) for 15 min and evaporated under diminished pressure (oil pump) under anhydrous conditions below 40°. The residual oil is triturated with dry ether (200 ml), and the resulting crystalline solid is collected by filtration under dry nitrogen, washed with dry ether, and dried *in vacuo* over phosphorus pentoxide: weight, 51 g (70%); m.p. 119° to 120°; ν_{max}^{KBr} 1740, 1775, 1805 (sh), and 1815 cm^{-1}. The filtrate is evaporated to dryness, and the residue is retreated with acetic anhdyride (200 ml) as just described. An additional 14.5 g (20%) of *2* is isolated; m.p. 118° to 120°.

(±)-(1,4/2)-4-Amino-3-hydroxycyclopentanemethanol (7a) and (±)-(1,4/3)-3-Amino-4-hydroxcyclopentanemethanol (7b) from 2 without Separation of Intermediate Isomers

A solution of *2* (51 g) in anhydrous methanol (790 ml; distilled from sodium) is stirred for 18 hr at room temperature in a stoppered flask and then evaporated under reduced pressure at room temperature. The residual colorless syrup[a] (*3a* + *3b*) is dissolved in benzene (850 ml), and ≃25 ml of benzene is removed by distillation to eliminate traces of water. Thionyl chloride (86 ml) is added to the hot solution (protected from atmospheric moisture),

---

[a]The IR spectrum (film) of the mixture of isomers includes CO bands at 1710 (sh) and 1730 cm^{-1}.

the mixture is stirred for 5 min, and pyridine (2 drops) is added; vigorous evolution of hydrogen chloride immediately occurs. When the reaction has subsided, *N*,*N*-dimethylformamide (2 drops) is added, and the mixture is boiled under reflux for 2 hr, kept overnight at room temperature, and evaporated *in vacuo* to remove the benzene and the excess of thionyl chloride. The residual oil (*4a* + *4b*) is dissolved in dry benzene, and anhydrous ammonia is bubbled into the solution for 0.5 hr. The mixture is purged with nitrogen for 10 min, heated to boiling, and treated with activated charcoal. The suspension is filtered through a pad of diatomaceous silica,[b] and the filtrate is evaporated to dryness *in vacuo*. The residue (60 g) is shown by TLC on silica gel H[c] with 19:1 chloroform—methanol (detection by spraying with alkaline potassium permanganate and by exposure to iodine vapor) to consist of *5a* plus *5b*, together with traces of impurities.

A suspension (protected from atmospheric moisture) of lithium borohydride (22.5 g) in freshly purified tetrahydrofuran (THF[d]; 3 l) is boiled under reflux for 1 hr and cooled to 40°. A

---

[b]Celite, Johns—Manville Products Corp., Lompoc, California.

[c]Silica Gel H for TLC (type 60) by E. Merck, distributed by Brinkmann Instruments, Inc., Westburg, New Jersey.

[d]The THF used in this reaction is dried and purified as follows. A mixture of commercial THF and calcium hydride is stirred overnight and filtered. Sodium bis(2-methoxyethoxy)aluminum hydride is added to the filtrate until bubbling ceases, and the THF is then distilled.

solution of the crude mixture of *5a* and *5b* in freshly purified THF[d] (200 ml) is added dropwise, with stirring, and the mixture is boiled under reflux for 2 hr, cooled to 5°, and diluted by dropwise addition of water (1 l). Amberlite CG-120 (H^+) cation-exchange resin (≈500 g) is added slowly, and the mixture is stirred overnight. The resin is separated by filtration and washed with water (1 l). The filtrate and washings are combined and treated with Dowex 50, 1 × 8, (OH^-) anion-exchange resin (≈700 g) to adjust the pH to 7. The resin is removed by filtration and washed with water (1 l), and the filtrate and washings are combined and evaporated under reduced pressure at 40° to a syrup. Two 150-ml portions of methanol are added to and evaporated from the syrup, which is then stirred with THF (150 ml) until it crystallizes. The white solid (*6a* + *6b*) is collected by filtration and dried *in vacuo*; yield, 34 g (83% from *2*). Thin-layer chromatography (80 μg applied to a plate of silica gel H[c]) with 3:1 chloroform—methanol (detection by spraying with alkaline potassium permanganate) showed only one spot; *6a* is not separated from *6b* by this system (or any other system tried).

To an aqueous solution of barium hydroxide, prepared by heating a mixture of water (6.9 l) and barium hydroxide octahydrate (340 g) to 40° to dissolve the hydroxide and then filtering to remove barium carbonate, is added bromine (37.7 g) at 5° to 10°.

As soon as the bromine has dissolved, a solution of the mixture *6a* and *6b* (34 g) in water (100 ml) is added in one portion. The solution is allowed to warm to room temperature, heated for 1 hr at 65° to 70°, refrigerated overnight, and made acid with 3M sulfuric acid (to pH 2). The mixture is stirred for 1 hr at 10° and filtered through a pad of diatomaceous silica[b] to remove barium sulfate, and the filtrate is allowed to drip through a column of Amberlite CG-120 (H^+) cation-exchange resin (420 g). The column is washed successively with water (4 l), 0.05M aqueous ammonia (until *7b* has been completely eluted), and 2M aqueous ammonia to elute *7a*. The 2M ammonia eluate is evaporated under reduced pressure to a syrup, which is dissolved in ethanol, the solution filtered to remove particles of resin, and the filtrate evaporated to a pale-yellow syrup[e]: yield of *7a*, 8.2 g (24.3% from *2*); TLC (80 μg applied to a plate of silica gel H[c]) with 4:1 ethanol—15M aqueous ammonia, and detection by spraying with alkaline potassium permanganate and with ninhydrin solutions, showed 1

---

[e] Compounds *7a* and *7b* readily absorb moisture. A small specimen of *7a* for analysis is dried *in vacuo* for 2 hr at 78° and handled as a hygroscopic compound during analytical procedures. Analysis calculated for $C_6H_{13}NO_2$: C, 54.93; H, 9.99; N, 10.68. Found: C, 54.65; H, 9.94; N, 10.46.

spot; ν_{max}^{film}[f] 3325, 3265, 3175, 3100, 2940, 2915, 2860, 2750, 1580, 1440, 1370, 1330, 1225, 1160, 1070, 1020, 960, 930, and 820 cm^{-1}; mass spectrum: *m/e* 131 (M^+), 114 $(M-OH)^+$, and 100 $(M-CH_2OH)^+$.

By the same procedure, *7b* is obtained from the 0.05<u>M</u> ammonia eluate as a red-brown syrup[g]; yield, 6.8 g (20.1% from *2*); TLC as above for *7a* showed 1 spot; ν_{max}^{film}[f] 3330, 3270, 3170, 2925, 2860, 2750, 1590, 1470, 1440, 1375, 1315, 1240, 1175, 1080, 1050, 1030, 965, 900, and 800 cm^{-1}; mass spectrum: *m/e* 131 (M^+), 113 $(M-H_2O)^+$, and 100 $(M-CH_2OH)^+$. The yield of *7a* + *7b* from *2* is 44.4%.

REFERENCE

1. Y. F. Shealy and C. A. O'Dell, *Tetrahedron Lett.*, 2231 (1969).

---

[f]The IR spectra of *7a* and *7b* consist of broad bands and shoulders. The most distinctive differences are the presence of a medium band at 1160 to 1150 cm^{-1} and a weak band at 820 to 810 cm^{-1} in the spectrum of *7a* and the absence of these distinct bands in the spectrum of *7b*.

[g]Short-path distillation of a specimen of *7b* at 25 μtorr gives a pale-yellow syrup. Found: C, 54.58; H, 9.80; N, 10.52.

[33] SYNTHESIS OF CRYSTALLINE PENTOFURANOSYL HALIDES SUITABLE FOR THE SYNTHESIS OF NUCLEOSIDES AND THEIR ANALOGS

Hassan S. El Khadem

Department of Chemistry and Chemical Engineering, Michigan Technological University, Houghton, Michigan 49931

INTRODUCTION

Pentofuranosyl halides are valuable synthetic intermediates for the preparation of nucleosides and nucleoside analogs. Crystalline halides are preferred as they can be purified and stored. Described here is the synthesis of three crystalline pentofuranosyl bromides: (a) tri-*O*-(*p*-nitrobenzoyl)-β-D-ribofuranosyl bromide (*3*), (b) tri-*O*-(*p*-nitrobenzoyl)-D-xylofuranosyl bromide (*9*), and (c) 3-deoxy-2,5-di-*O*-(*p*-nitrobenzoyl)-α-D-*threo*-pentofuranosyl bromide (*16*).

Bromides are better leaving groups than chlorides and fluorides and are, therefore, more suited for nucleophilic replacement. They are also more stable than iodides. Substitution on the benzoyl groups used for protection of pentofuranosyl halides often yields crystalline derivatives by increasing the polarity of the benzene ring in the order NO_2>OMe>Cl>Me. The *p*-nitrobenzoyl group is by far the most favorable to inducing crystallization.

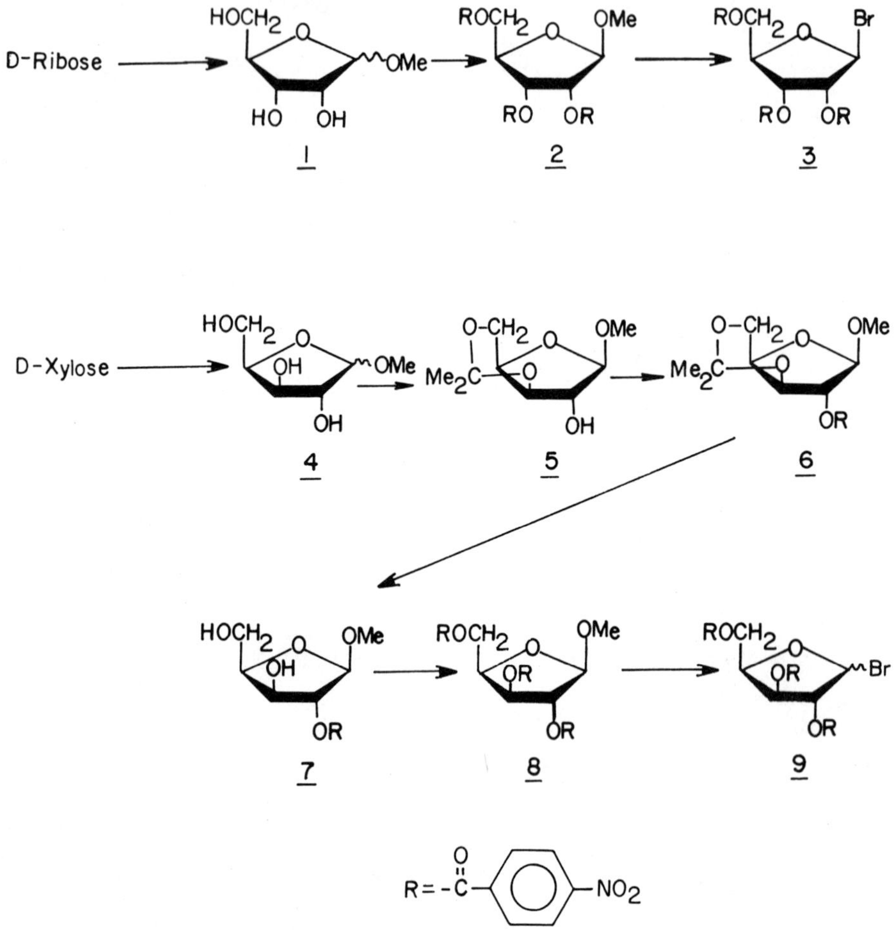

Although *p*-nitrobenzoyl groups on *O*-2 tend to destabilize the (cyclic) orthoester carbonium ion that is the intermediate in a nucleophilic replacement, there is evidence that such intermediates are formed, and the three pentofuranosyl bromides described here react with nucleophiles to give anomerically pure products having the *trans* configuration between the aroyl group on *O*-2 and

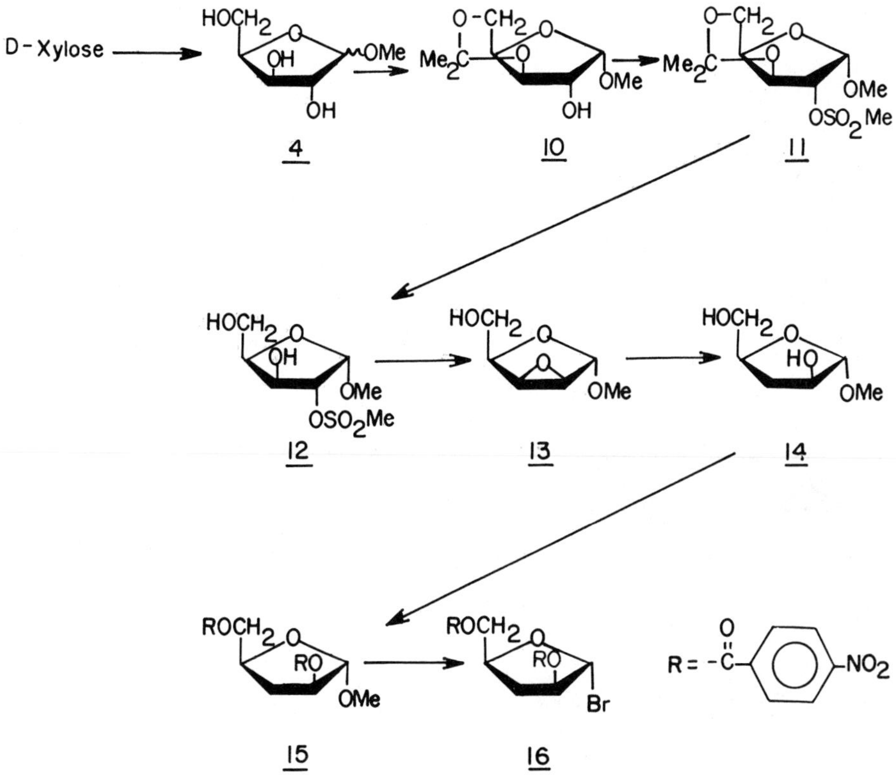

the entering nucleophile.

The procedures herein given for the synthesis of the three pentofuranosyl bromides start with D-ribose and D-xylose and follow the scheme outlined.

TRI-*O*-(*p*-NITROBENZOYL)-β-D-RIBOFURANOSYL BROMIDE (*3*)

Background

At present the most widely used D-ribofuranosyl halides are syrups: (a) tri-*O*-benzoyl-D-ribofuranosyl chloride[1] and

(b) bromide.[2] Both are obtained from a crystalline precursor, 1-*O*-acetyl-2,3,5-tri-*O*-benzoyl-β-D-ribofuranose.[2] Attempts to prepare crystalline *p*-substituted-benzoyl-D-ribofuranosyl halides such as the *p*-toluoyl and *p*-anisoyl[3] derivatives resulted in the formation of syrupy products. Two crystalline D-ribofuranosyl halides have recently been described; tri-*O*-(*p*-chlorobenzoyl)-β-D-ribofuranosyl bromide was obtained by Szerker and Bardos[4] in five steps from methyl D-ribofuranoside, and 2,3-*O*-isopropylidene-5-*O*-trityl-D-ribofuranosyl chloride was prepared by Fox and co-workers[5] for use in reactions where neighboring-group participation is not desired. Lately, Earl and Townsend[6] have succeeded in crystallizing Todd's tri-*O*-acetyl-β-D-ribofuranosyl chloride.[7]

Procedure

Methyl 2,3,5-tri-O-(p-nitrobenzyl)-β-D-ribofuranoside[8] *(2)*. To a solution of D-ribose (12.5 g) in anhydrous methanol (250 ml) is added anhydrous methanol (25 ml) containing hydrogen chloride (1.11 g). The solution is kept for 90 min at room temperature and the reaction is quenched with pyridine (25 ml). The solution is evaporated to a yellow syrup, which is treated with pyridine (25 ml), and the evaporation is repeated. A solution of the yellow syrup[2] (*1*) in pyridine (300 ml) and chloroform (100 ml) is stirred and cooled to 0°, and *p*-nitrobenzoyl chloride (50 g) is added in portions during 90 min below 10°; the mixture is kept for 15 hr at room temperature and then cooled to 0°. Ice (300 g) is added, and the aqueous layer is extracted with two 25-ml portions of dichloromethane. The extracts are combined and evaporated under diminished pressure to remove most of the pyridine. The solid residue is dissolved in dichloromethane (500 ml) and the solution successively washed with a saturated solution of sodium hydrogen carbonate (4 × 200 ml), water (4 × 200 ml), ice-cold 1.5M sulfuric acid (2 × 150 ml), and water (2 × 200 ml), dried (anhydrous magnesium sulfate), and evaporated to dryness under diminished pressure. The solid product is triturated with anhydrous ether, and the resulting crystals are recrystallized from acetone—petroleum ether; yield: 44.5 g (86%); m.p. 169.5° to 170°, $[\alpha]_D^{20}$ +79° (*c* 1.38, chloroform).

2,3,5-Tri-O-(p-nitrobenzoyl)-β-D-ribofuranosyl bromide[8] *(3)*. A solution of compound *2* (7.25 g) in anhydrous dichloromethane (40 ml) is cooled to 0°, and treated with acetyl bromide (5 ml) and a saturated solution of hydrogen bromide in acetic acid (75 ml) purified by treating it with $B(OAc)_3$ according to the method prescribed by Jolley.[9] The mixture is kept for five days at 4° and then evaporated under diminished pressure to an orange syrup. Slow coevaporation with 25-ml portions of dry benzene and then with dry benzene (20 ml) plus dry toluene (20 ml) affords the bromide as light-yellow crystals. These are washed with anhydrous diethyl ether and dried overnight under high vacuum. The yield is 6.19 g (79%) and recrystallized from dichloromethane it gives needles, m.p. 100° to 105°, $[\alpha]_D^{20}$ +55.4° (*c* 1.58, chloroform).

TRI-*O*-(*p*-NITROBENZOYL)-D-XYLOFURANOSYL BROMIDE (9)

Background

The first tri-*O*-acyl-D-xylofuranosyl halide to be obtained was the chloride, prepared by Baker and Schaub[10] from 2,3,5-tri-*O*-benzoyl-D-xylofuranose and hydrogen chloride; they also described tri-*O*-benzoyl-D-xylofuranosyl bromide, later prepared from 2,3,5-tri-*O*-benzoyl-D-xylofuranose by Fox and co-workers,[11] who also obtained Baker's chloride from 1-*O*-acetyl-2,3,5-tri-*O*-benzoyl-D-xylofuranose. A number of D-xylofuranosyl halides having good leaving groups, such as the tosyloxy and mesyloxy groups, on C-3 were prepared by Baker and co-workers,[12,13] Finally tri-*O*-acetyl-D-xylofuranosyl bromide[14] and chloride[15] were prepared from tetra-*O*-acetyl-D-xylofuranose. All of these D-xylofuranosyl bromides and chlorides as well as two D-xylofuranosyl fluorides recently described[16,17] were syrups.

Procedure

Methyl-3,5-O-isopropylidene α-and β-D-xylofuranoside[18] *(4)*. D-Xylose (110 g, 73 mmol) in 0.5% hydrogen chloride in methanol (3 l) is stirred for 5 hr at room temperature and then neutralized with DOWEX-1 X-8 (OH^-) anion-exchange resin (600 ml). The filtrate and washings are combined and evaporated to dryness, giving a mixture of methyl α- and β-D-xylofuranoside, which is mixed with acetone (1.5 l), *p*-toluenesulfonic acid (0.60 g), and anhydrous copper(II) sulfate (225 g) and stirred for 60 hr at room temperature. The solid is removed by filtration and washed with acetone (3 × 200 ml); the filtrate and washings are combined, made neutral with concentrated ammonia (2.5 ml), and evaporated under diminished pressure at 40°. A solution of the residue in water (100 ml) is extracted with dichloromethane (5 × 150 ml), and the extracts are combined, dried (magnesium sulfate) and evaporated under diminished pressure at 40° to yield a mixture of methyl 3,5-*O*-isopropylidene-α- and β-D-xylofuranoside (130 g). Distillation of the mixture at 25 mtorr (micro-Vigreux column) gives the α anomer (*10*) at 66° to 67° (yield 45.2 g), which is used in Part III. Subsequent distillation at 25 mtorr (Claisen head) gives the β anomer (*5*) at 83° to 85° (yield 42.9 g).

Methyl 3,5-O-isopropylidene-2-O-(p-nitrobenzoyl)-β-D-xylofuranoside[8] (*6*). A solution of methyl 3,5-*O*-isopropylidene-β-D-xylofuranoside[18] (*5*, 42.9 g) in pyridine (250 ml) is cooled in an ice bath, *p*-nitrobenzoyl chloride (43.1 g) is added, and the mixture is kept for 30 min in the ice bath and stirred for 24 hr at room temperature. Ice (400 g) is added, and the mixture kept for 1 hr at room temperature to hydrolyze the excess of *p*-nitrobenzoyl chloride. Dichloromethane (500 ml) is added and the suspension filtered to remove the *p*-nitrobenzoic acid that separates. The filtrate is successively washed with water (250 ml) and a cold saturated solution of sodium hydrogen carbonate (3 × 500 ml), dried (anhydrous magnesium sulfate), and evaporated to dryness under diminished pressure at 40°. Toluene (3 × 100 ml) is added to and evaporated from the residue to remove the last traces of pyridine, and the product is recrystallized from methanol (66.2 g, 89.2%), m.p. 94° to 95°.

*Methyl 2,3,5-tri-*O-(p-*nitrobenzoyl)*-β-D-*xylofuranoside*[8] *(8)*. A solution of compound *8* in 70% acetic acid (23.0 ml) is kept for 2 hr at 50°, evaporated to dryness under diminished pressure at 50°, and the residue dried by adding and evaporating toluene (25 ml) and then ethanol (15 ml). A solution of the resultant greenish syrup (7) in pyridine (125 ml) is cooled in an ice bath and treated with *p*-nitrobenzoyl chloride (11.6 g); the mixture is stirred under anhydrous conditions for 30 min in an ice bath and for 20 hr at room temperature. Ice (100 g) is added, the mixture is kept until the ice has melted, and dichloromethane (125 ml) is added. The suspension is filtered, and the filtrate is transferred to a separatory funnel. The aqueous layer is removed, and the dichloromethane layer is successively washed with a saturated aqueous solution of sodium hydrogen carbonate (3 × 150 ml) and water (150 ml), dried (anhydrous magnesium sulfate), and evaporated to dryness. The residue is dissolved in the minimal volume of boiling acetone, and petroleum ether is added until the ratio of acetone to petroleum ether is 2:3. The resulting crystals are removed by filtration and recrystallized twice from acetone—petroleum ether; yield, 15.3 g of chromatographically pure crystals, m.p. 144°. The mother liquors afford another 1.2 g of crystals (overall yield, 95.3%).

2,3,5-Tri-O-(p-nitrobenzoyl)-α,β-D-xylofuranosyl bromide[8] *(9)*. A solution of compound *8* (3 g) in dry dichloromethane (60 ml) presaturated with hydrogen bromide is kept for 24 hr at 0° to 3° under anhydrous conditions. The mixture is evaporated under diminished pressure at 40°, and then dry toluene (3 × 25 ml) is added to and evaporated from the product. The resulting solid is crystallized from chloroform—hexane (yield, 2.2 g, 68%). Recrystallization from chloroform—hexane gives an analytical sample of the anomers of the bromide. Integration of the NMR spectrum of the mixed bromides (9) shows the α and β anomers to be present in the ratio of 5:2.

3-DEOXY-2,5-DI-*O*-(*p*-NITROBENZOYL)-α-D-*THREO*-PENTOFURANOSYL BROMIDE (*16*)

Background

This synthesis uses, as the starting material, methyl 3,5-*O*-isopropylidene-α-D-xylofuranoside (*10*). Mesylation, deblocking, and treatment with methoxide affords *13*, which is then catalytically hydrogenated exclusively at O-3 to give methyl 3-deoxy-α-D-*threo*-pentofuranoside (*14*). *p*-Nitrobenzoylation and treatment with hydrogen bromide affords the desired 3-deoxy-2,5-di-*O*-(*p*-nitrobenzoyl)-α-D-*threo*-pentofuranosyl bromide (*16*).

Procedure

Methyl 3,5-O-isopropylidene-2-O-(methylsulfonyl)-α-D-xylofuranoside (11). To a solution of methyl 3,5-*O*-isopropylidene-α-D-xylofuranoside[18] (37.4 g) in pyridine (105 ml) cooled to -5° is added methanesulfonyl chloride (25 ml) dropwise. The mixture is kept for 24 hr at 20°, poured onto crushed ice, stirred for 15 min, and then extracted with dichloromethane (4 × 250 ml). The extracts are combined and evaporated, and residual pyridine is removed by repeated co-evaporation with toluene—ethanol. The orange-colored residue crystallizes from a mixture of carbon tetrachloride (50 ml) and hexane (10 ml); yield 43.7 g (84.5%), m.p. 85° to 86°. After recrystallization from carbon tetrachloride, the crystals have m.p. 86° to 87°, $[\alpha]_D$ +103.8° (c 1, methanol) (lit.[18] for syrupy material, $[\alpha]_D$ +65.7°); R_F 0.15 and 0.43 in hexane—acetone (3:1 and 1:1).

Methyl 2-O-(methylsulfonyl)-α-D-xylofuranoside[18] (12).

A solution of compound *11* (40 g) in a mixture of acetic acid (120 ml) and water (60 ml) is heated for 2.5 hr at $50 \pm 1^\circ$, evaporated to dryness, and the residue coevaporated with toluene (3×10 ml); a yellow residue is obtained that crystallizes from ether in colorless plates; yield 32.8 g (95%), m.p. 89°, $[\alpha]_{\underline{\underline{D}}}^{22}$ +166.7° (*c* 1, methanol); R_F 0.11 and 0.22 in hexane—acetone (3:1 and 1:1).

Methyl 2,3-anhydro-α-D-lyxofuranoside[18] *(13)*. To a solution of compound *12* (30.5 g, 125 mmol) in methanol (150 ml) cooled to -10° is added a freshly prepared solution of 2M sodium methoxide (70 ml) in methanol with stirring during 1 hr. After 2 hr at -10° the mixture is kept for 3 days at $0^\circ \pm 1^\circ$ under dry nitrogen. Filtration through Celite, followed by washing with methanol (5×70 ml), removes most of the sodium methanesulfonate. The filtrate and washings are combined, neutralized with acetic acid (3.5 ml), and evaporated to dryness. The residue is dissolved in water (30 ml) and the solution extracted with dichloromethane (5×50 ml). The extracts are combined, dried, and evaporated to a light-yellow solid, which is recrystallized from hot benzene (20 ml) plus hexane (2 ml); yield 17.4 g (94.5%), m.p. 81° to 82°, lit.[18] m.p. 80° to 82°; R_F 0.22 in 13:7 hexane—acetone.

*Methyl 3-deoxy-α-D-*threo-*pentofuranoside*[19] *(14)*. Compound *13* (5 g, 35 mmol) in ethanol (100 ml) containing Raney nickel (two teaspoonsfuls) is hydrogenated for 21 hr at 60° to 65°/6C lb·in^{-2}. The suspension is filtered through Celite and the catalyst washed with ethanol (5 × 50 ml). The filtrate and washings are combined and evaporated to a syrup (5 g), which is chromatographed on a column (3 × 48 cm) of silica gel with 4:1 hexane—acetone for elution. Fractions (15 ml each) are collected and examined by TLC; fractions 38 to 50 (2.59 g) are pure, whereas fractions 34 to 36 and 51 to 56 contain traces of impurities that can be removed by one recrystallization from ethanol. Total yield of *4* (after purification): 4.6 g (88.9%), m.p. 47°, $[\alpha]_{\underline{\underline{D}}}^{22}$ +134.8° (c 1, methanol); R_F 0.169 in 13:7 hexane—acetone.

Methyl 3-deoxy-2,5-di-O-(p-nitrobenzoyl)-α-D-threo-pentofuranoside[19] *(15).* To a solution of compound *14* (22.2 g, 150 mmol) in pyridine (600 ml), cooled to -5°, is added *p*-nitrobenzoyl chloride (84.6 g, 450 mmol). After 2 hr at -5° the mixture is stirred for 24 hr at 20°, poured onto crushed ice (3.5 l), and kept for 15 hr at 5°. The solid that separates is removed by filtration, washed with cold water, dissolved in hot benzene (500 ml), and the solution passed through a column (2 × 30 cm) of silica gel which is then eluted with benzene (1.5 l). The eluate is concentrated to 150 ml, hexane (20 ml) is added, and the product crystallizes slowly as light-yellow needles; yield: 44.6 g (66.6%), m.p. 135° to 136°, $[\alpha]_{\underline{\underline{D}}}^{22}$ +49.8° (c 1, chloroform); R_F 0.67 in

13:7 hexane—acetone.

*3-Deoxy-2,5-di-O-p-nitrobenzoyl-α-*D*-*threo*-pentofuranosyl bromide*[19] *(16)*. A suspension of compound *15* (10 g, 25 mmol) in acetic acid (45 ml), acetyl bromide (3 ml), and dichloromethane (45 ml) is gently warmed until dissolution occurs and then cooled to 5°. Cold, anhydrous hydrogen bromide in acetic acid (33% w/v, 45 ml) is added, and the mixture is stirred for 1.25 hr at 0° ± 2°. Co-evaporation with toluene (3 × 30 ml) gives a yellow solid, which is recrystallized from dichloromethane (20 ml) plus ether (15 ml) to afford light-yellow plates; yield: 8.6 g (77.0%), m.p. 152° to 154°, $[\alpha]_D^{22}$ +107.8° (c 1, chloroform); R_F 0.210 in 3:1 hexane—acetone.

REFERENCES

1. H. M. Kissman, C. Pidacks, and B. R. Baker, *J. Am. Chem. Soc.*, 77, 18 (1955).
2. F. Weigand and F. Wirth, *Chem. Ber.*, 85, 100 (1952).
3. H. Zinner and L. Balan, *J. Prakt. Chem.*, 18, 79 (1962).
4. G. L. Szekes and T. J. Bardos, *J. Med. Chem.*, 13, 708 (1970).
5. R. S. Klein, H. Ohrui, and J. J. Fox, *J. Carbohydr. Nucleos. Nucleot.*, 1, 265 (1974).
6. R. A. Earl and L. B. Townsend, *J. Carbohydr. Nucleos. Nucleot.* 1, 177 (1976).
7. G. A. Howard, B. Lythgoe, and A. R. Todd, *J. Chem. Soc.*, 1052 (1947).

8. H. S. El Khadem, T. D. Andichya, D. A. Niemeyer, and J. Kloss, *Carbohydr. Res.*, 47, 233 to 240 (1976).

9. W. Jolly, *The Synthesis and Characterization of Inorganic Compounds*, Prentice-Hall, Englewood Cliffs, New Jersey, 1970, p. 116.

10. B. R. Baker and R. E. Schaub, *J. Am. Chem. Soc.*, 77, 5900 (1955).

11. J. J. Fox, N. Yung, J. Cavoll, and G. B. Brown, *J. Am. Chem. Soc.*, 78, 2117 (1956).

12. A. Benitez, O. P. Crews, Jr., L. Goodman, and B. R. Baker, *J. Org. Chem.*, 25, 1946 (1960).

13. C. D. Anderson, L. Goodman, and B. R. Baker, *J. Am. Chem. Soc.*, 81, 3967 (1959).

14. O. P. Crews, Jr., and L. Goodman, in W. W. Zorbach and R. S. Tipson (eds.), *Synthetic Procedures in Nucleic Acid Chemistry*, Wiley, New York, Vol. I, 1968, pp. 139 to 142.

15. A. Magnani and Y. Mikuriya, *Carbohydr. Res.*, 28, 158 (1973).

16. L. D. Hall, P. R. Steiner, and C. Pedersen, *Can. J. Chem.*, 48, 1155 (1970).

17. S. Jacobsen, S. R. Jenson, and C. Pedersen, *Acta Chem. Scand.*, 26, 1561 (1972).

18. B. R. Baker, R. E. Schaub, and J. H. Williams, *J. Am. Chem. Soc.*, 77, 7 (1955).

19. H. S. El Khadem, T. D. Audichya, and M. J. Withee, *Carbohydr. Res.*, 33, 329 (1974).

[34] METHYL 2-DEOXY-4-THIO-β-D-*ERYTHRO*-PENTOFURANOSIDE AND ITS α ANOMER

The Displacement of the p-*Chlorophenylsulfonyloxy Group with the Thioacetate Anion*

Yun-Lung Fu and Miroslav Bobek

Department of Experimental Therapeutics and Grace Cancer Drug Center, Roswell Park Memorial Institute, New York State, Department of Health, Buffalo, New York 14263

INTRODUCTION

Anomeric methyl 2-deoxy-4-thio-D-ribofuranosides are useful as intermediates in the preparation of 2'-deoxy-4'-thionucleosides. The title compounds (*12* and *13*), which have been synthesized by Nayak and Whistler,[1] can be prepared more conveniently from methyl 2-*O*-benzoyl-3-*O*-*p*-tolylsulfonyl-L-arabinopyranoside (*1*) via selective epoxide opening with lithium aluminum hydride and nucleophilic displacement of the *p*-chlorophenylsulfonyloxy group with the thioacetate anion.[2]

PROCEDURE

Methyl 2-O-Benzoyl-3-O-p-tolylsulfonyl-β-L-arabinopyranoside (1)

To a solution of methyl 2-*O*-benzoyl-β-L-arabinopyranoside[3] (60 g, 224 mmol) in anhydrous pyridine (200 ml) cooled to below -10° in

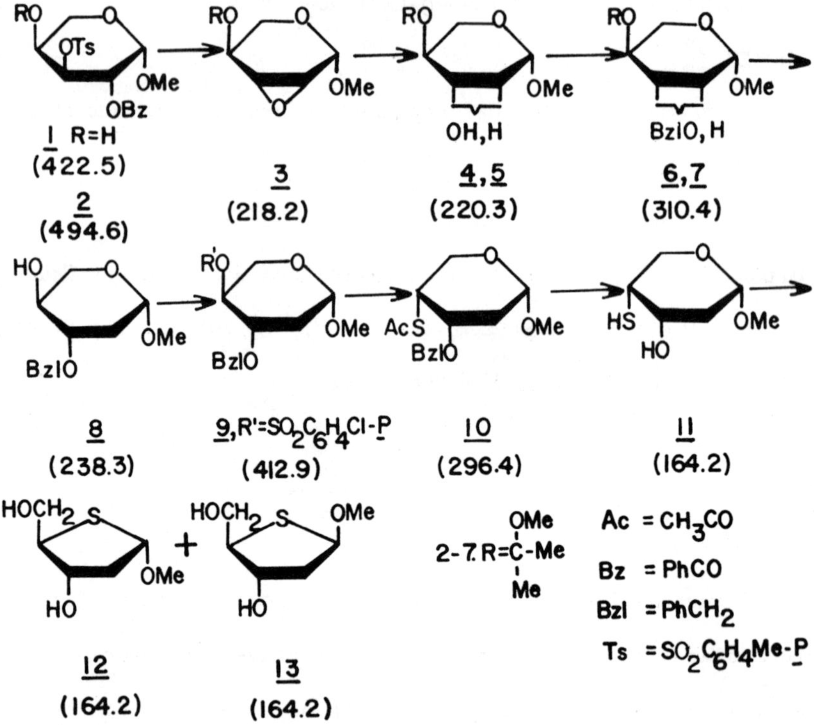

an ice-bath is added dropwise, with stirring and continuous cooling below -10°, a precooled (0°) solution of *p*-toluenesulfonyl chloride (45.2 g, 236 mmol) in anhydrous pyridine (200 ml). The mixture is stirred for 16 hr at 0° and then for 7 hr at room temperature. Ice water (800 ml) is added dropwise, with stirring, and after 16 hr at room temperature the precipitate (*A*, which largely contains methyl 2-*O*-benzoyl-3,4-di-*O*-*p*-tolylsulfonyl-β-L-arabinoside and some *1*) is removed by filtration and the filtrate is extracted with chloroform (5 × 200 ml). The extracts are combined and evaporated to a syrup, which is dissolved in chloroform

(600 ml). The solution is washed successively with cold (0°) 1M hydrochloric acid (2 × 100 ml), water (2 × 100 ml), saturated sodium hydrogen carbonate solution (3 × 100 ml), and water (until the aqueous layer is neutral) and is dried (sodium sulfate) and evaporated[a] to a syrup that is dissolved in hot methanol (200 ml), and the solution evaporated[a] to a viscous residue. This is cooled in ice for 3 hr, and is then kept at room temperature (for crystallization). Methanol (30 ml) is added and the crystals collected by filtration. Recrystallization from methanol gives 48 g of *1*; m.p. 112.5° to 112.7° (lit.[4] m.p. 111° to 113°).

A solution of the precipitated disulfonate by-product (A) in chloroform (200 ml) is washed with water (2 × 100 ml), dried (sodium sulfate), and evaporated[a] to a viscous residue which is dissolved in boiling methanol (300 ml), and the solution kept at room temperature for crystallization. The 3,4-di-*p*-toluenesulfonate crystallizes in the form of needles. The crystals are removed by filtration and washed with methanol (2 × 20 ml), m.p. 162° to 163°. The filtrates are combined and evaporated[a] to a syrup, which is dissolved in boiling methanol (50 ml). The solution is cooled to 25°, and nucleated with a few crystals of *1*. After being kept overnight at room temperature, the crystals are collected by filtration to give an additional 6 g (m.p. 112° to

---

[a]Evaporation is performed under diminished pressure at a bath temperature of 30° to 40°.

113°) of compound *1*; total yield 44 g (57.5%).

Methyl 2-O-Benzoyl-4-O-(methoxyisopropyl)-3-O-(p-tolylsulfonyl)-β-L-lyxoside (2)

A solution of compound *1* (41 g, 97 mmol) in alcohol-free chloroform[b] (100 ml) is diluted with anhydrous benzene[c] (100 ml). To this solution is added 2-methoxypropene[5] (50 g) followed by 5 drops of a saturated solution of hydrogen chloride in chloroform, and the mixture is kept under nitrogen for 1 hr at room temperature. To neutralize the hydrogen chloride and to keep the solution alkaline,[d] triethylamine (1 ml) is added to the mixture while the flask is swirled. The resulting mixture is evaporated[a] at a bath temperature of 30° to a syrup. On addition of absolute alcohol (25 ml), compound *2* crystallizes readily. The crystals are collected by filtration; yield 47 g (98%), m.p. 118° to 119°, $[\alpha]_D^{25}$ +198.3° (c 2.9, chloroform); NMR data ($CDCl_3$): δ 1.41 (s, 6 H, 2 $C-CH_3$), 2.29 (s, 3 H, tosyl CH_3), 3.30 and 3.33 (two s, 6 H, two $O-CH_3$), and 6.98 to 7.88 (m, 9 H, aromatic H). If used for the preparation of *3*, compound *2* need not be isolated.[d]

Methyl 2,3-Anhydro-4-O-(1-methoxyisopropyl)-β-L-lyxoside (3)

A solution of compound *2* (51 g, 103 mmol) in anhydrous benzene

---

[b] Alcohol-free chloroform is prepared by passing reagent-grade chloroform through a column of activated alumina.

[c] Refluxed with, and distilled from, metallic sodium.

[d] The 1-methoxyisopropyl group is very labile; it is readily hydrolyzed in the presence of water and a catalytic amount of acid.

(500 ml) is cooled to 0° in an ice-water bath, and freshly prepared methanolic sodium methoxide [from sodium (2.5 g, 109 mmol) in anhydrous methanol (125 ml)] precooled to 0° is then added. The mixture is kept for 5 hr at room temperature, and the precipitate (sodium *p*-toluenesulfonate) is removed by filtration and washed with benzene (4 × 50 ml, each portion containing a few drops of triethylamine). The filtrates are combined and evaporated to a syrup,[a] which is dissolved in benzene (1 l). The solution is washed with water (50 ml), saturated with sodium chloride, and evaporated[a] to a syrup. Methyl benzoate is removed at 40°/0.01 torr, and compound *3* crystallizes as white needles during the process; yield 21.7 g (96.5%), m.p. 48.5° to 49°, $[\alpha]_D^{25}$ +56.5° (*c* 3.3, chloroform), NMR data ($CDCl_3$): δ 1.42 (s, 6 H, 2 C-CH_3), 3.26 and 3.45 (2 s, 6 H, 2 O-CH_3), and 4.93 (d, 1 H, $J_{1,2}$ 2.2 Hz, H-1).

*Methyl 2-Deoxy-4-O-(1-methoxyisopropyl)-β-L-*threo*-pentopyranoside (4) and Methyl 3-Deoxy-4-O-(1-methoxyisopropyl)-β-L-*threo*-pentofuranoside (5)*

Into a 2-l, 3-necked, round-bottomed flask, equipped with a magnetic stirrer, a dropping funnel, and a reflux condenser (drying tube) are introduced anhydrous ethyl ether (500 ml) and lithium aluminum hydride (9 g). To the stirred suspension is added, dropwise, a solution of *3* (21.8 g) in anhydrous ethyl ether (500 ml), in such a way as to produce gentle refluxing. After 3 hr the excess of hydride is decomposed by dropwise addition of ethyl acetate (30 ml) followed by water (50 ml). The white precipitate is removed by filtration and washed thoroughly with ethyl ether until all of the sugar derivative has been removed. Sodium chloride (20 g) is added to the combined filtrates, the mixture is extracted with benzene (4×100 ml), and the extracts are combined and evaporated[a] to give a crude, viscous mixture of compounds *4* and *5* (20.8 g, 94.4%), that can be used without purification for the next reaction.

For preparation of analytical samples, the mixture (2 g) is chromatographed on a column of dry silica gel (60 to 200 mesh) by eluting with 10:1:1 benzene—ethyl acetate—triethylamine to give syrupy compounds *4* (1.8 g) and *5* (0.15 g). The NMR data (acetone-d_6) are: for *4*, δ 1.35 (s, 6 H, 2 C-CH_3), 3.21 and 3.35 (2 s, 6 H, 2 O-CH_3), 4.48 (d of d, $J_{1,2,2'}$ 2.7 and 7.4 Hz, H-1); for *5*, δ 1.30 (s, 6 H, 2 C-CH_3), 3.2 and 3.35 (2 s, 6 H, 2 O-CH_3), 4.52 (d, $J_{1,2,2'}$ 3 Hz, H-1).

*Methyl 3-O-Benzyl-2-deoxy-β-L-*threo*-pentopyranoside (8)*

The syrupy mixture of compounds *4* and *5* (26.6 g) is dissolved in freshly distilled dimethyl sulfoxide (200 ml) with stirring under nitrogen. Sodium hydride (12 g)[e] is added to the solution in small portions, so that no vigorous evolution of hydrogen occurs, and the mixture is stirred and cooled in an ice bath[f] for 4 hr.[g] α-Bromotoluene (10 ml) is added dropwise to the cooled[f] mixture during 45 min, and the mixture is stirred for 6 to 8 hr at room temperature. Sodium hydride (10 g) and α-bromotoluene (10 ml) are added to the mixture, with cooling in an ice bath as just described, and the mixture is stirred for 6 to 8 hr at room temperature; this procedure is repeated twice. Triethylamine (40 ml) is now added dropwise to the mixture, with cooling and stirring, to convert the excess of α-bromotoluene into an ammonium salt. The excess of sodium hydride is decomposed by dropwise addition of water (50 ml) while the mixture is kept cold and well stirred. The mixture is then saturated with sodium chloride (50 g), and extracted with chloroform (4 × 300 ml), and the extracts are evaporated[a] to remove chloroform; dimethyl sulfoxide[h] is removed

---

[e]Obtained from 25 g of a 50% dispersion in oil, washed with anhydrous ethyl ether (4 × 100 ml), decanted, and dried under diminished pressure just before using.

[f]It is cooled in such a way that it can be stirred freely; thus, when the reaction mixture freezes on the walls of the reaction flask, the ice bath is removed, and, when it melts, the flask is recooled in the ice bath.

[g]Or, when the color of the reaction mixture turns a deep green before the time period stated, α-bromotoluene should be added.

[h]If dimethyl sulfoxide is not completely removed, it will lower the yield of compound *8*.

at 70°/0.01 torr, to give a syrup that is dissolved in benzene (500 ml). The solution is washed with water (50 ml) containing triethylamine (5 ml) and evaporated to give a syrupy mixture of compounds *6* and *7*. This mixture is dissolved in 4:1 (v/v) acetic acid—water (50 ml), and the solution is stirred for 10 min at room temperature, and evaporated to a syrup which is co-evaporated several times with portions (10 ml) of dry toluene until all of the acetic acid has been removed. The syrupy residue is dissolved in a boiling mixture of benzene (10 ml) and cyclohexane (50 ml), and the hot solution is covered with 300 ml of boiling petroleum ether (b.p. 38° to 50°). The solution is then cooled and kept at 4° to effect crystallization of compound *8* as long white needles. The crystals are collected by filtration and washed with 1:10 (v/v) cyclohexane—petroleum ether (3 × 20 ml); yield 20.7 g (72%), m.p. 73° to 74°; NMR data ($CDCl_3$): δ 1.65 and 2.24 (2 m, 2 H, H-2), 3.48 (s, 3 H, OCH_3), 4.39 (d of d, 1 H, $J_{1,2,2'}$ 2.8 and 8.2, H-1), 4.62 (d of d, 2 H, *J* 12 Hz, benzylic H), and 7.34 (br, s, 5 H, aromatic H).

Methyl 3-O-Benzyl-4-O-(p-chlorophenylsulfonyl)-2-deoxy-β-L-threo-pentoside (9)

To a solution of compound *8* (6 g, 25 mmol) in a mixture of alcohol-free chloroform (3 ml) and dry pyridine (50 ml) is added a solution of *p*-chlorobenzenesulfonyl chloride (30 g, 142 mmol) dry pyridine (100 ml), and the solution is stirred for 16 hr at 80°, cooled, and poured into ice water (500 ml), and the mixture is stirred for 16 hr at room temperature. It is then extracted with chloroform (4 × 100 ml), and the extracts are combined, successively washed with cold (0°) 2M hydrochloric acid (3 × 100 ml), water (2 × 100 ml), saturated sodium hydrogen carbonate solution (2 × 100 ml), and water (100-ml portions, until neutral), dried (sodium sulfate), and evaporated[a] to give syrupy *9*, which crystallizes from 95% ethanol as small needles; yield 11.3 g (98%), m.p. 126° to 127°, $[\alpha]_D^{25}$ + 56.7° (*c* 1.0, chloroform); NMR data ($CDCl_3$): δ 1.68 and 2.22 (2 m, 2 H, H-2), 3.44 (s, 3 H, OCH_3), 4.34 to 4.62 (3 H, benzylic and H-1, overlaps), and 7.12 to 7.88 (m, 9 H, aromatic H).

*Methyl 4-S-Acetyl-3-O-benzyl-2-deoxy-4-thio-α-D-*erythro*-pentoside (10)*

Compound *9* (20 g, 48 mmol) is stirred for 16 hr with freshly recrystallized potassium thiolacetate (20 g, 175 mmol) in dry *N,N*-dimethylformamide (200 ml) at 75° to 80° (oil-bath temperature) under a current of dry nitrogen. The mixture is then cooled to 0° and poured, with stirring, into dry xylene (2 l). After 16 hr at room temperature under dry nitrogen the precipitated salts are removed by filtration and washed with dry xylene (3 × 100 ml). The filtrates are combined and evaporated (oil pump) below 40° (bath temperature), and the residue is extracted with heptane (4 × 150 ml). The extracts are combined and evaporated[a] to a syrup that is dissolved in dry pyridine (100 ml), and the solution cooled to 0°. To this solution is added acetic anhydride (50 ml), and the mixture is stirred for 16 hr at room temperature, poured with stirring into ice water (1 l), and stirred for 16 hr at room temperature. The mixture is extracted with chloroform (4 × 100 ml), and the extracts are combined, washed with water (2 × 150 ml), dried (sodium sulfate), and evaporated[a] to a syrup, which on crystallization from 95% ethanol gives 9.9 g (70%) of compound *10*; m.p. 69° to 70°, $[\alpha]_{D}^{25}$ +13° (*c* 1.0, chloroform); NMR data ($CDCl_3$): δ 1.86 (m, 2 H, H-2), 2.38 (s, 3 H, thioacetyl H), 3.46 (s, 3 H, OCH_3), 4.42 (d of d, 1 H, $J_{1,2,2'}$ 7.0 and 3.5 Hz, H-1), 4.58 (s, 2 H, benzylic H), and 7.36 (br, s, 5 H, aromatic H).

*Methyl 2-Deoxy-4-thio-α-D-*erythro*-pentopyranoside (11)*

Liquid ammonia (200 ml, freshly distilled from sodium) is added to a solution of *10* (3.25 g, 11 mmol) in 1,2-dimethoxyethane (20 ml). While the solution is stirred under a current of dry nitrogen, freshly cut sodium is added in small pieces (70 to 100 mg) one at a time until the blue color of the solution persists for 30 min. The excess of sodium is then carefully decomposed with a slight excess of solid ammonium chloride, and the ammonia is allowed to evaporate in a current of dry nitrogen. Chloroform (50 ml) is added to the residue, and nitrogen is bubbled through the mixture to remove the last traces of ammonia. The mixture is then extracted with chloroform (3 × 50 ml) and filtered, and the solid is washed with chloroform (2 × 10 ml). The filtrate and washings are combined and evaporated[a] to syrupy *11*; yield 1.3 g; ν_{max}^{film} 2570 (SH) and 3500 cm^{-1} (OH).

*Methyl 2-Deoxy-4-thio-α- and -β-D-*erythro-*pentofuranosides (12 and 13)*

A solution (500 ml) of compound *11* (6.5 g) in 0.1% methanolic hydrogen chloride solution is boiled under reflux under a nitrogen atmosphere for 90 min, cooled to room temperature, and made neutral with solid lead carbonate. The precipitate is removed by filtration and washed with methanol (4 × 50 ml). The filtrate and washings are combined and evaporated to give a syrupy mixture of compounds *12* and *13*. This mixture is separated by chromatography on a column of dry silica gel (60 to 200 mesh) with 10:1 (v/v) chloroform—methanol as the eluant, the effluents being monitored by TLC on silica gel with 10:1 (v/v) chloroform—methanol. The fraction containing the component having R_F 0.31 is evaporated to yield syrupy *12*, 2.39 g (37%). The fraction containing the component having R_F 0.19 is evaporated, to give syrupy *13*, 3.86 g (59.5%). For *12*, $[\alpha]_D^{25}$ +315.6° (*c* 1.08, chloroform)[i]; NMR data ($CDCl_3$): δ 2.30 (m, 2 H, H-2), 3.26 (s, 3 H, OCH_3), 4.95 (d of d, 1 H, $J_{1,2,2'}$ 2.5 and 5.0 Hz, H-1); for *13*, $[\alpha]_{\underline{\underline{D}}}^{25}$ 278° (c 1.02, chloroform)[i]; NMR data ($CDCl_3$): δ 2.33 (m, 2H, H-2), 3.30 (s, 3 H, OCH_3), 4.98 (d of d, 1 H, $J_{1,2,2'}$ 2.5 and 4.5 Hz, H-1).

REFERENCES

1. U. G. Nayak and R. L. Whistler, *Justus Liebigs Ann. Chem.*, 741, 131 (1970).

2. Y. L. Fu and M. Bobek, *J. Org. Chem.*, 41, 3831 (1976).

3. M. A. Oldham and J. Honeyman, *J. Chem. Soc.*, 986 (1946).

4. E. J. Reist, L. V. Fisher, and D. E. Gueffroy, *J. Org. Chem.*, 31, 226 (1966).

5. M. S. Newman and M. C. V. Zwan, *J. Org. Chem.*, 38, 2910 (1973).

---

[i] For *12*, lit.[1] $[\alpha]_{\underline{\underline{D}}}^{25}$ +314° (c 1, chloroform); for *13*, $[\alpha]_{\underline{\underline{D}}}^{25}$ −278°.

[35] METHYL 2,3-ANHYDRO-5-DEOXY-α-D-RIBOFURANOSIDE

John A. Montgomery, H. Jeanette Thomas, and Sarah D. Clayton

Kettering-Meyer Laboratory, Southern Research Institute, Birmingham, Alabama 35205

INTRODUCTION

Methyl 2,3-anhydro-5-deoxy-α-D-ribofuranoside (*6*), an important intermediate for the synthesis of 5-deoxyisonucleosides,[1] as well as other 2- and 3-substituted 5-deoxyglycosides, has been prepared by Kuzuhara and Emoto[2] from 1,2-*O*-isopropylidene-α-D-xylofuranose (*1*).[a] An improved synthesis of *6* has been achieved starting from 1,2-*O*-isopropylidene-3,5-di-*O*-*p*-tolylsulfonyl-α-D-xylofuranose[3] (*2*).

PROCEDURE

1,2-O-Isopropylidene-3,5-di-O-p-tolylsulfonyl-α-D-xylofuranose[3] *(2)*

A solution of 1,2-*O*-isopropylidene-α-D-xylofuranose (*1*)[a] (104 g, 547 mmol) in pyridine (2 l), dried over 4A molecular sieve[b] is stirred and cooled to 5°. A solution of *p*-toluenesulfonyl

---

[a] Pfanstiehl Laboratories, Inc.

[b] Beads, size 8 × 12, from Linde Division, Union Carbide Corporation, 3839 West Alabama Street, Houston, Texas 77027.

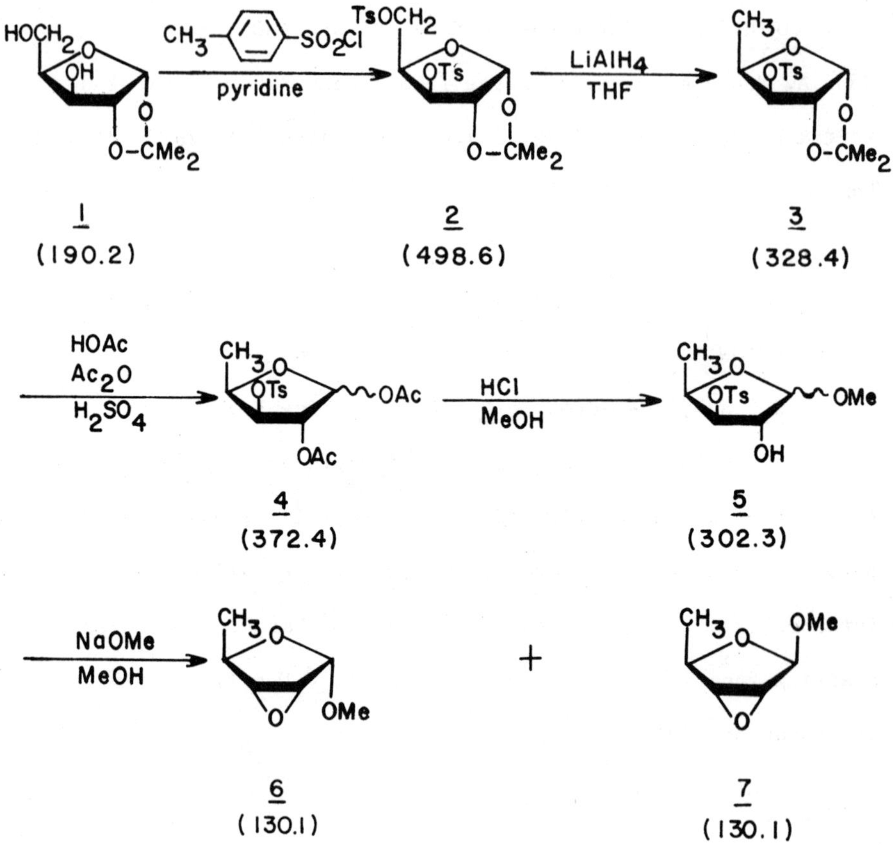

chloride (210 g, 1.094 mol) in chloroform (500 ml) is added at such a rate that the temperature remains below 10°, and then the mixture is stirred overnight at room temperature. Thin-layer chromatography[c] with 1:1 benzene—ethyl acetate reveals the presence of monotosyl compound; therefore, the mixture is cooled to 5° before the addition of *p*-toluenesulfonyl chloride (110 g,

---

[c]Analtech Silica Gel GF plates, 250-μm layer.

547 mmol) in chloroform (250 ml). This procedure is repeated on the 4th day. On the 5th day only a small amount of monotosyl compound remains; therefore, a further portion (50 g, 263 mmol) of *p*-toluenesulfonyl chloride in chloroform (100 ml) is added, resulting in complete reaction. The mixture is chilled to 5°, water (160 ml) is added dropwise, and the mixture is stirred for 0.5 hr and then poured into ice and water (5 l). The product is extracted into chloroform (3 × 2.5 l), and the extracts are combined, freed of pyridine by washing with ice-cold 1.5M sulfuric acid (3 × 2.5 l), followed by water (3 × 2 l) until the washings are neutral, dried (magnesium sulfate), and evaporated to dryness. The residue is recrystallized from ethyl alcohol (1.3 l), yield 254.7 g (93.5%), m.p. 96° to 98° (capillary). The product is homogeneous by TLC with 1:1 benzene—ethyl acetate.

5-Deoxy-1,2-O-isopropylidene-3-O-p-tolylsulfonyl-β-D-xylofuranose (3)

A mixture of compound *2* (254.7 g, 0.51 mol) and lithium aluminum hydride (80 g) in tetrahydrofuran (2 l) is boiled overnight under reflux (TLC with 3:1 benzene—ethyl acetate indicated complete reaction) and then treated successively with water (80 ml), 15% sodium hydroxide solution (80 ml), and water (240 ml). The gelatinous solid is removed by filtration and washed with ether. The filtrates are combined and evaporated to dryness, the residue is dissolved in chloroform, and the solution is successively washed with M sodium hydroxide and water, dried (magnesium sulfate), and evaporated to dryness, to yield a chromatographically pure product; yield 137.9 g (82%). This material, identical with that prepared by the *p*-toluenesulfonylation of 5-deoxy-1,2-*O*-isopropylidene-α-D-xylofuranose,[2] is used in the next step without further purification.

1,2-Di-O-acetyl-5-deoxy-3-O-p-tolylsulfonyl-D-xylofuranose[2] (*4*)

A solution of compound *3* (137.9 g, 0.42 mol) in acetic acid (1.86 l) and acetic anhydride (211 ml) is cooled to 10°, and concentrated sulfuric acid (111 ml) is added dropwise below 20°. The solution is stirred overnight at room temperature, poured into ice and water (6 l), and extracted with chloroform (5 × 1 l). The extracts are combined, washed with saturated sodium hydrogen carbonate solution until the wash remains basic, and then with water until it is neutral. The chloroform solution is dried (magnesium sulfate) and evaporated to dryness to yield a syrup (chromatographically pure by TLC in 3:1 benzene—ethyl acetate); yield 145.8 g, 93.5%.

Methyl 5-Deoxy-3-O-p-tolylsulfonyl-D-xylofuranoside (*5*)

A solution of compound *4* (145.8 g, 0.39 mol) in 1% methanolic hydrogen chloride (2.4 l) is stirred for 20 hr at room temperature, chilled, and made neutral by addition of an excess of solid sodium hydrogen carbonate. The suspension is filtered, and the filtrate is evaporated under diminished pressure at 30° to 35°; the residue is suspended in chloroform (1.8 l), the suspension is dried (magnesium sulfate) and filtered, and the filtrate is evaporated to dryness. The resulting syrup (106.5 g, 90%) is an essentially pure mixture of the two anomers.[d] Separation of the anomers is accomplished by dry-column chromatography, followed by recrystallization. The syrup is dissolved in chloroform and deposited on a dry column of silica gel[e] (300 g) by evaporation with a rotary evaporator; the mixture is then placed on the top of a nylon column (7.5 cm) containing dry-column silica gel (1,700 g) preequilibrated with the elutant (99:1 chloroform—methanol).[f] Based on their ultraviolet absorption the bands are cut and separately eluted with ethyl acetate. Each eluate is evaporated, and the residue is recrystallized from benzene—hexane.[g] The yield of the β anomer is 21.5 g (18%), m.p. 70° to

---

[d]In another preparation of approximately the same size the crude mixture was used for preparing a mixture of the anomers of methyl 2,3-anhydro-5-deoxy-D-ribofuranoside, which were then separated by distillation: β, 19.7 g (44.4%), b.p. 46° to 62°/13.5 torr; α, 10.4 g (23.4%), b.p. 74° to 80°/13.5 torr.

[e]ICN Pharmaceuticals, Woelm silica gel, activity 111, containing a fluorescent indicator.

[f]This column must be supported in a glass column because chloroform attacks the nylon.

[g]These compounds are well separated in TLC with 3:1 benzene—ethyl acetate.

73° (capillary); ^{1}H NMR data ($CDCl_3$—Me_4Si): δ 4.75 (d, $J_{1,2}$ 2.0 Hz, H-1). The yield of the α anomer is 11.8 g (10%), m.p. 70° to 72° (capillary); ^{1}H NMR data ($CDCl_3$—Me_4Si): δ 4.90 (d, $J_{1,2}$ 5.0 Hz, H-1).

Methyl 2,3-Anhydro-5-deoxy-α(and β)-D-ribofuranoside[2] *(6 and 7)*

A solution of compound 5 (103 g, 0.34 mol) in dry[h] methanol (220 ml) is cooled to 0°, and a solution (at 0°) of sodium methoxide (21.1 g, 0.39 mol) in dry methanol (185 ml) is added. The mixture is then kept for 4 days at 4°, mixed with Celite (13 g) to facilitate filtration, and the mixture is stirred for a few minutes and filtered. The filtrate is rendered neutral in the cold with glacial acetic acid and carefully evaporated at atmospheric pressure (to avoid loss of the β anomer). The residue is suspended in ether, the suspension is dried (magnesium sulfate) and filtered, and the filtrate is evaporated at atmospheric pressure. The residue is distilled in a Vigreux column (20 cm) with receivers cooled in dry ice. The yield of the β anomer is 19.7 g (44.4%), b.p. 46° to 62°/13.5 torr. The yield of the α anomer is 10.4 g (23.4%), b.p. 74° to 80°/13.5 torr.[i]

---

[h]Dried over Linde 3A molecular sieve, 0.469-cm pellets.

[i]If this reaction is conducted with pure α anomer, the yield is 96% of satisfactory material. However, the yield of pure methyl 5-deoxy-3-*O*-*p*-tolylsulfonyl-α-D-xylofuranoside is 10%.

REFERENCES

1. J. A. Montgomery, S. D. Clayton, and H. J. Thomas, *J. Org. Chem.*, 40, 1923 (1975).

2. H. Kuzuhara and S. Emoto, *Agric. Biol. Chem.*, 27, 689 (1963).

3. P. Karrer and A. Boettcher, *Helv. Chim. Acta*, 36, 837 (1953).

[36] CRYSTALLINE 2,3,5-TRI-*O*-ACETYL-β-D-RIBOFURANOSYL CHLORIDE

Robert A. Earl and Leroy B. Townsend

Department of Biopharmaceutical Sciences and Department of Chemistry, University of Utah, Salt Lake City, Utah 84112

INTRODUCTION

Noncrystalline 2,3,5-tri-*O*-acetyl-β-D-ribofuranosyl chloride (*2*) has been prepared[1] by treating a cold, ethereal solution of 1,2,3,5-tetra-*O*-acetyl-β-D-ribofuranose (*1*) with anhydrous hydrogen chloride. However, it has been shown[2] that this procedure yields a product containing several unidentified impurities.

An improved procedure[2] utilizes dichloromethane as the solvent and yields a highly pure form of *2* that crystallizes readily. Crystalline *2* may be stored for months at -16°, under strictly anhydrous conditions, without showing evidence of decomposition.

Therefore, it is no longer necessary to use "freshly prepared" 2,3,5-tri-*O*-acetyl-β-D-ribofuranosyl chloride whenever *2* is needed for a reaction. Crystalline *2* has been used successfully[3] for the preparation of a pyrazolo[3,4-*d*]pyrimidine nucleoside possessing significant biological activity.

PROCEDURE

Syrupy *1* is prepared in quantitative yield by dissolving dry

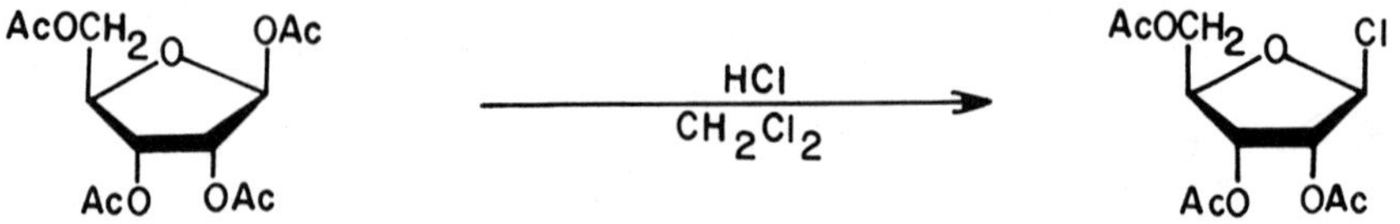

1 (318.3) 2 (294.7)

1,2,3,5-tetra-*O*-acetyl-β-D-ribofuranose[4] (*1*) (50 g) in cold (0°) dry[a] dichloromethane (320 ml) and rapidly saturating the solution with anhydrous hydrogen chloride (passed through anhydrous calcium chloride). The solution is kept for 2.5 hr at 0° while dry hydrogen chloride gas is slowly bubbled through the solution in order to maintain saturation. The solvent is then removed *in vacuo* (bath temperature, 25°), and the residue is coevaporated with dry xylene (3 × 50 ml). The colorless syrup is stored for 18 hr at -16° under vacuum and then allowed to warm slowly to room temperature, when crystallization usually commences.[b] Recrystallization is carried out by dissolving the solid in warm, dry peroxide-free isopropyl ether (freshly distilled from lithium aluminum hydride), cooling the solution to ≃ +20°, and introducing a nucleating crystal. The solution is then slowly cooled to +5° and

---

[a] In order to obtain a crystalline product it is absolutely essential that glassware and solvents be dry. Dichloromethane may be dried by refluxing it over calcium hydride, followed by distillation, and then storing the distillate over 3A molecular sieves.

[b] Spontaneous crystallization has been observed to occur during this stage of several different preparations. Occasionally several freeze—thaw cycles are required for this to occur.

kept for 1 hr at +5°. The supernatant liquor is decantated while the flask is flushed with dry nitrogen. The solid mass of crystals is then washed with cold (+5°) isopropyl ether (2 × 20 ml) in the same manner. The crystals are dried *in vacuo* for 1 hr at 25°; weight 40.8 g (88%), m.p.[c] 46° to 50°. A sample recrystallized from isopropyl ether has m.p. 50° to 51°, $[\alpha]_{D}^{27}$ -25.9° (*c* 1.45, ethanol-free chloroform); NMR data ($CDCl_3$): δ 6.06 (sharp singlet, H-1').

REFERENCES

1. H. Zinner, *Chem. Ber.*, 83, 153 (1950).

2. R. A. Earl and L. B. Townsend, *J. Carbohydr. Nucleos. Nucleot.*, 1, 177 (1974).

3. R. A. Earl and L. B. Townsend, *J. Heterocycl. Chem.*, 11, 1033 (1974).

4. R. D. Guthrie and S. C. Smith, *Chem. Ind.* (Lond.), 547 (1968); *Biochem. Prep.*, 13, 1 (1971).

---

[c]Samples of material with m.p. 46° to 50° have been stored over Drierite at -16° for nine months with no change in appearance, m.p. or optical rotation. Frozen samples should be thawed in a dessicator and the crystals kept under dry nitrogen as much as possible during transfer operations.

[37] 2-AMINO-1-β-D-RIBOFURANOSYLIMIDAZOLE

Reduction of the 2-Nitro Group of D-Ribosylazomycin

Ganapathi R. Revankar and Roland K. Robins

ICN Pharmaceuticals, Inc., Nucleic Acid Research Institute, Irvine, California 92715

INTRODUCTION

The isolation of a 4-amino-D-ribosylimidazole-5-carboxamide from the culture medium of sulfonamide-inhibited *Escherichia coli*,[1] and the enzymic synthesis of a 5-amino-D-ribosylimidazole[2] focused the attention of the present authors on imidazole nucleosides having an amino group attached to C-2 of the imidazole residue as in compound *2*. The synthesis of *2* described here uses the 1-β-D-ribofuranosyl derivative[3] (*1*) of the antibiotic azomycin.[4,5] Catalytic reduction of the D-ribosylazomycin provided an excellent yield of reasonably stable *2*.

PROCEDURE

2-Amino-1-β-D-ribofuranosylimidazole (*2*)

To a suspension of 2-nitro-1-β-D-ribofuranosylimidazole[3] (*1*, D-ribosylazomycin; 3.0 g, 12.2 mmol) in absolute methanol (300 ml) is added platinum oxide (0.30 g), and compound *1* is hydrogenated in a Parr hydrogenator at 40 lb·in^{-2} for 1.5 hr at 25° to

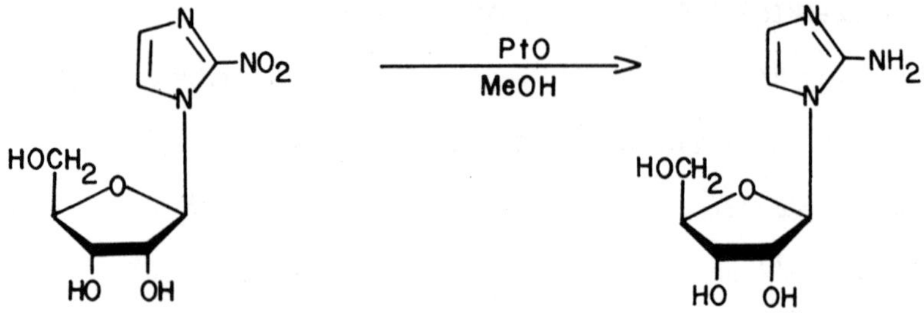

1 (245.2) 2 (215.2)

28°.[a] The suspension is removed by filtration through a Celite pad, and the filter cake is thoroughly washed with methanol (5 × 25 ml). The filtrate and washings are combined, flushed with nitrogen, and evaporated[b] to dryness under reduced pressure. The residual gum is chromatographed on a dry column (100 g, 3 × 50 cm) of silica gel[c] with 5:1 (v/v) methanol—chloroform as the eluant. The fractions containing the product are pooled, and evaporated[b] to dryness. A solution of the residue in ethanol (10 ml) is slowly poured into cold anhydrous ether (200 ml) with efficient stirring. The mixture is stirred for 2 to 3 hr at room temperature and then kept overnight in a refrigerator (0° to 4°). The

---

[a]At the end of this time TLC on silica gel F-254 (E. Merck) with the upper phase 4:2:1 (v/v) ethyl acetate—water—propyl alcohol, showed one major product, R_F 0.035 (starting material, R_F 0.66).

[b]The evaporation is conducted in a rotary evaporator under diminished pressure (bath temperature <30°).

[c]ICN-Woelm silica gel (70 to 230 mesh) is used.

almost white solid is removed by filtration, washed with anhydrous ether (2 × 25 ml), and dried *in vacuo* over phosphorus pentaoxide at room temperature; yield 2.14 g (81.3%), m.p. 135° to 137°, $[\alpha]_D^{25}$ -67.4° (*c* 1.0, water); λ_{max} in nm (pH 1 and 7) 214 (ε_{mM} 8.00); $\lambda_{max}^{pH\ 11}$ 228 nm (ε_{mM} 3.40); ^1H NMR data (Me_2SO-d_6): δ 6.78 (d, *J* 1.5 Hz, H-5), 6.42 (d, *J* 1.5 Hz, H-4), 5.45 (d, *J* 5.0 Hz, H-1'), and 5.30 (broad s, NH_2, and other sugar protons).

Analysis calculated for $C_8H_{13}N_3O_4$(215.21): C, 44.64; H, 6.09; N, 19.53. Found: C, 44.71; H, 6.39; N, 19.25.

REFERENCES

1. G. R. Greenberg, *J. Am. Chem. Soc.*, 74, 6307 (1952).
2. B. Levenberg and J. M. Buchanan, *J. Biol. Chem.*, 224, 1005 (1957).
3. R. J. Rousseau, R. K. Robins, and L. B. Townsend, *J. Heterocycl. Chem.*, 4, 311 (1967).
4. K. Maeda, T. Osato, and H. Umezawa, *J. Antibiot., Ser. A*, 6, 182 (1953).
5. S. Nakamura, *Chem. Pharm. Bull.*, 3, 379 (1955).

[38] DIMETHYL 1-(2,3,5-TRI-*O*-BENZOYL-β-D-RIBOFURANOSYL)IMIDAZOLE-4,5-DICARBOXYLATE AND DIMETHYL 1-β-D-RIBOFURANOSYLIMIDAZOLE-4,5-DICARBOXYLATE

Ribosylation of a Substituted Imidazole via the Stannic Chloride Procedure

P. Dan Cook and Roland K. Robins

ICN Pharmaceuticals, Inc., Nucleic Acid Research Institute, 2727 Campus Drive, Irvine, California 92715

INTRODUCTION

Dimethyl 1-β-D-ribofuranosylimidazole-4,5-dicarboxylate (*5*) and its acyl derivatives, such as dimethyl 1-(2,3,5-tri-*O*-benzoyl-β-D-ribofuranosyl)imidazole-4,5-dicarboxylate (*4*), are potentially valuable intermediates for the synthesis of purine nucleosides (such as xanthosine), purine nucleoside analogs, and analogs of 5-amino-1-β-D-ribofuranosylimidazole-4-carboxamide riboside (AICAR). Syrupy dimethyl 1-(2,3,5-tri-*O*-acetyl-β-D-ribofuranosyl)imidazole-4,5-dicarboxylate has been prepared by the condensation of the silver salt of dimethyl imidazole-4,5-dicarboxylate (*1*) with 2,3,5-tri-*O*-acetyl-β-D-ribofuranosyl chloride.[1] The stannic chloride-catalyzed condensation of dimethyl 1-(trimethylsilyl)-imidazole-4,5-dicarboxylate (*2*) with 1-*O*-acetyl-2,3,5-tri-*O*-

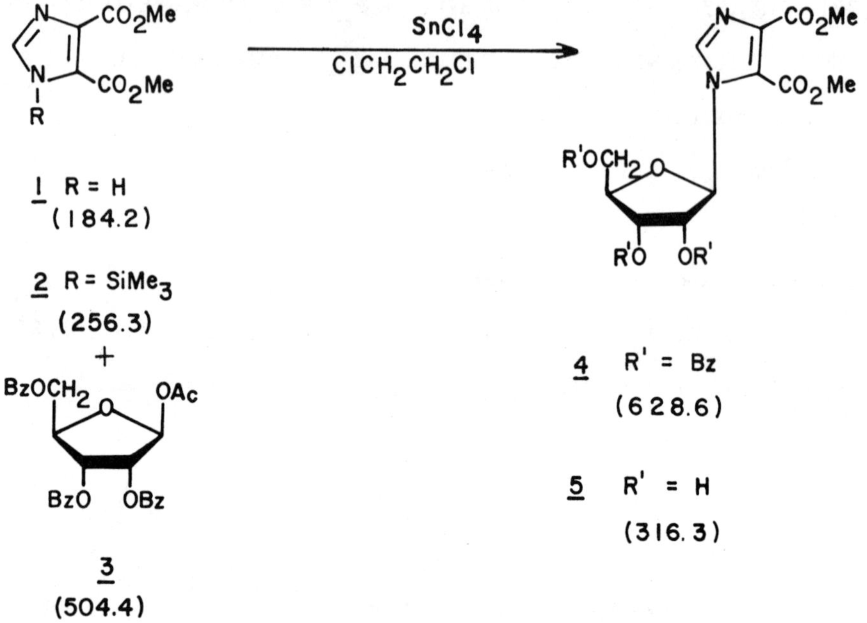

benzoyl-β-D-ribofuranose (*3*) provides crystalline *4* and is a significant improvement over the silver salt procedure.[2] Protected nucleoside *4* was converted into xanthosine according to the procedure of Todd and co-workers,[1] thus verifying its structure.

PROCEDURE

Dimethyl 1-(2,3,5-Tri-O-benzoyl-β-D-ribofuranosyl)imidazole-4,5-dicarboxylate (4)

A mixture of dimethyl imidazole-4,5-dicarboxylate (*1*, 11.04 g, 0.06 mol), is boiled under reflux, with stirring, under anhydrous conditions for 8 hr with hexamethyldisilazane (150 ml) and ammonium sulfate (0.160 g). The excess of hexamethyldisilazane is removed under diminished pressure,[a] providing the trimethylsilyl derivative *2* as a colorless oil, which is dissolved in dry 1,2-dichloroethane (150 ml). 1-*O*-Acetyl-2,3,5-tri-*O*-benzoyl-β-D-ribofuranose (*3*, 30.24 g, 0.06 mol) is added, followed by direct addition[b] of anhydrous stannic chloride (10.08 ml, 86.4 mmol). Thin-layer chromatography of a methanolized aliquot on silica gel with 10:1 chloroform—methanol indicated almost complete conversion of the sugar and base into the product after 15 min at room temperature.[c] The solution, which gradually darkens, is stirred at room temperature for 6 to 12 hr, diluted with chloroform (150 ml), the solution poured with vigorous stirring into 5% sodium hydrogen carbonate solution (1.2 l), and the mixture filtered through a bed of Celite. The organic layer is separated, washed with water (50 ml), dried (magnesium sulfate), and evaporated under reduced pressure to a beige foam (37.0 g).[d]

---

[a]The vacuum from a water aspirator is intermittently applied directly to the refluxing solution. Caution is taken to avoid loss of product from foaming. The heating mantle, or oil bath, is not shut off or turned down until the excess of hexamethyldisilizane has been removed.

[b]Relatively small volumes of stannic chloride (20 ml or less) are measured in, and delivered from, a syringe having an 18-gauge needle. Larger volumes are measured in a graduated cylinder and added rapidly.

[c]The temperature of the solution rises from ≃30° to between 45° and 55° on addition of the stannic chloride and then falls to between 25° and 28° within 1.5 hr.

[d]According to TLC this material is quite pure and it constitutes a 99% yield.

Crystallization of the foam from ethanol provides nucleoside *4* as white microcrystals (33.9 g, 90%), m.p. (after being dried for 3 hr at 80°) 121° to 122°, $[\alpha]_{\underline{\underline{D}}}^{25}$ -11.0 (*c* 1, methanol); PMR data (Me_2SO-d_6): δ 3.76 (s, 3, CH_3), 3.82 (s, 3, CH_3), 6.63 (d, 1, *J* 6 Hz, H-1'), and 8.41 (s, 1, H-2).

Analysis calculated for $C_{33}H_{28}N_2O_{11}$ (628.57): C, 63.06; H, 4.49; N, 4.46. Found: C, 63.00; H, 4.44; N, 4.29.

Dimethyl 1-β-D-Ribofuranosylimidazole-4,5-dicarboxylate (5)

A solution of compound *4* (5.0 g, 7.96 mmol) in sodium methoxide (from 20 mg of sodium in 10 ml of dry methanol), and dry methanol (150 ml) is boiled under reflux for 45 min, cooled to room temperature, and treated with Amberlite IRC-50 ion-exchange resin (10 ml). The suspension is stirred for 30 min, the resin is removed by filtration, and the filtrate is evaporated *in vacuo* to a semi-solid residue, which is triturated with anhydrous ether (25 ml, in two portions). The solid is recrystallized from methanol to provide *5* (2.14 g, 85%) as small white needles, m.p. (after being dried at 100° for 2 hr) 134° to 135°, $[\alpha]_{\underline{\underline{D}}}^{25}$ -4.41° (*c* 1, water); λ_{max} in nm ($\varepsilon_{\underline{mM}}$), at pH 1, 242 (6,748); at pH 7, 252 (7,362); at pH 11, 250 (7,699); PMR data (Me_2SO-d_6): δ 3.84 (s, 3, CH_3), 3.90 (s, 3, CH_3), 5.92 (d, 1, *J* 4 Hz, H-1'), and 8.41 (s, 1, H-2).

Analysis calculated for $C_{12}H_{16}N_2O_8$ (316.26): C, 45.57;

H, 5.10; N, 8.86. Found: C, 45.61; H, 5.23; N, 8.88.

REFERENCES

1. G. A. Howard, A. C. McLean, G. T. Newbold, F. S. Spring, and A. R. Todd, *J. Chem. Soc.*, 232 to 234 (1949).

2. This procedure has been used for the synthesis of: (a) pyrimidine nucleosides [see U. Niedballa and H. Vorbruggen, *J. Org. Chem.*, 39, 3654 (1974) and references cited therein], (b) purine and purine analog nucleosides [see F. W. Lichtenthaler, P. Voss, and A. Heerd, *Tetrahedron Lett.*, 2141 (1974)], and (c) imidazole nucleosides [see P. D. Cook, R. J. Rousseau, A. M. Mian, P. Dea, R. B. Meyer, Jr., and R. K. Robins, *J. Am. Chem. Soc.*, 98, 1492 (1976).

[39] 5-FLUORO-1-β-D-RIBOFURANOSYLIMIDAZOLE-4-CARBOXAMIDE

Synthesis of Fluoroimidazole Nucleosides via Coupling by the Stannic Chloride Method

John C. Reepmeyer, Kenneth L. Kirk, and Louis A. Cohen

Laboratory of Chemistry, National Institute of Arthritis, Metabolism, and Digestive Diseases, National Institutes of Health, Bethesda, Maryland 20014

INTRODUCTION

5-Fluoro-1-β-D-ribofuranosylimidazole-4-carboxamide (*3*) has been found to have broad-spectrum antiviral activity and to inhibit polynucleotide biosynthesis in cell culture.[1] The coupling of benzoylated D-ribofuranosyl chloride with the bis(trimethylsilyl) derivative (*1a*) of 4-fluoroimidazole-5-carboxamide (*1*), in the presence of mercuric acetate[2] or molecular sieve,[3] provides *2* along with its anomer or N-3 isomer, or both.[4] It was subsequently found that condensation with stannic chloride as the catalyst[5] leads to *2* exclusively; accordingly, the latter procedure is the method of choice for this precursor of *3*.

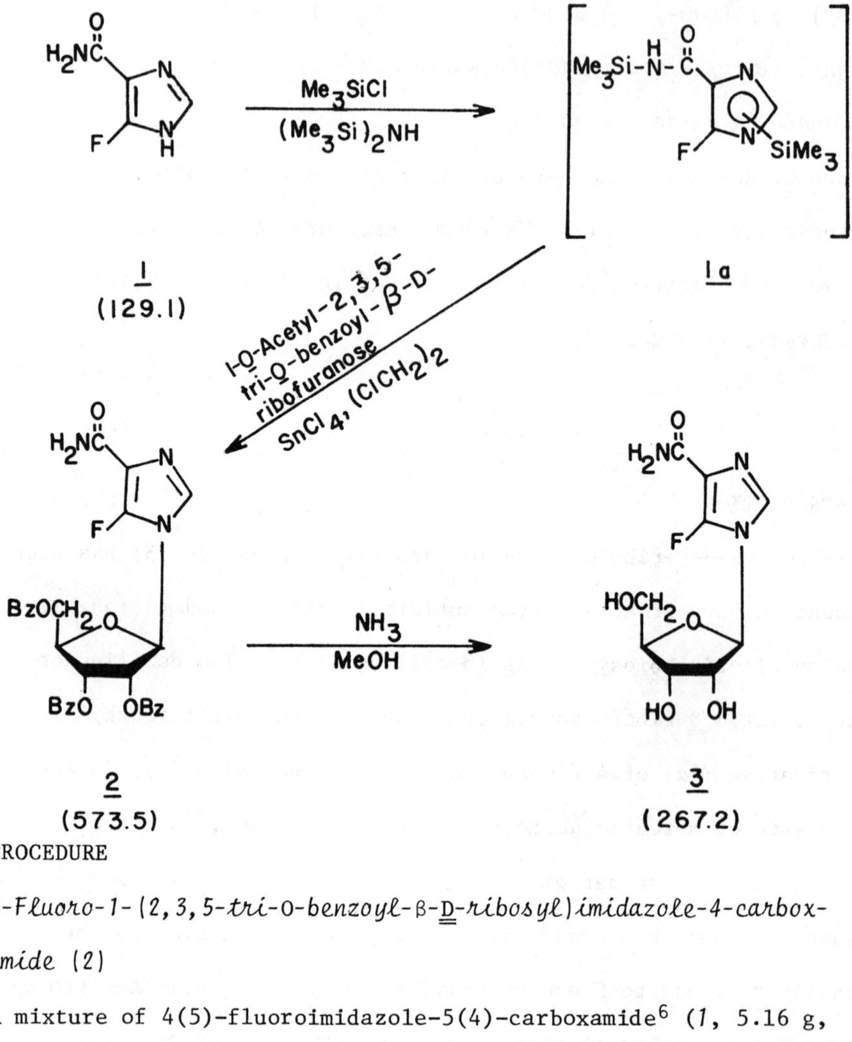

PROCEDURE

5-Fluoro-1-(2,3,5-tri-O-benzoyl-β-D-ribosyl)imidazole-4-carboxamide (2)

A mixture of 4(5)-fluoroimidazole-5(4)-carboxamide[6] (*1*, 5.16 g, 0.04 mol), chlorotrimethylsilane (5 ml, 0.04 mol) and hexamethyldisilazane (50 ml) is stirred for 1 hr in a flask (with reflux condenser) submerged to its neck in an oil bath kept at 130°. During the reaction, ammonium chloride sublimes into the condenser leaving a clear solution that is evaporated at atmospheric pressure, the last traces being removed by adding and evaporating dry toluene; residual toluene is removed under diminished pressure. Without purification, the product 5-fluoro-bis(trimethylsilyl)-imidazole-4-carboxamide[a] is dissolved in dry 1,2-dichloroethane (25 ml), and is added to a solution of 1-*O*-acetyl-2,3,5-tri-*O*-benzoyl-β-D-ribose[b] (20.16 g, 0.04 mol) in 1,2-dichloroethane (200 ml). The mixture is stirred and cooled in an ice bath, and a solution of freshly distilled stannic chloride (4.6 ml, 0.04 mol) in 1,2-dichloroethane (25 ml) is added dropwise. The clear mixture is stirred for 13 hr at room temperature and is then poured into saturated sodium hydrogen carbonate solution (400 ml). The resulting emulsion is filtered through a layer of Celite—sand, and the filter aid is washed thoroughly with 1,2-dichloroethane. The organic layer is separated, dried (magnesium sulfate), and evaporated under diminished pressure. The oily residue is triturated several times with cyclohexane,[c] giving 22.03 g (96%) of a white amorphous solid; TLC shows one major

---

[a]Formation of the bis(trimethylsilyl) derivative is demonstrated by the mass spectrum, *m/e* 273 (M^+), 258 ($M^+ - 15$, base peak). Owing to the susceptibility of the silyl compound to hydrolysis, every precaution is taken to exclude atmospheric moisture during its preparation and use.

[b]Obtained from Sigma Chemical Co.

[c]The desired compound is contaminated with a compound of much higher R_F. Extraction of the white solid with cyclohexane is continued until this impurity is eliminated.

component, R_F 0.21 (silica gel, 1:1 ethyl acetate—benzene); ν_{max}^{KBr} 1724 (C=O of benzoic ester) and 1675 cm^{-1} (C=O of amide). Without further purification, the product is debenzoylated.

5-Fluoro-1-β-D-ribofuranosylimidazole-4-carboxamide (3)

A solution of the tribenzoate (*2*) (22.03 g) in methanol (600 ml) is saturated with ammonia at 0°, kept for 22 hr at room temperature, and evaporated under diminished pressure to a viscous, pale-yellow oil. This is washed several times with ether (to remove methyl benzoate and other impurities) and the gummy residue is crystallized from 95% ethanol to give a colorless solid (6.64 g) m.p. 159° to 161°. Recrystallization of this and concentration of the mother liquors give a total yield of 6.03 g (60%) of *3*,[d] m.p. 164° to 165°, $[\alpha]_D^{19}$ -48.2° (*c* 1.02, water); λ_{max}^{MeOH} 226 nm (ε_{mM} 11.70) ; NMR data (Me_2SO-d_6): δ 7.77 (s, 1, H-2),[f] 7.22 (br s, 2, -$CONH_2$), 5.52 (d, 1, H-1', J_{HH} 6 Hz), ν_{max}^{KBr}

---

[d]The structure and stereochemistry of *3* are shown by the conversion[4] of the compound into 5-amino-1-β-D-ribofuranosylimidazole-4-carboxamide by methanolic ammonia during 13 days at 60° to 65°. The chromatographic properties, optical rotation, and UV and NMR spectra of this product are identical to those of an authentic sample.

[e]The chemical-ionization mass-spectrum of *3* shows a peak of 30% intensity at *m/e* 262 (MH^+), the base peak (*m/e* 130) corresponding to *1* as a readily formed fragmentation product.

[f]Whereas the H-2 signal of *3* appears as a singlet, the H-2 signal of the isomeric 4-fluoro-1-β-D-ribofuranosylimidazole-5-carboxamide (Me_2SO-d_6) is a doublet (δ 8.10, J_{HF} 2 Hz). This difference in coupling behavior has been observed for other pairs of *N*-alkyl-fluoroimidazoles, and may have general applicability in the elucidation of structure.[4]

1675 (amide carbonyl stretching), 1650, 1613, and 1590 cm^{-1}.[e]

REFERENCES

1. E. DeClercq, M. Luczak, J. C. Reepmeyer, K. L. Kirk, and L. A. Cohen, *Life Sci.*, 17, 187 (1975).

2. K. J. Ryan, E. M. Acton, and L. Goodman, *J. Org. Chem.*, 31. 1181 (1966).

3. M. P. Mertes and M. T. Shipchandler, *J. Heterocycl. Chem.*, 8, 133 (1971).

4. J. C. Reepmeyer, K. L. Kirk, and L. A. Cohen, *Tetrahedron Lett.*, 4107 (1975).

5. U. Niedballa and H. Vorbrüggen, *J. Org. Chem.*, 39, 3654 (1974).

6. K. L. Kirk and L. A. Cohen, *J. Am. Chem. Soc.*, 95, 4619 (1973).

[40] 2'-AZIDO-2'-DEOXYCYTIDINE

Amination of 2'-Azido-2'-deoxyuridine

J. Hobbs and Fritz Eckstein

Max-Planck-Institut für Experimentelle Medizin, Abteilung Chemie, Göttingen, Germany

INTRODUCTION

2'-Azido-2'-deoxycytidine[1] is prepared from 3',5'-di-*O*-acetyl-2'-azido-2'-deoxyuridine[2] by treatment with thionyl chloride-*N*,*N*-dimethylformamide and subsequent reaction with methanolic ammonia.[3] It serves as the starting material for the 5'-diphosphate, which specifically inhibits ribonucleotide reductases[4] and can be polymerized by polynucleotide phosphorylase.[1] The 2'-azido group in the azidonucleoside, as well as the azidonucleotide, can be reduced by hydrogen in the presence of platinum dioxide.[1]

PROCEDURE

2'-Azido-2'-deoxycytidine

2'-Azido-2'-deoxy-3',5'-di-*O*-acetyluridine (1.0 g) is dissolved in 13.9 ml of ethanol-free chloroform, dry dimethylformamide (0.139 ml) and thionyl chloride (2.2 ml) are then added. The solution is heated at reflux temperature for 6.5 hr, cooled to

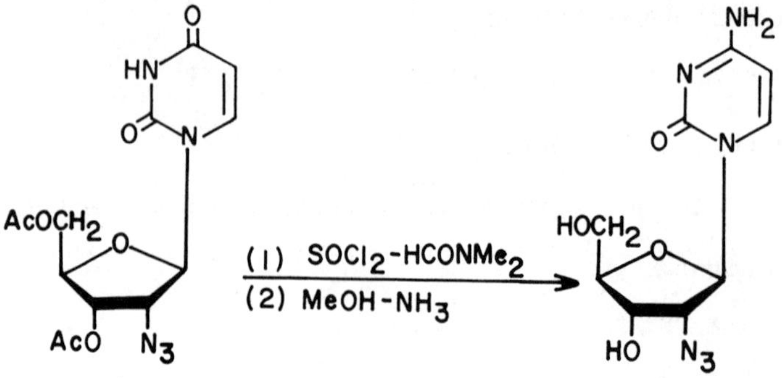

1 (353.3) 2 (268.2)

room temperature, evaporated, and dissolved in methanol (60 ml) which is 50% saturated with ammonia. The solution is stirred at room temperature for 5 days. TLC on SiO_2 (methanol—chloroform, 40:60, v/v) shows the major product to have an R_F of 0.62. The solution is evaporated, and the product separated by preparative TLC[a] in the above system, the desired band is eluted with methanol, evaporated, the residue dissolved in water and applied to a column of Dowex 1 × 2 (OH^-) (1.7 × 21.5 cm). The column is washed with water and the required product eluted with methanol—water, 30:70, v/v). The solvent is evaporated, and the remaining gum dissolved in a little ethanol and stored at 5° to give white crystals. The yield (determined spectrophotometrically) is 47%.

---

[a]Silica plates (2 mm thick) for preparative chromatography obtained from E. Merck, Darmstadt, Germany are developed once in analytical-grade methanol and dried prior to use in this preparation.

REFERENCES

1. J. Hobbs, H. Sternbach, M. Sprinzl, and F. Eckstein, *Biochemistry*, 12, 5138 (1973).

2. J. P. H. Verheyden, D. Wagner, and J. G. Moffatt, *J. Org. Chem.*, 36, 250 (1971).

3. J. Zemlička and F. Šorm, *Collect. Czech. Chem. Commun.*, 30, 2052 (1965).

4. L. Thelander, B. Larsson, J. Hobbs, and F. Eckstein, *J. Biol. Chem.*, 251, 1398 (1976).

[41] 2'-DEOXY-5-METHYLCYTIDINE

Amination of Thymidine

Helmut Vorbrüggen and Konrad Krolikiewicz

Schering A. G., 1000 Berlin 65, West Germany

INTRODUCTION

Uridine derivatives have usually been transformed into the corresponding cytidine derivatives by using the classical method of Fox and co-workers[1] that involves acylation of the hydroxyl groups of the D-ribosyl group, thiation of the 4-carbonyl group, and, finally, amination. Our silylation—amination procedure[2] effects this transformation in one step, although the conversion of *1* into *2* is more difficult, owing to steric hindrance by the 5-methyl group and the instability of *1*, than the amination of uridine to afford cytidine.[2]

PROCEDURE

2'-Deoxy-5-methylcytidine (2)

A mixture of thymidine[a] (*1*), hexamethyldisilazane[b] (31.4 ml,

---

[a]Pharma Waldhof GmbH, 4000 Düsseldorf 11, West Germany; Aldrich Chem. Co.

[b]Standard quality; Dow Chem. Co.; Wacker Chemie, GmbH, 8000 München 22, West Germany; redistilled, b.p. 127° to 128°.

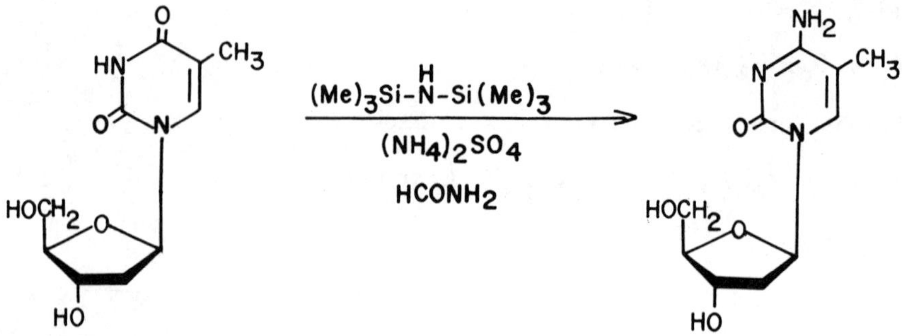

1 (242.2)   2 (241.3)

150 mmol), absolute formamide[c] (0.8 ml, 20 mol), and ammonium sulfate (0.132 g, 1 mmol) is stirred magnetically in a 100-ml, antimagnetic autoclave,[d] and heated for 72 hr at 140° (oil-bath temperature); the temperature[e] inside the autoclave is 130° to 135°. The autoclave is then cooled, the pressure is released, and the dark liquid (containing dark flakes) is boiled under reflux with methanol (400 ml) for 6 hr and filtered. The filter is washed with methanol, and the filtrate and washings are combined and evaporated. The residue is extracted with ethanol (4 × 120 ml), the extract is filtered, and the filtrate is evaporated to give a brownish foam (2.8 g). This is treated with hot ethanol (60 ml), leaving 300 mg of dark, insoluble material. The

---

[c]E. Merck, Darmstadt.

[d]Dr. Thiedig Co., 1000 Berlin 36, West Germany.

[e]The temperature of the reaction should not exceed 140°, as pertrimethylsilylated thymidine decomposes to thymine above this temperature.[2]

extract is cooled, and chromatographed on a column of silica[f] (250 g) by elution with ethanol. The first 600 ml elutes ≈0.2 to 0.3 g of unreacted *1*, which is contaminated by a yellowish impurity. The next 400 ml elutes 1.8 g (75%) of slightly yellowish *2*, which is homogenous in TLC.[g] A solution of *2* (1.8 g) in ethanol (20 ml) is cooled to 10°, and a slight excess of hydrogen chloride in absolute ether is added. Evaporation (30°/12 torr) and recrystallization[h] from ethanol (20 ml) gives, in two crops, 1.69 g (61%) of the pure hydrochloride of *2*, m.p. 153° to 154° (dec.) [lit.[1] 154° to 155° (dec.)].

REFERENCES

1. J. J. Fox, D. van Praag, I. Wempen, I. L. Doerr, L. Cheong, J. E. Knoll, M. L. Eidinoff, A. Bendich, and G. B. Brown, *J. Am. Chem. Soc.*, 81, 178 (1959); J. J. Fox, N. Miller, and I. Wempen, *J. Med. Chem.*, 9, 101 (1966).

2. H. Vorbrüggen and U. Niedballa, *Angew, Chem. Int. Ed. Engl.*, 10, 657 (1971); H. Vorbrüggen, K. Krolikiewicz, and U. Niedballa, *Justus Liebigs Ann. Chem.*, 988 (1975).

---

[f]E. Merck, Darmstadt, (70 to 230 mesh).

[g]On plates of silica F_{254} (Merck) with ethanol; *1*, R_F 0.75; *2*, R_F 0.39.

[h]As the hydrochloride of *2* decomposes on extended heating in ethanol, all solutions should be heated only briefly, and should be evaporated *in vacuo*. The dark impurities formed on longer heating may, however, be removed by treatment with charcoal.

[42] 6-METHYLCYTIDINE AND 2'-DEOXY-6-METHYLCYTIDINE

Synthesis of 6-Methylpyrimidine Nucleosides by the Trimethylsilyl Method

Michael W. Winkley and Roland K. Robins

Wyeth Laboratories, Radnor, Pennsylvania 19010 and ICN Nucleic Acid Research Institute, Irvine, California 92664

INTRODUCTION

6-Methylpyrimidine nucleosides are simple, unnatural nucleosides that exist in the unusual *syn* conformation in both the crystal state[1,2] and in solution.[3–10] These nucleosides also display an unusual susceptibility to acid hydrolysis.[11] The introduction of the trimethylsilyl method[12] has made possible the first synthesis[13] of this interesting group of nucleosides.

PROCEDURE

6-Methylcytidine[13] (*6*)

To the crystalline trimethylsilyl derivative[a] (*2*, 10 g) is added 2,3,5-tri-*O*-benzoyl-D-ribofuranosyl bromide (from 25 g of the

[a]Compound *2* is prepared from *1* by the general method of Wittenburg.[14] The pyrimidine (*1*) in an excess of hexamethyldisilazane with a catalytic quantity of ammonium sulfate is heated under reflux under anhydrous conditions until complete dissolution is achieved. The excess of hexamethyldisilazane is removed by distillation under vacuum (aspirator). The residual oil (*2*), which crystallizes on standing, is used without further purification.

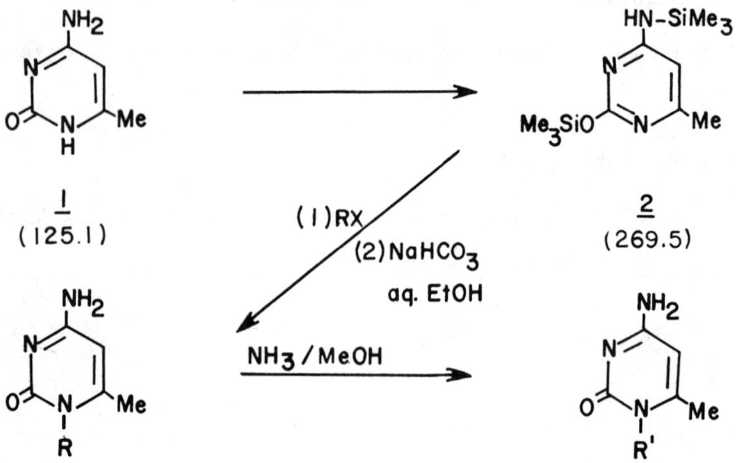

3, R = 2,3,5-tri-O-benzoyl-β-D-ribofuranosyl (569.6)

4, R = 2-deoxy-3,5-di-O-p-toluoyl-β-D-*erythro*-pentofuranosyl- (477.5)

5, R = 2-deoxy-3,5-di-O-p-toluoyl-α-D-*erythro*-pentofuranosyl- (477.5)

6, R' = β-D-ribofuranosyl- (257.2)

7, R' = 2-deoxy-β-D-*erythro*-pentofuranosyl- (241.2)

8, R' = 2-deoxy-α-D-*erythro*-pentofuranosyl- (241.2)

1-*O*-acetyl derivative[b] in dry acetonitrile (150 ml), and the mixture is stirred under anhydrous conditions until dissolution occurs. After 3 days at room temperature the solution is evaporated to a syrup that is treated with sodium hydrogen carbonate, water, and ethanol. The mixture is evaporated to dryness, and

---

[b]This compound is prepared by the procedure described for 2,3,5-tri-*O*-acetyl-D-ribofuranosyl bromide described in reference 13.

the last traces of moisture are removed from the residue by successive evaporations with absolute ethanol. The residue is extracted with dichloromethane, and the extract is evaporated to dryness; the extraction is repeated, and the extract is evaporated, to give crude *3* (31.0 g).

A portion (22.4 g) of crude *3* is dissolved in a small volume of benzene, and the solution is applied to a column (5.0 × 46 cm) of alumina (Merck) prepacked in benzene, and the material is washed on with benzene (1 l). Elution with 4:1 benzene—ethyl acetate (2 l), followed by ethyl acetate (3 l), removes the sugar component. The eluant is then changed to 19:1 ethyl acetate—methanol (4.3 l) to elute the major nucleosidic component (*3*). The fractionation is monitored by TLC on alumina HF 254 (Brinkmann) with 4:1 ethyl acetate—methanol as the developer and on SilicAR 7GF (Mallinkrodt) with 9:1 ethyl acetate—methanol as the developer. The solution containing *3* is evaporated to give *3* as a syrup (10.3 g).[c] Further elution with 9:1 ethyl acetate—methanol (2 l) removes additional *3* contaminated with a nucleosidic component of undetermined structure. The PMR spectrum of ($CDCl_3$) of *3* exhibits peaks at δ 2.34 (s, Me-6), 5.91 (s, H-5), and 6.12 (s, $J_{1',2'} < 1$ Hz, H-1').

---

[c]This compound has been obtained[15] in crystalline form, m.p. 190° to 192°.

A solution of syrupy *3* (10.3 g) in methanol (200 ml) presaturated with ammonia at 0° is kept in a pressure bottle for 3 days at room temperature. The suspension is filtered, and the filtrate is concentrated to a small volume; crystallization occurs during the concentration. Isopropyl alcohol is added, and the volume is further diminished. The yield of white crystals of essentially pure *6* is 3.7 g (40% from 1-*O*-acetyl-2,3,5-tri-*O*-benzoyl-D-ribose), m.p. 232° to 236° (dec.). Compound *6* may be recrystallized from water—ethanol—isopropyl alcohol. Pure *6* has $[\alpha]_D^{30}$ -41.5° (*c* 1.55, water); $\lambda_{max}^{pH\ 1}$ in nm 278 (ε_{mM} 14.70), $\lambda_{min}^{pH\ 1}$ 241 (1.50), $\lambda_{max}^{pH\ 14}$ 273 (9.90), and $\lambda_{min}^{pH\ 14}$ 252 (5.80); PMR data (D_2O): δ 2.37 (s, Me-6), 5.73 (d, $J_{1',2'}$ 3.5 Hz, H-1'), and 5.88 (s, H-5).

Other workers[15,16] have reported more recent syntheses of *6* by different procedures. Their results confirm the data and the structure[13] given for *6*.

2'-Deoxy-6-methylcytidine (*7*)

2-Deoxy-3,5-di-*O*-*p*-toluoyl-D-*erythro*-pentofuranosyl chloride (35 g, prepared[17] from 50 g of methyl 2-deoxy-3,5-di-*O*-*p*-toluoyl-D-*erythro*-pentofuranoside) is added to *2* (prepared from 15 g of *1*). Molecular sieves (type 4A, 1.59-mm pellets; 20 g) and dry acetonitrile (450 ml) are added, and the mixture, protected from moisture, is stirred for 4 days at room temperature, diluted with dichloromethane, and filtered through Celite. The filtrate is evaporated to dryness, and the residue is treated with sodium hydrogen carbonate, water, and ethanol. The mixture is evaporated to dryness, and the final traces of moisture are removed from the residue by successive evaporations with absolute ethanol. The residue is extracted with chloroform, and then with ethyl acetate. The extracts are combined and evaporated to a syrup that is dissolved in benzene, and the solution applied to a column (5.0 × 43 cm) of alumina (Merck) prepacked in benzene. The material is washed on with benzene (500 ml). Elution with 4:1 benzene—ethyl acetate (1 1) followed by ethyl acetate (6 1) removes the sugar component. The eluant is then changed to 49:2 ethyl acetate—methanol (1 1) followed by 10:1 ethyl acetate—methanol (1 1) to remove the major nucleosidic component *4*. The fractionation is monitored by TLC on SilicAR 7GF with 4:1 ethyl acetate—methanol as the developer. (Fractions obtained by further elution with 10:1 ethyl acetate—methanol contain a mixture of *4* and *5*.) Fractions containing only *4* are pooled and evaporated to give syrupy *4* (8.5 g). Compound *4* is dissolved in methanol (300 ml) presaturated at 0° with ammonia, and the solution is kept in a pressure bottle for 3 days at room temperature. The suspension is filtered, the filtrate is evaporated to dryness, and the residue is partitioned between chloroform and water. The aqueous phase is washed thrice with chloroform and evaporated to dryness. Evaporation with absolute ethanol removes the final traces of water from syrupy *7*. Crystallization from ethanol—isopropyl alcohol gives 3.0 g (14%) of *7*, m.p. 159° to 162°. A solution of this slightly yellow product in methanol is decolorized with charcoal and evaporated to a syrup that is crystallized from methanol—isopropyl alcohol to yield 2.70 g, m.p. 164° to 166°; this process is repeated to raise the m.p. to 173° to 176°. Pure *7* has $[\alpha]_{D}^{34}$ +32.2° (*c* 1, water); $\lambda_{max}^{pH\ 1}$ in nm 278 (ε_{mM} 14.90), $\lambda_{min}^{pH\ 1}$ 241 (0.96), $\lambda_{max}^{pH\ 14}$ 273 (9.80), $\lambda_{min}^{pH\ 14}$ 252 (5.60); PMR data (D_2O): δ 2.40 (s, Me-6), 5.87 (s, H-5), 6.20 (t, width 13.5, $J_{1',2'}$ 6.7 Hz, H-1').

The α-D anomer (*8*) can also be isolated as described,[13] for use in an unequivocal anomeric assignment.

REFERENCES

1. D. Suck, W. Saenger, and A. Vorbrüggen, *Nature*, 235, 333 (1972).

2. D. Suck and W. Saenger, *J. Am. Chem. Soc.*, 94, 6520 (1972).

3. D. W. Miles, M. J. Robins, R. K. Robins, M. W. Winkley, and H. Eyring, *J. Am. Chem. Soc.*, 91, 824, 831 (1969).

4. D. W. Miles, W. H. Inskeep, M. J. Robins, M. W. Winkley, R. K. Robins, and H. Eyring, *J. Am. Chem. Soc.*, 92, 3872 (1970).

5. M. P. Schweizer, J. T. Witkowski, and R. K. Robins, *J. Am. Chem. Soc.*, 93, 277 (1971).

6. M. P. Schweizer, E. B. Banta, J. T. Witkowski, and R. K. Robins, *J. Am. Chem. Soc.*, 95, 3770 (1973).

7. M. P. Schweizer and G. P. Kreishmann, *J. Magn. Reson.*, 9, 334 (1973).

8. A. Holý, *Collect. Czech. Chem. Commun.*, 38, 3912 (1973).

9. A. Holý, *Collect. Czech. Chem. Commun.*, 39, 3374 (1974).

10. I. Kulakowska, A. Rabczenko, and D. Shugar, *Biochem. Biophys. Res. Commun.*, 48, 65 (1972).

11. R. E. Notari, D. T. Witiak, J. L. DeYoung, and A. J. Lin, *J. Med. Chem.*, 15, 1207 (1972).

12. See I. Iwai, T. Nishimura, and B. Shimizu, in W. W. Zorbach and R. S. Tipson (eds.), *Synthetic Procedures in Nucleic Acid Chemistry*, Wiley-Interscience, New York, 1968, Vol. 1, pp. 388 to 394 and references cited therein.

13. M. W. Winkley and R. K. Robins, *J. Org. Chem.*, 33, 2822 (1968).

14. E. Wittenburg, *Z. Chem.*, 4, 303 (1964).

15. M. Prystaš and F. Šorm, *Collect. Czech. Chem. Commun.*, 34, 331, 2316 (1969).

16. R. S. Klein and J. J. Fox, *J. Org. Chem.*, 37, 4381 (1972).

17. See C. C. Bhat, in W. W. Zorbach and R. S. Tipson (eds.), *Synthetic Procedures in Nucleic Acid Chemistry*, Wiley-Interscience, New York, 1968, Vol. 1, pp. 521 to 522 and reference cited therein.

[43] *O*'-ALKYL DERIVATIVES OF 1-β-D-ARABINOFURANOSYLCYTOSINE (AraC)

Dialkyl Sulfate Alkylation of Sugar Hydroxyl Groups of Cytosine Nucleosides

Edward Darzynkiewicz and David Shugar

Department of Biophysics, Institute of Experimental Physics, University of Warsaw, 02-089 Warszawa; and Institute of Biochemistry and Biophysics, Academy of Sciences, 02-532 Warszawa, Poland

INTRODUCTION

The relative resistance to alkylation of N-3 of the cytosine residue and the exocyclic amino group of 1-substituted cytosines in strongly alkaline, *aqueous* media,[1] may be profitable to alkylate the glycosyl hydroxyl groups of *any* cytosine nucleoside.[2] In contrast to treatment with a diazoalkane in the presence of a Lewis acid such as stannous chloride ($SnCl_2$), which gives high yields of the 2'- and 3'-*O*-alkylnucleosides when their glycosyl group has *cis*-hydroxyl groups (e.g., D- or L-ribose[3]; contribution 132 of this volume, or D-lyxose[4]) alkylation in strongly alkaline medium provides a mixture of all seven of the possible mono-, di- and tri-*O*'-alkyl analogs, which are readily separated by column

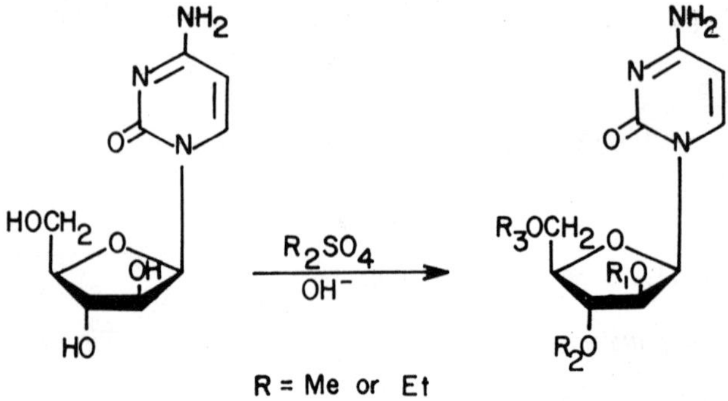

AraC R_1, R_2, R_3 = H, Me or Et

Examples of nomenclature: (a) if R_1 = Me, R_2 = H, R_3 = Me the resulting product is 2',5'-di-*O*-methyl-AraC (2',5'-Me_2AraC); (b) if R_1 = H, R_2 = Et, R_3 = H, the product is 3'-*O*-ethyl-AraC (3'-EtAraC).

chromatography.[2,4] The presence of *cis*-hydroxyl groups is not a requirement. Application of this procedure to the preparation of the methyl (or ethyl) ethers of the therapeutically important AraC and, by subsequent deamination, of the corresponding *O*'-alkyl derivatives of AraU[4] is described.

PROCEDURE

O'-Methylation of AraC

To a solution of AraC[a] (5.0 g, 20.6 mmol) in M potassium hydroxide (500 ml) is added dimethyl sulfate (40 ml, 400 mmol) in 1-ml portions with continuous stirring during 10 hr. Simultaneously, to

---

[a]AraC is readily available commercially; it is also readily prepared in the laboratory, most conveniently by the procedure of Kanai and co-workers.[5]

maintain the alkalinity of the mixture, 10M potassium hydroxide (50 ml) is added in 10-ml portions over the same period. The reaction is terminated by addition of concentrated sulfuric acid to approximate neutrality,[b] the solution is evaporated almost to dryness under diminished pressure,[c] and the residue is extracted portionwise with methanol (300 ml). The extracts are combined and filtered to remove inorganic salts. The extract ($A_{270}^{pH\ 7}$ 171,600 units) is evaporated to dryness under diminished pressure, and the oily residue is taken up in water (700 ml) and loaded on a column (35 × 2.5 cm) of Dowex 50-W X-2 (H^+) ion-exchange resin (200 to 400 mesh). The column is washed with water until the effluent is neutral, and elution then performed with M potassium hydroxide; the nucleoside products are present in the first 150 ml of effluent ($A_{270}^{pH\ 7}$ 160,000 units). To this solution is added water (350 ml), and the solution is passed through a column (35 × 3.5 cm) of Dowex 50-W X-2 (NH_4^+) ion-exchange resin (20 to 50 mesh); the column is washed with water until the effluent exhibits no UV absorption. The eluate is evaporated to dryness,[d] and

---

[b]At this point TLC on Merck HF_{254} silica gel with 4:1 (v/v) chloroform—methanol shows the presence of four spots, having R_F values (and approximate relative intensities under a dark UV lamp) as follows: 0.15 (40%), unreacted AraC; 0.30 (25%), mono-*O*'-methyl-AraC; 0.50 (25%), di-*O*'-methyl-AraC; 0.65 (10%), tri-*O*'-methyl-AraC.

[c]A rotary evaporator is used.

[d]Particular attention should be paid to removal of ammonia.

the residue is dissolved in water (25 ml) ($A_{270}^{pH\ 7}$ 158,000 units) and loaded on a column (60 × 6 cm) of Dowex-1 X-2 (OH^-) ion-exchange resin (200 to 400 mesh) prewashed with water.[6] Elution, initially with water, and subsequently with increasing concentrations of methanol—water, causes resolution of all seven of the possible *O*'-methyl derivatives of AraC (see Figure 1).

The combined eluates for each peak are evaporated to dryness, and each residue is crystallized from anhydrous ethanol by slow evaporation over phosphorus pentaoxide in a desiccator at atmospheric pressure. Where crystallization does not proceed readily, or only in low yield, the hydrochloride is obtained in crystalline form as follows; the pH of the ethanolic solution is brought to 2 with 2*M* hydrochloric acid, the solution is evaporated to dryness, and the excess of hydrochloric acid is removed by several evaporations with ethanol. The resulting hydrochloride is crystallized from anhydrous ethanol as described for the nucleoside (see Table 1 for yields and melting points of products).

O'-Ethylation of AraC

To a solution of AraC (2.5 g, 10.3 mmol) in *M* potassium hydroxide (500 ml) stirred continously is added diethyl sulfate (40 ml, 268 mmol) in 1-ml portions during 12 hr, with concurrent addition of 10*M* potassium hydroxide (50 ml) in 5-ml portions over the same period. The reaction is terminated by addition of concentrated

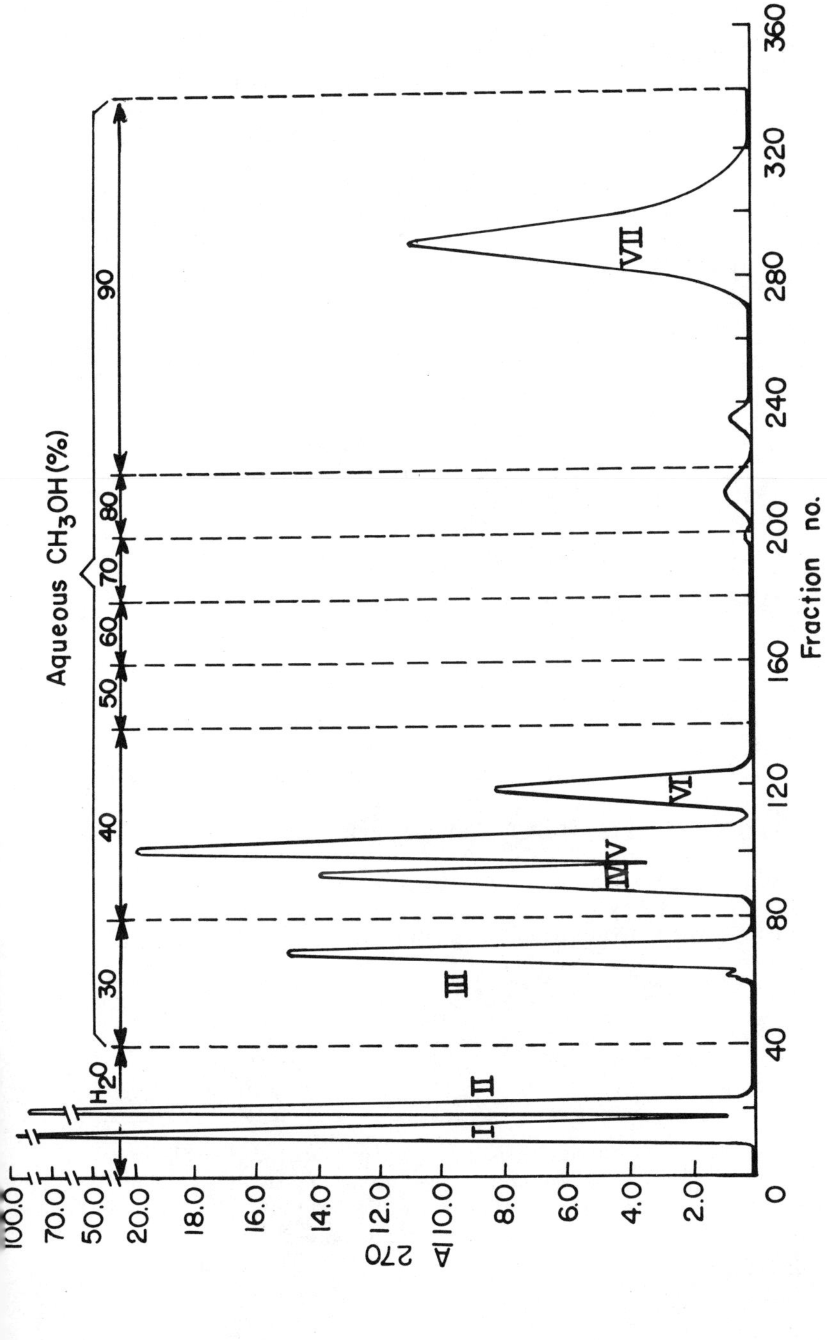
Aqueous CH3OH (%)
H2O
30
40
50
60
70
80
90
I
II
III
IV
V
VI
VII
A 270
100.0
70.0
50.0
20.0
18.0
16.0
14.0
12.0
10.0
8.0
6.0
4.0
2.0
0
40
80
120
160
200
240
280
320
360
Fraction no.

FIGURE 1

Figure 1. Elution Pattern, on a Column (60 × 6.5 cm) of Dowex-1 X-2 (OH^-) Ion-Exchange Resin (200 to 400 mesh), of Products of Methylation of AraC with Dimethyl Sulfate in Alkaline Medium. Fractions (100 ml each) are collected at 10-min intervals. Elution is conducted with increasing concentrations of aqueous methanol as indicated. Material giving $A_{270}^{pH\ 7}$ 158,000 units is deposited on the column: peak I (2',3',5'-Me_3AraC, 21,000), II 2',3'-Me_2AraC, 21,000), III (3',5'-Me_2AraC, 6200), IV (2',5'-Me_2AraC, 6500), V (2'-MeAraC, 11,600), VI (5'-MeAraC, 6400), and VII (3'-MeAraC, 15,000 units).

sulfuric acid to neutrality.[e] Subsequent processing and column separation of products are conducted as for the *O*'-methyl derivatives. The yields and melting points of products, following crystallization, are given in Table 1.

Identification of Products

Identification of the mono-*O*'-methyl derivatives is facilitated by the fact that unambiguous syntheses have been reported for 5'-MeAraC[7] and 2'-MeAraC.[8,9] However, all of the methyl and ethyl derivatives may readily be indentified on the basis of their PMR spectra, profiting from the magnetic anisotropy of the O'-Me bonds and the additivity of the effects of one or more methyl groups on a given, non-exchangeable proton of the glycosyl group, as illustrated in Table 2.

---

[e]At this point, TLC on Merck HF_{254} silica gel with 4:1 (v/v) chloroform—methanol shows the presence of four spots, having R_F values (and relative intensities under a dark UV lamp) as follows: 0.15 (40%), unreacted AraC; 0.35 (35%), mono-*O*'-ethyl-AraC; 0.50 (20%), di-*O*'-ethyl-AraC; and 0.70 (5%), tri-*O*'-ethyl-AraC.

TABLE 1. Melting Points and Overall Yields (Following Crystallization[a]) of *O*'-Methyl and *O*'-Ethyl Derivatives of AraC from Typical Alkylations with Dialkyl Sulfate in Aqueous Alkaline Medium

AraC Derivative	Crystal Form	m.p. (°C)	Yield, mg (%)
2'-Me	needles (HCl salt)	195 to 197	390 (7.7%)
3'-Me	small prisms	264 to 266	170 (3.4%)
	needles (HCl salt)	217 to 221	364 (7.2%)
5'-Me	platelets	192 to 193	190 (3.8%)
2',3'-Me_2	large prisms	211 to 213	545 (11.0%)
2',5'-Me_2	irregular prisms	163 to 166	110 (2.2%)
3',5'-Me_2	platelets	203 to 205	177 (3.5%)
2',3',5'-Me_3	prisms	229 to 231	146 (3.0%)
2'-Et	parallelopipeds	194 to 195	143 (5.6%)
3'-Et	prisms	242 to 243.5	171 (6.2%)
5'-Et	long needles	205 to 205.5	245 (9.5%)
2',3'-Et_2	prisms (HCl salt)	179 to 182	126 (4.5%)
2',5'-Et_2	prisms (HCl salt)	192 to 195	67 (2.5%)
3',5'-Et_2	prisms (HCl salt)	183 to 186	105 (4.0%)
2',3',5'-Et_3	prisms (HCl salt)	202 to 204	130 (4.6%)

[a]Products crystallized from anhydrous ethanol, with the exception of 2',3'-Et_2AraC and 2',3',5'-Et_3AraC, which were crystallized from ethyl acetate—ethanol.

Table 2. Chemical Shifts (in PPM *versus* Internal DSS[a]) of the H-1', Methyl, and Methylene Protons of the Various *O*'-Methyl and *O*'-Ethyl Derivatives[b] of AraC in D_2O at pD 3

AraC and *O*'-Alkyl Derivatives of AraC	H-1'	2'-CH_3	3'-CH_3	5'-CH_3	2'-CH_2		3'-CH_2		5'-CH_2
					H_A	H_B	H_A	H_B	$H_A \equiv H_B$
AraC	6.20								
2'-Me	6.30	3.37							
3'-Me	6.11		3.50						
5'-Me	6.19			3.46					
2',3'-Me_2	6.19	3.37	3.51						
2',5'-Me_2	6.30	3.38		3.45					
3',5'-Me_2	6.11		3.50	3.47					
2',3',5'-Me_3	6.20	3.37	3.51	3.45					
2'-Et	6.30	1.08			3.68	3.51			
3'-Et	6.13		1.23				3.77	3.71	
5'-Et	6.20			1.22					3.67
2',3'-Et_2	6.23	1.08	1.23		3.69	3.45	3.76	3.72	
2',5'-Et_2	6.30	1.08		1.22	3.66	3.52			3.66
3',5'-Et_2	6.13		1.22	1.22			3.77	3.70	3.68
2',3',5'-Et_3	6.23	1.08	1.22	1.22	3.67	3.46	3.77	3.70	3.66

[a] 2,2-Dimethyl-2-silapenten-5-sulfonate, sodium salt.

[b] For the 2'- and 3'-*O*-ethyl groups, the signals corresponding to the methylene protons (denoted as H_A and H_B) have been assigned on the basis of a conformational analysis of the ethyl substituent. For the 5'-*O*-ethyl substituent, the A and B methylene protons are indistinguishable.

O'-*Alkyl Derivatives of AraU*

Each of the foregoing *O*'-alkylAraC analogs may be converted into the corresponding *O*'-alkylAraU derivative by simple deamination with nitrous acid or, preferably, with bisulfite.[2,10,11]

For those AraC analogs having a free 2'-hydroxyl group,[b] buffer anion-catalyzed deamination is by far the simplest procedure.[12] A solution of the sample in M acetic acid (≃20 mg/ml) is heated in a sealed ampoule in a boiling-water bath for 7 hr. The solution is then evaporated to dryness, the residue is dissolved in water, and the solution reevaporated to dryness. The residue is kept over pellets of potassium hydroxide overnight under vacuum (to remove residual acetic acid), dissolved in water, the solution passed through a small column of Dowex 50-W (H^+) ion-exchange resin (to remove any traces of unreacted *O*'-alkyl-AraC), and the effluent evaporated to dryness. Most of the products are obtained as colorless oils.

REFERENCES

1. J. T. Kuśmierek and D. Shugar, *Acta Biochim. Polon.*, 18, 413 (1971).

2. J. T. Kuśmierek, J. Giziewicz, and D. Shugar, *Biochemistry*, 12, 194 (1973).

3. M. J. Robins, S. R. Naik, and A. S. K. Lee, *J. Org. Chem.*, 39, 1891 (1974).

---

[b]For example, 5'-MeAraC, or 3',5'-AraC.

4. E. Darzynkiewicz and D. Shugar, *Acta Biochim. Polon.*, 21, 305 (1974).

5. T. Kanai, T. Kijima, O. Maruyama, and M. Ichino, *Chem. Pharm. Bull.*, 18, 2569 (1970).

6. C. A. Dekker, *J. Am. Chem. Soc.*, 87, 4027 (1965).

7. J. Giziewicz, J. T. Kuśmierek, and D. Shugar, *J. Med. Chem.*, 15, 839 (1972).

8. J. Giziewicz and D. Shugar, *Acta Biochim. Polon.*, 20, 73 (1973).

9. J. A. Montgomery and A. G. Laster, *J. Med. Chem.*, 17, 360 (1974).

10. R. Shapiro, R. E. Servis, and M. Welcher, *J. Am. Chem. Soc.*, 92, 422 (1970).

11. H. Hayatsu, Y. Wataya, and K. Kai, *J. Am. Chem. Soc.*, 92, 724 (1970).

12. R. E. Notari, M. L. Chin, and R. Wittebort, *J. Pharm. Sci.*, 61, 1189 (1972).

[44] 4-AMINO-1-β-D-*ARABINO*-PENTOFURANOSYL-2(1*H*)-PYRIMIDINONE (1-β-D-ARABINOFURANOSYLCYTOSINE)

One-Step Synthesis of 1-β-D-Arabinofuranosylcytosine

Jiří Beránek, Thomas J. Delia,[a] and Pavel Drašar

Institute of Organic Chemistry and Biochemistry, Czechoslovak Academy of Sciences, 16610 Prague, Czechoslovakia

INTRODUCTION

The cancerostatic 1-β-D-arabinofuranosylcytosine (*2*) displays its primary clinical activity against the acute myeloblastic form of leukemia.[1] A comparative study of the known synthesis of *2* was recently made.[2] The present procedure is a convenient, one-step synthesis of 1-β-D-arabinofuranosylcytosine[3] (*2*) from the naturally occurring nucleoside cytidine (*1*) by reaction with diphenyl carbonate (see contribution 170); an anion-exchange resin (borate form) is used for separation of *2* from *1*. A similar reaction was described in a preliminary communication.[4] More recently two important methods have been developed for the preparation of *2* via the 2,2'-anhydro derivative, namely, by reaction of cytidine (*1*) with phosphoryl chloride,[5] or with thionyl chloride.[6]

---

[a]Visiting scientist. Present address: Central Michigan University, Mt. Pleasant, Michigan 48858, U.S.A.

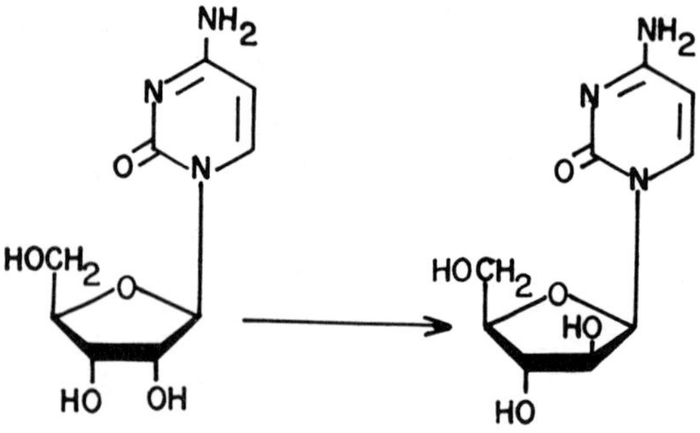

1
(243.2)

2
(243.2)

PROCEDURE

1-β-D-Arabinofuranosylcytosine (2) *and its Hydrochloride Salt*

A mixture[b] of cytidine (*1*) (600 g, 2.56 mol), diphenyl carbonate (560 g, 1.19 equiv.), *N,N*-dimethylformamide (750 ml), and water[c] (45 ml, 2.5 mol) is heated[d] to 120° with continuous stirring. Additional portions of diphenyl carbonate (3 × 40 g, a further 0.25 equiv.) are added at 1.5 hr, 2.5 hr, and 3.5 hr. After 6 hr,[e] the mixture is cooled, and slowly poured, with efficient

---

[b]The reaction is performed in a 6-l flask fitted with a reflux condenser, a mechanical stirrer, a thermometer, and a stoppered side neck.

[c]Water can be replaced by sodium hydrogen carbonate (12 g).

[d]The bath temperature is maintained at ≃130°.

[e]The solution turns dark brown during this period of time. The reaction can be conducted for a shorter time (30 min) at a higher temperature (150°), according to the original procedure for the preparation of a 2,2'-anhydronucleoside.[7] In experiments on a larger scale, the lower temperature and the longer heating are recommended.

stirring[f] into ether (2.5 l). The precipitate is removed by filtration, dried,[g] and dissolved in 60% aqueous methanol (5 l), the solution filtered with charcoal, and the filtrate evaporated.[h] The residue is dissolved in water (2.5 l), and the solution is passed through a column[i] of 9 kg of Zerolite FF (14 to 52 mesh), borate form.[j] The column is eluted with water.[k] Fraction 1, a brown liquid, is collected until the effluent turns yellow (≃4.6 l).[l] The yellow fraction 2 (≃105 l)[m,n,o] is evaporated,[h]

---

[f]"Vibromischer" is used.

[g]On a sintered-glass funnel.

[h]A rotary evaporator, a water pump (20 Torr), and a bath temperature of 35° to 40° are used.

[i]Diameter 10 cm, length 130 cm.

[j]An aqueous solution of sodium tetraborate (≃5 kg) is passed through a column of a strong, anion-exchange resin (OH^-), and the column is washed with water until the pH value of the effluent is about 7.5. The resulting "borate column" separates the *arabino* derivative *2* from the *ribo* derivative *1* (which stays on the column).

[k]The effluent is checked by TLC on plates of Silufol[R] silica gel (Kavalier Glassworks, Votice, Czechoslovakia) in 9:1 (v/v) methanol—M ammonium acetate saturated with sodium tetraborate [compound (R_F value): *1* (0.16), *2* (0.52), 1-β-D-arabinofuranosyluracil (0.62), uridine (0.22)].

[l]Fraction 1 gives several spots in TLC, but mainly a spot for *2*.

[m]Fraction 2 contains compound *2* as the sole, UV-absorbing material, and some boric acid. This fraction is used to isolate *2* as the free nucleoside.

[n]To isolate the unreacted cytidine, the column is eluted with 5% acetic acid.[o] The effluent (≃30 l), containing cytidine as the sole, UV-absorbing material, boric acid, and acetic acid, is concentrated[h] to ≃5 l, and the boric acid precipitate is removed by filtration. The filtrate is concentrated[h] to 1 l and then to 500 ml; boric acid is removed by filtration each time. The

and the residue is repeatedly evaporated with methanol (12 × 1 l) to remove the boric acid.[p] The residue[q] is crystallized from methanol to give 104 g of *2* (as the free nucleoside), m.p. 205° to 216°. On recrystallization from 50% aqueous ethanol, 85.8 g of *2* is obtained; m.p. 215° to 217°, $[\alpha]_D^{25}$ +155.2° (*c* 0.7, water); $\lambda_{max}^{H_2O}$ 275 nm (log ε 3.90), $\lambda_{min}^{H_2O}$ 254 nm (log ε 3.73). The combined mother liquors from the crystallizations are added to fraction 1 (4.6 l) from the "borate column," and the whole is evaporated.[h] The oily residue is dissolved in water (800 ml), and the solution is passed through a column of 1.5 kg of Dowex 50WX2 (H^+) resin (50 to 100 mesh). The column is washed with water (≃8 l).[o,r]

---

residual brown solution is passed through a column of Dowex 50 WX2 (H^+) resin (1 kg; 50 to 100 mesh). The column is eluted with water[o,r] (7 l), and the cytidine (*1*) is then eluted with 5% aqueous ammonium hydroxide[o] (≃5 l). The brown effluent is evaporated,[h] the residue is dissolved in 30% aqueous methanol (1 l), and the solution is passed through a column of 1 kg of Dowex-2 X-4 (OH^-) resin. The column is eluted with 30% aqueous methanol, and then with 0.1M aqueous ammonium hydrogen carbonate. The fraction containing[k] cytidine (*1*) is evaporated,[h] and the residue is coevaporated with water (to remove ammonium hydrogen carbonate). The residue is crystallized from 50% aqueous ethanol to yield 198.45 g of cytidine (*1*).

[o]Until the end of UV absorption.

[p]A sample of the methanolic solution gives a green flame when borate is present.

[q]All of the 1-β-D-arabinofuranosylcytosine could be isolated in the form of its hydrochloride. However, it was intended to obtain both the free nucleoside *2* and its hydrochloride in crystalline form, as described.

[r]To remove the uracil derivatives.

The product (*2*) is then eluted with 4% aqueous ammonium hydroxide (≈15 1),[o] and the eluate is evaporated.[h] The oily residue (212 g) which contains compound *2* as the sole, UV-absorbing material,[k,s] is dissolved in methanol (500 ml), and 30% methanolic hydrogen chloride is added until the pH value of a sample diluted with water is ≈2 to 3. The solution is evaporated,[h] and the residue is coevaporated twice with methanol and crystallized from 50% aqueous ethanol to give 135 g of the hydrochloride of *2* which on recrystallization from 50% aqueous ethanol yields 121.1 g of 1-β-D-arabinofuranosylcytosine hydrochloride, m.p. 188° to 193°; $[\alpha]_D^{25}$ +131.0° (*c* 0.7, water); $\lambda_{max}^{H_2O}$ 276 nm (log ε 4.00), $\lambda_{min}^{H_2O}$ 246 nm (log ε 3.59 The overall yield of the procedure (based on the free nucleoside is 190.85[g] (31.9%) of *2* (i.e., 219.2 g of the hydrochloride). The yield of *2* with respect to the recovered cytidine[n] is 47.5%.

REFERENCES

1. G. P. Bodey, E. J. Freireich, R. W. Monto, and J. S. Hewlett, *Cancer Chemother. Rep.*, 53, 59 (1969).

2. J. Brokeš and J. Beránek, *Collect. Czech. Chem. Commun.*, 39. 3100 (1974).

---

[s]For better isolation the 1-β-D-arabinofuranosylcytosine (*2*) contained in the syrupy residue is converted into its hydrochloride.

3. J. Beránek and T. J. Delia, in G. K. Daikos (ed.), *Progress in Chemotherapy* (Proceedings of the 8th International Congress of Chemotherapy, Athens, 1973), Vol. III., Hellenic Society for Chemotherapy, Athens, 1974, pp. 832 to 834; Czechslovakian Patent 164,401 (1971); *C.A.* 79, 19043W (1973).

4. K. K. Ogilvie, *Carbohydr. Res.*, 24, 210 (1972).

5. K. Kikugawa and M. Ichino, *J. Org. Chem.*, 37, 284 (1972).

6. T. Sowa and K. Tsunoda, *Bull. Chem. Soc. Japan*, 48, 505 (1975).

7. A. Hampton and A. W. Nichol, *Biochemistry*, 5, 2076 (1966).

[45] 1-β-D-ARABINOFURANOSYLCYTOSINE (Ara-C) AND "VIRAZOLE"

The Synthesis of Ara-C via β-D-Xylofuranosylcytosine Derivatives and of "Virazole" by the Trimethylsilyl Ether "Solution" Method

Bunji Shimizu and Akio Saito

Central Research Laboratories, Sankyo Co. Ltd., Tokyo, Japan

INTRODUCTION

1-β-D-Arabinosylcytosine (*5*) has antileukemic activities as well as other biological properties.[1] Compound *5* has been synthesized by various procedures.[2] Treatment of protected D-xylofuranosyl chloride (*2*) with N^4-acetylbis(trimethylsilyl)cytosine (*3*) in the presence of mercuric bromide leads to the formation of protected 1-β-D-xylofuranosylcytosine (*4*). Subsequent hydrolysis with aqueous sodium hydroxide yields[3] *5*.

"Virazole" (1-β-D-ribofuranosyl-1,2,4-triazole-3-carboxamide) (*9*) is a synthetic nucleoside having significant activity against a broad spectrum of DNA and RNA viruses.[4] Compound *9* was first prepared by the silylation—glycosylation procedure and by the acid-catalyzed fusion procedure.[5] A more convenient procedure involves treatment of 2,3,5-tri-*O*-benzoyl-D-ribofuranosyl bromide (*7*) with the trimethylsilyl derivative of methyl 1,2,4-triazole-3-carboxylate (*6*) in the presence of mercuric bromide, to give

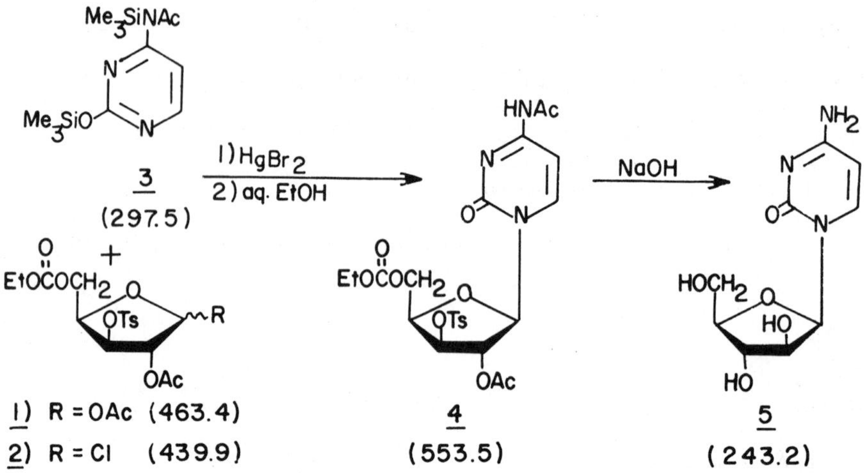

Ts = - $SO_2C_6H_4Me$-p

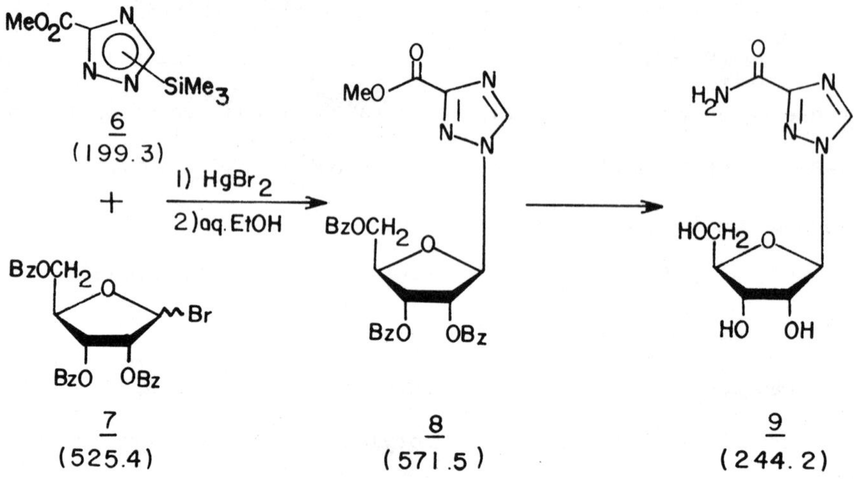

1 - OAc (504.5)   Bz = PhCO

the protected D-ribofuranosyltriazole ester (*8*). Debenzoylation of *8* with methanolic ammonia affords[6] over 90% of *9*. No by-product was obtained in this prodcedure.

PROCEDURE

N^4-*Acetyl*-N,2-*bis(trimethylsilyl)cytosine*[7] *(3)*

To a suspension of dry, powdered N^4-acetylcytosine (15.3 g, 0.10 mol) and chlorotrimethylsilane (33 g, 0.31 mol) in dry benzene (350 ml) is added dropwise with stirring a solution of triethylamine (31 g, 0.31 mol) in dry benzene under anhydrous conditions at room temperature, and stirring is continued for 5 hr. The precipitated mixture of triethylamine hydrochloride and N^4-acetylcytosine is removed by filtration and washed with three 30-ml portions of dry benzene. The filtrate and washings are combined and evaporated to dryness under diminished pressure. The resulting viscous oil is distilled at 139° to 141°/6 torr to give 21.4 g (75%, based on the N^4-acetylcytosine consumed) of a colorless oil (*3*), and the product crystallizes on being kept at room temperature; m.p. 62° to 68°.

N^4-*Acetyl-1-*[2-O-*acetyl*-5-O-*(ethoxycarbonyl*-3-O-p-*tolylsulfonyl*-β-D-*xylofuranosyl)*]*cytosine (4)*

To a solution of N^4-acetyl-bis(trimethylsilyl)cytosine[7] (*3*) (2.14 g, 7.2 mmol) and *2* [obtained from the reaction of 2.76 g (6 mmol) of 1,2-di-*O*-acetyl-5-*O*-(ethoxycarbonyl)-3-*O*-*p*-tolylsulfonyl-D-xylofuranose (*1*) with dry ethereal hydrogen chloride] in dry benzene (35 ml) is added mercuric bromide (2.16 g, 6 mmol). The mixture is stirred overnight at room temperature; the crystals gradually dissolve, giving a clear solution that is evaporated to dryness under diminished pressure. The residue is dissolved in ethanol and the solution is evaporated to dryness. The residue is dissolved in chloroform, and the solution is successively washed with 20% potassium iodide solution (200 ml) and water (100 ml), dried, and evaporated to dryness under diminished pressure. The residue (3.4 g) is chromatographed on silica gel (50 g). The fraction eluted with 9:1 benzene—acetone gives the protected nucleoside *4*; 3.1 g (92% from *3*), amorphous, $[\alpha]_{\underline{\underline{D}}}^{20}$ -0.5° (*c* 4.0, chloroform); λ_{max}^{EtOH} in nm 227, 248, and 297.

1-β-D-Arabinofuranosylcytosine (5)

To a vigorously stirred solution of *4* (3.31 g, 6 mmol) in methanol (140 ml) and water (26 ml) is added, dropwise, 4% aqueous sodium hydroxide (7.12 ml) during 6 hr. The mixture is evaporated to dryness under diminished pressure; the residue is dissolved in water, and the solution is applied to the top of a column of Dowex 1 X-4 (OH^-) ion-exchange resin (30 ml). The column is washed with water, and the product (*5*) is eluted with 50% methanol, the fractions being monitored by their UV absorption. Fractions absorbing at 270 nm are collected, combined, and evaporated to dryness under diminished pressure. The residue is recrystallized from 50% aqueous ethanol to give *5* (1.08 g; 75%, based on *4*), m.p. 213° to 214°, $[\alpha]_{\underline{\underline{D}}}^{20}$ +146° (*c* 1.04, water); $\lambda_{max}^{pH\ 2}$

212 and 278 nm, $\lambda_{max}^{pH\ 12}$ 225 (infl.) and 272 nm.

Methyl 1-(2,3,5-Tri-O-benzoyl-β-D-ribofuranosyl)-1,2,4-triazole-3-carboxylate (8)

To a solution of the trimethylsilyl derivative[5] (*6*) of methyl 1,2,4-triazole-3-carboxylate (4.0 g, 20 mmol) and 2.3,5-tri-*O*-benzoyl-D-ribofuranosyl bromide (*7*) [obtained from 10.08 g (20 mmol) of 1-*O*-acetyl-2,3,5-tri-*O*-benzoyl-D-ribose[a] in dry benzene (60 ml)] is added mercuric bromide (7.2 g, 20 mmol), and the mixture is stirred overnight at room temperature. The crystals gradually dissolve, giving a clear solution; this is evaporated to dryness under diminished pressure, and the residue is dissolved in ethanol, and the solution evaporated to dryness. The residue is dissolved in chloroform, and the solution is successively washed with 20% potassium iodide solution (200 ml) and water (100 ml), dried, and evaporated to dryness under diminished pressure. The residue is crystallized from 99% ethanol to give 10.52 g (92%) of *8*, m.p. 136° to 138° (lit.[5] m.p. 137° to 139°).

1-β-D-Ribofuranosyl-1,2,4-triazole-3-carboxamide ("Virazole") (9)

A solution of *8* (5.72 g, 10 mmol) in methanol (200 ml) (presaturated with dry ammonia at 0°) is kept in a sealed flask for two days at room temperature, and evaporated under diminished pressure; the residue is crystallized from ethanol—ether, to give *9*,

---

[a]Cyclo Chemical Corporation, Los Angeles, Calif. 90001.

2.4 g (98%), m.p. 173° to 174°, $[\alpha]_{\underline{\underline{D}}}^{25}$ -36.5° (*c* 1.0, water) [lit.[5] m.p. 174° to 176°, $[\alpha]_{\underline{\underline{D}}}^{25}$ -36.5° (*c* 1.0, water)].

REFERENCES

1. S. S. Cohen, *Progr. Nucleic Acid Res. Mol. Biol.*, 5, 2 to 88 (1966).

2. E. R. Walwick, W. K. Roberts, and C. A. Dekker, *Proc. Chem. Soc.*, 84 (1959); U. S. Patent 3,116,282 (1963); T. Y. Shen, H. M. Lewis, and W. V. Ruyle, *J. Org. Chem.*, 30, 835 (1965); H. P. M. Fromageot and D. B. Reese, *Tetrahedron Lett.*, 3499 (1966); L. E. Orgel and R. A. Sanchez, *Chem. Eng. News*, 30 (Oct. 6, 1969); K. Kikugawa and M. Ichino, *Tetrahedron Lett.*, 867 (1970); T. Kanai, T. Kojima, O. Maruyama, and M. Ichino, *Chem. Pharm. Bull.*, 18, 2569 (1970); M. Kaneko and B. Shimizu, ibid., 20, 1050 (1972).

3. B. Shimizu and F. Shimizu, *Chem. Pharm. Bull.*, 18, 1060 (1970).

4. J. H. Huffman, R. W. Sidwell, G. P. Khare, J. T. Witkowski, L. B. Allen, and R. K. Robins, *Antimicrob. Agents Chemother.*, 3, 235 (1973); G. P. Khare, R. W. Sidwell, J. T. Witkowski, L. N. Simon, and R. K. Robins, ibid., 3, 517 (1973).

5. J. T. Witkowski, R. K. Robins, R. W. Sidwell, and L. N. Simon, *J. Med. Chem.*, 15, 1150 (1972).

6. B. Shimizu and A. Saito, Japan Patent Prov. Publ., 49-86372 (1974).

7. T. Nishimura and I. Iwai, *Chem. Pharm. Bull.*, 12, 352 (1964).

[46] 1-(5-AZIDO-5-DEOXY-β-D-ARABINOFURANOSYL)CYTOSINE AND 1-(5-AMINO-5-DEOXY-β-D-ARABINOFURANOSYL)CYTOSINE

A Novel Synthesis of 5'-Amino and 5'-Azido Analogs of D-Arabinosylcytosine

Tai-Shun Lin and William H. Prusoff

Department of Pharmacology, Yale University School of Medicine, New Haven, Connecticut 06510

INTRODUCTION

The 5'-amino analog of IdUrd [i.e., 5'-amino-2',5'-dideoxy-5-iodouridine (AIU)] has been found[1] to exert a marked antiviral effect against *Herpes simplex* virus type 1, not only in cell culture but, more important, in rabbits with experimental *Herpes keratitis*. Furthermore, the antiviral activity is accompanied by an unusual lack of toxicity to the host cell as well as to experimental animals. It was therefore of interest to determine whether a similar modification of other antiviral nucleosides would result not only in retention of antiviral activity but also in a concomitant decrease in toxicity to the host cells. This procedure describes the synthesis of the 5'-amino analog of D-arabinosylcytosine (ara-C) via the 5'-*O*-tosyl and 5'-azido intermediates.

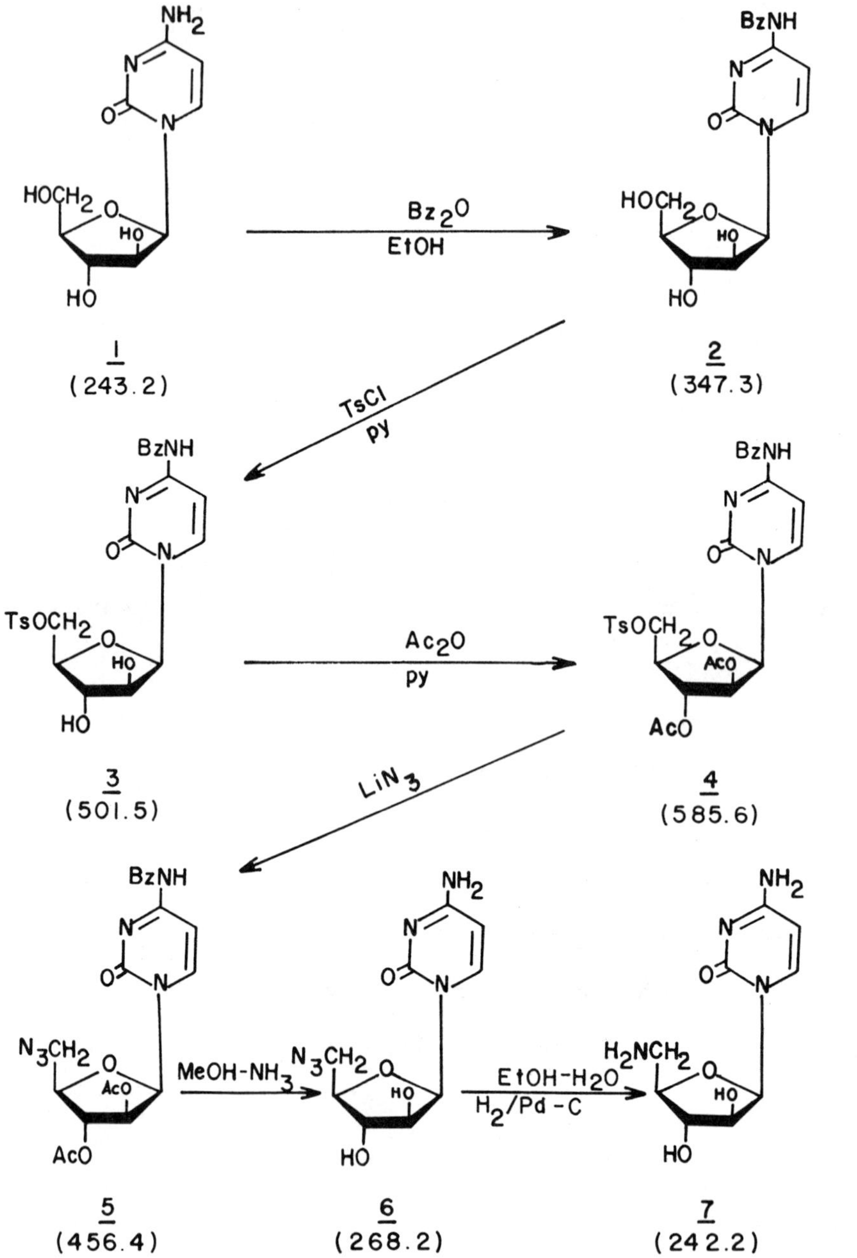
Bz$_2$O
EtOH
TsCl
py
Ac$_2$O
py
LiN$_3$
MeOH-NH$_3$
EtOH-H$_2$O
H$_2$/Pd-C
1
(243.2)
2
(347.3)
3
(501.5)
4
(585.6)
5
(456.4)
6
(268.2)
7
(242.2)

PROCEDURE

N-*Benzoyl-1-β-*D*-arabinofuranosylcytosine*[2] (*2*)

To a stirred, refluxing solution of 1-β-D-arabinofuranosylcytosine (*1*, 8.00 g, 32.90 mmol) in absolute ethanol (400 ml) is added benzoic anhydride (8.09 g, 32.90 mmol). During the refluxing, three further portions (8.00 g each) of the anhydride are added hourly. After the final addition, the solution is boiled under reflux for 1 hr, cooled, and concentrated almost to dryness under diminished pressure. The syrupy residue is triturated with ether, to afford crystals. The suspension is filtered and the solid extracted with ether (10 × 300 ml). The crystalline product remaining weighs 10.30 g (90%); m.p. 188° to 189°; λ_{max}^{MeOH} 259 (ε_{mM} 24.03) and 305 nm (ε_{mM} 11.53).

N-*Benzoyl-1-(5-O-p-tolylsulfonyl-β-*D*-arabinofuranosyl)cytosine*[2](*3*)

To a solution of *N*-benzoyl-1-β-D-arabinofuranosylcytosine (*2*, 9.50 g, 27.40 mmol) in anhydrous pyridine (100 ml) at 0° (ice bath) is added *p*-toluenesulfonyl chloride (7.80 g, 41.00 mmol), and the mixture is stirred for 2 hr while undergoing cooling in an ice bath. The solution is then refrigerated (3°) for a further 22 hr and evaporated under diminished pressure at room temperature. The residue is thoroughly washed with ether and dried *in vacuo* overnight to afford a glass that is used immediately, without further purification, for the next step.

N-Benzoyl-1-(2,3-di-O-acetyl-5-O-p-tolylsulfonyl-β-D-arabinofuranosyl)cytosine[2] *(4)*

A solution of compound *3* and acetic anhydride (30 ml, 32.4 g, 0.32 mol) in anhydrous pyridine (50 ml) is kept for 24 hr at room temperature. The excess of acetic anhydride and pyridine are removed *in vacuo*, and the residue is taken up in ethanol (120 ml). The suspension is clarified with Norit (decolorizing carbon) and filtered through a Celite pad, and the filtrate is evaporated to dryness, affording a glass that is crystallized from ethanol—ether—petroleum ether (b.p. 30° to 60°). The crystals are collected by filtration, washed thoroughly with 10:1 ether—ethanol, and dried, to give 10.7 g (66% based on *2*) of compound *4* as white needles, m.p. 169° to 170°; λ_{max}^{MeOH} 261 (ε_{mM} 23.88) and 303 nm (ε_{mM} 9.33).

N-Benzoyl-1-(2,3-di-O-acetyl-5-azido-5-deoxy-β-D-arabinofuranosyl)cytosine[2] *(5)*

A solution of *4* (8.20 g, 14.00 mmol) and lithium azide (2.10 g, 42.00 mmol) in dry *N,N*-dimethylformamide (70 ml) is stirred magnetically for 2 hr at 85° to 90° and evaporated under diminished pressure at 50°. The residue is coevaporated several times with ethanol and then triturated with ice-cooled water (150 ml). The azido derivative *5* crystallizes as white crystals, which are collected by filtration, successively washed thoroughly with ice-water and ether, and dried. The product weighs 4.75 g (74%); m.p. 165° to 166°; λ_{max}^{MeOH} 260 ($\varepsilon_{\underline{mM}}$ 24.22) and 303 nm ($\varepsilon_{\underline{mM}}$ 9.83).

1-(5-Azido-5-deoxy-β-D-arabinofuranosyl)cytosine[2] *(6)*

A solution of *5* (4.36 g, 9.56 mmol) in methanol (200 ml) presaturated with ammonia is kept for 22 hr in a glass pressure bottle at 3° with stirring and then evaporated under diminished pressure. Ether is added to the resulting syrup, and compound *6* crystallizes. The product is collected by filtration, washed thoroughly with ether, and dried. One recrystallization from ethanol gives 2.25 g (88%) of an analytically pure sample as fine, white needles; m.p. 197° to 198° (dec.); $\lambda_{max}^{H_2O}$ 272 nm ($\varepsilon_{\underline{mM}}$ 9.02), $\lambda_{min}^{H_2O}$ 249 nm.

1-(5-Amino-5-deoxy-β-D-arabinofuranosyl)cytosine[2] *(7)*

A mixture of compound *6* (2.00 g, 7.46 mmol) and 10% palladium-on-charcoal (1.10 g) in 1:1 ethanol—water (120 ml) is hydrogenated for 3 hr at room temperature and a pressure of 37 lb·in^{-2}. The catalyst is removed by filtration through a Celite pad, and the filtrate is concentrated to a small volume. To the concentrate is added ether, and compound 7 crystallizes as fine white crystals, which are collected by filtration, washed with a small amount of ice-cooled 1:1 ethanol—water, and ether, and dried to yield 1.42 g (78%) of product, m.p. 201° to 202° (dec.). One recrystallization from ethanol gives an analytically pure sample, m.p. 203° to 204° (dec.); $\lambda_{max}^{H_2O}$ 272 nm ($\varepsilon_{m\underline{M}}$ 6.92), $\lambda_{min}^{H_2O}$ 250 nm.

REFERENCES

1. Y. C. Cheng, J. P. Neenan, B. Goz, D. C. Ward, and W. H. Prusoff, *J. Virol.*, 1284 (1975).

2. T. S. Lin and W. H. Prusoff, *J. Carbohydr. Nucleos. Nucleot.*, 2, 185 (1975).

[47] 1-CYTOSIN-1-YL-1-*S*-ETHYL-1-THIO-D-XYLITOL

(1R) and (1S) Epimers of an Acyclic-Sugar Pyrimidine Nucleoside Analog Prepared by the Trimethylsilyl Fusion Method

Derek Horton and Satish S. Kokrady

Department of Chemistry, The Ohio State University, Columbus, Ohio 43210

INTRODUCTION

Tetra-*O*-acetyl-D-xylose diethyl dithioacetal[1] (*1*) is converted into the unstable syrupy 2,3,4,5-tetra-*O*-acetyl-1-bromo-1-*S*-ethyl-1-thio-D-xylitol (*2*) by a procedure improved in our laboratory.[2] This unstable bromide is used immediately in the next step. On fusion with 2,4-bis(trimethylsilyl)cytosine[3] (*3*) a mixture of the (1*R*) (*4a*) and (1*S*) (*4b*) epimers of 2,3,4,5-tetra-*O*-acetyl-1-(cytosin-1-yl)-1-*S*-ethyl-1-thio-D-xylitol is obtained.[4,5] These epimers are separated by fractional recrystallization, and converted into their *O*-deacetylated counterparts (*5a* and *5b*) by saponification with methanolic ammonia.

PROCEDURE

2,3,4,5-Tetra-O-acetyl-1-bromo-1-S-ethyl-1-thio-D-xylitol (2)

To a solution of tetra-*O*-acetyl-D-xylose diethyl dithioacetal[1] (*1*) (4.24 g, 10 mmol) in anhydrous ether (50 ml) is added a solution

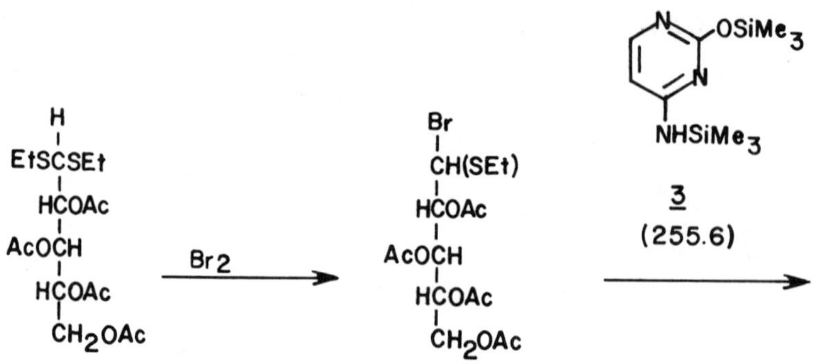

1 (424.5) 2 (443.3) 3 (255.6)

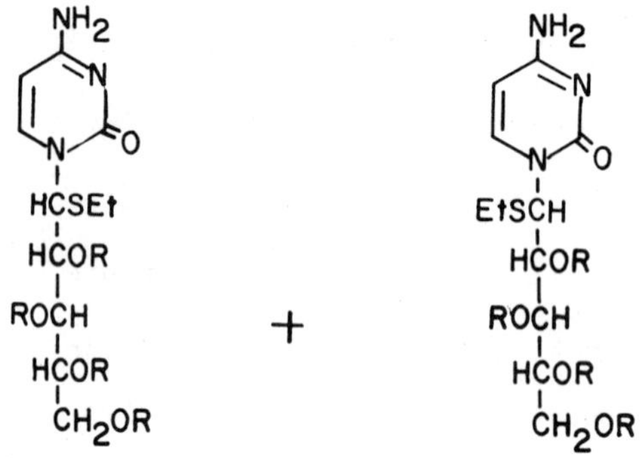

4a (R=Ac) (473.7) 4b (R=Ac) (473.7)

5a (R=H) (305.5) 5b (R=H) (305.5)

of bromine (1.6 g, 10 mmol) in ether (25 ml). After 15 min at -25°, the solution is concentrated under diminished pressure to 5 ml, and then carbon tetrachloride (20 ml) is added and the mixture is evaporated to remove volatile by-products and the excess of bromine. This evaporation procedure is repeated four times. The resulting bromide[2] is very unstable and is used immediately in the next step.

2,4-Bis(trimethylsilyl)cytosine (3)

A magnetically stirred mixture of cytosine (1.65 g, 15 mmol) and hexamethyldisilazane (10 ml) is boiled under reflux until a clear solution results (1 hr). The excess of hexamethyldisilazane is removed under diminished pressure, and the resulting semi-solid product[3] (3) is used without further purfication.

(1R)-2,3,4,5-Tetra-O-acetyl-1-(cytosin-1-yl)-1-S-ethyl-1-thio-D-xylitol (4a) and its (1S) Epimer (4b)

The syrupy bromide (2) (4.43 g, 10 mmol) is mixed with 2,4-bis-(trimethylsilyl)cytosine (3, 3.83 g, 15 mmol) in a flask that is then evacuated with a water pump. The flask is closed, and the mixture is slowly heated to fusion (140°) and kept molten for 30 min; the mixture is then cooled and evaporated to dryness. The dark-colored residue is treated with 80% methanol (20 ml), and the mixture is evaporated to dryness. The residue is extracted with hot chloroform (250 ml) and the extract filtered. The filtrate is washed with water, dried (sodium sulfate), and evaporated to a thick glass (4.8 g) which, on TLC (Silica Gel G, Merck) with 9:1 chloroform—methanol as developer, shows a major

(R_F 0.36) and a minor nucleoside component (R_F 0.63). Trituration of this glassy material with ether gives an amorphous solid that is dissolved in chloroform, and ether is added to incipient turbidity. On keeping the solution, compound *4a* separates as a sharp-melting product; yield 1.5 g (32%), m.p. 111° to 112°, $[\alpha]_D^{22}$ +90° (*c* 0.35, chloroform). Recrystallization from acetone affords pure *4a*; yield 1.1 g (23%), m.p. 125° to 126°, $[\alpha]_D^{22}$ +92° (*c* 0.3, chloroform); R_F 0.39 (9:1 chloroform—methanol); λ_{max}^{MeOH} 277 nm (ε_{mM} 11.25); NMR (100 MHz, $CDCl_3$): δ 7.82 (doublet, $J_{5,6}$ 7.3 Hz, H-6 of pyrimidine), 6.47 (doublet, H-5 of pyrimidine), 6.27 (doublet, $J_{1,2}$ 7.7 Hz, H-1), 5.85 to 5.44 (multiplet, H-2,3,4), 4.28 (quartet, $J_{4,5b}$ 4.0 Hz, $J_{5a,5b}$ 12.1 Hz, H-5b), 4.12 (quartet, $J_{4,5a}$ 5.5 Hz, H-5a), 2.54 (quartet, J 7.0 Hz, CH_2 of Et), 2.14, 2.11, 2.12, 1.98 (singlets, OAc), and 1.22 (triplet, CH_3 of Et). The product, although sharp-melting, gave a diffuse X-ray powder diffraction pattern.

The mother liquor from the isolation of *4a* is evaporated to a thick glass (1.3 g) that is dissolved in acetone. On keeping the solution, compound *4b* separates as a sharp-melting solid; yield 900 mg (19%), m.p. 94 to 96°, $[\alpha]_D^{21}$ -134° (*c* 0.8, chloroform); R_F 0.39 (9:1 chloroform—methanol); λ_{max}^{MeOH} 277 nm (ε_{mM} 9.70); NMR (100 MHz, $CDCl_3$): δ 7.71 (doublet, $J_{5,6}$ 7.3 Hz, H-6 of pyrimidine), 5.87 (doublet, $J_{1,2}$ 3.5 Hz, H-1), 5.65 to 5.35 (multiplet,

H-2,3,4), 4.04 (quartet, $J_{4,5a}$ 5.0 Hz, $J_{5a,5b}$ 12.0 Hz, H-5b), 3.74 (quartet, $J_{4,5a}$ 6.9 Hz, H-5a), 2.57 (quartet, J 7.5 Hz, CH_2 of Et), 2.17, 2.04, 2.02, 1.98 (singlets, OAc), and 1.23 (triplet, CH_3 of Et). The product, although sharp-melting, gave a diffuse X-ray powder diffraction pattern.

The dextrorotatory epimer is assigned as (1*R*), and the levorotatory one (1*S*) on the basis of the generalized heterocycle rule[6] and by comparison with related compounds[7] having structures confirmed by X-ray crystallography.

(1R)-1-(Cytosin-1-yl)-1-S-ethyl-1-thio-D-xylitol, (5a) and its (1S) Epimer (5b)

The acetate *4a* (260 mg) is dissolved in methanol (20 ml) presaturated with ammonia at 0°. The solution is kept overnight at -25°, and then concentrated until crystallization occurs. Recrystallization of the crude product from aqueous methanol affords pure *5a*; yield 160 mg (96%), m.p. 160 to 162°, $[\alpha]_D^{21}$ +160° (*c* 0.1, water), $\lambda_{max}^{H_2O}$ 275 nm (ε_{mM} 11.21); X-ray powder diffraction data: 8.54 m, 7.03 m, 4.47 w, 4.30 w, 3.99 vs (1,1), 3.90 m, 3.47 m, 3.04 w, 2.96 m, and 2.98 s.

The acetate *4b* (600 mg) is saponified with methanolic ammonia by the procedure just described. The product is recrystallized from aqueous methanol to give pure *5b*; yield 350 mg (92%), m.p. 160 to 162°, $[\alpha]_D^{20}$ -150° (*c* 0.2, water); $\lambda_{max}^{H_2O}$ 277 nm (ε_{mM} 9.8);

X-ray powder diffraction data (somewhat diffuse pattern): 11.04 s (2), 9.50 vw, 8.38 w, 7.53 m, 6.83 w, 6.37 s, 5.33 w, 5.06 s, 4.62 vs (1,1), 4.40 w, 4.09 w, 3.91 vw, and 3.73 vs (1,1).

REFERENCES

1. S. M. Olin, *Methods Carbohydr. Chem.*, 1, 148 to 151 (1962).

2. D. Horton and S. S. Kokrady, *Carbohydr. Res.*, 24, 333 to 342 (1972).

3. T. Nishimura and I. Iwai, *Chem. Pharm. Bull.*, 12, 352 to 356 (1964).

4. D. Horton, D. C. Baker, and S. S. Kokrady, *Ann. N. Y. Acad. Sci.*, 255, 131 to 150 (1975).

5. S. S. Kokrady, Ph.D. thesis (D. Horton, preceptor), The Ohio State University, 1972; D. Horton and S. S. Kokrady, *Diss. Abstr. Intern. B*, 34, 597 (1973).

6. H. S. El Khadem and Z. M. El Shafei, *Tetrahedron Lett.*, 1887 to 1889 (1963).

7. D. Horton, K. Blieszner, and R. A. Markovs, *Proc. Int. Conf. Transfer Ribonucleic Acids and Their Components*, Poznan, Poland, 1976, pp. 60 to 85; R. A. Markovs, Ph.D. Thesis (D. Horton, preceptor), The Ohio State University, 1975; *Diss. Abstr. Intern. B*, 34, 597 (1973).

[48] 2,5'-ANHYDRONUCLEOSIDES OF 2'-DEOXY-D-*ERYTHRO*-PENTOSYL-PYRIMIDINES

*Two-Step Synthesis of 2,5'-Anhydronucleosides From Thymidine, 2'-Deoxyuridine, and 2'-Deoxy-5-fluorouridine**

K. A. Watanabe, Uri Reichman, C. K. Chu, and Jack J. Fox

Laboratory of Organic Chemistry, Memorial Sloan-Kettering Cancer Center, Sloan-Kettering Institute, Sloan-Kettering Division of Graduate School of Medical Sciences, Cornell University, New York, New York 10021

INTRODUCTION

2,5'-Anhydronucleosides are potentially versatile intermediates in chemical syntheses. Although the Shibuya—Mitsunobu reaction[1] can be applied to the synthesis of 2,5'-anhydrides of D-ribofuranosylpyrimidines, this reaction gives an intractable mixture when applied to 2'-deoxynucleosides.[2] Selective tosylation of *1* by the method of Reist and co-workers[3] gives the crystalline 5'-*p*-toluenesulfonate (*2*) in ≃60% yield. Treatment of *2* with 1,5-diazabicyclo[5,4,0]undec-5-ene (DBU) in refluxing acetonitrile gave 2,5'-anhydronucleoside (*3*) in crystalline form. This method is much simpler than the procedures reported.[4,5]

---

*The authors acknowledge the support of the U.S. Public Health Service, Grant Nos. CA-08748 and CA-18601.

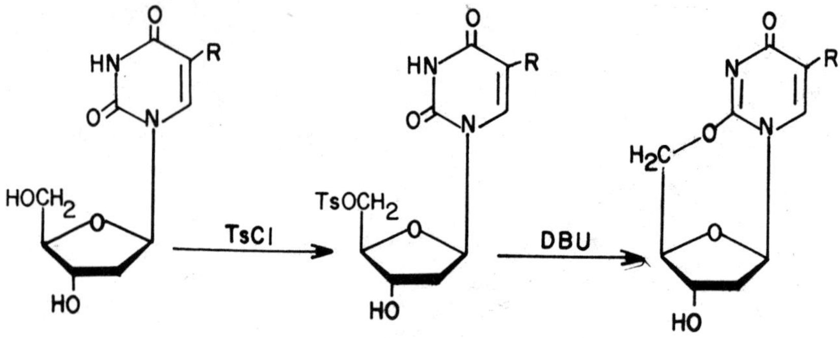

1		2	3
a,	R = CH_3 (242.2)	a, (394.3)	a, (224.2)
b,	R = H (228.2)	b, (380.3)	b, (210.2)
c,	R = F (246.2)	c, (398.3)	c, (228.2)

Secrist[6] has described the synthesis of several anhydronucleosides by treatment of mesylated nucleosides with DBU in dichloromethane.

PROCEDURE

General Method of Tosylation

A solution of nucleoside *1* (*a*, *b*, or *c*) (0.04 mol) in hot pyridine (100 ml) is cooled to 10°, *p*-toluenesulfonyl chloride (8.4 g, 0.044 mol) is added, and the mixture is stirred overnight at room temperature. Water (250 ml) is then added, and the mixture is extracted with chloroform (3 × 200 ml). The extracts are combined, washed successively with saturated sodium hydrogen carbonate solution (200 ml) and water (200 ml), dried (sodium sulfate), and evaporated to dryness. Traces of pyridine are removed by several coevaporations with ethanol, and the residue is processed as in the individual descriptions that follow.

5'-O-p-Tolylsulfonylthymidine (2a)

A suspension of the solid residue containing *2a* in ethanol (100 ml) is boiled under reflux for 5 min and filtered while hot. Compound *2a* is obtained as colorless needles (8.2 g, 53%), m.p. 171° to 176° (dec.) [lit.[3] m.p. 168° to 169° (dec.)]. From the mother liquor are obtained further quantities of *2a* (2.3 g, 15%) by evaporation, filtration of the precipitate, and recrystallization from methanol.

2'-Deoxy-5'-O-p-tolylsulfonyluridine (2b)

The residue containing *2b* is crystallized from methanol. Compound *2b* is obtained as colorless needles (10.5 g, 69%), m.p. 152° to 157° (dec.) [lit.[7] m.p. 157° to 158° (dec.)].

2'-Deoxy-5-fluoro-5'-O-p-tolylsulfonyluridine (2c)

The residue containing *3c* is crystallized from methanol. Compound *2c* is obtained as colorless needles (9.5 g, 60%), m.p. 153° to 158° (dec.) [lit.[8] m.p. 157° to 158° (dec.)].

General Method of Formation of Anhydronucleosides

A solution of the 5'-*O*-*p*-tolylsulfonylnucleoside *2* (5 mmol) in a mixture of acetonitrile (100 ml, dried over 4A molecular sieves) and DBU (0.84 g, 5.5 mmol) is boiled under reflux with stirring for 24 hr, during which time colorless crystals are precipitated.

The reaction mixture is treated as in the following individual descriptions.

2,5'-Anhydrothymidine (3a)

The mixture containing *3a* is filtered while hot, the insoluble material is discarded, and the filtrate is cooled. Compound *3a* (460 mg) is precipitated and is removed by filtration. The filtrate is evaporated to dryness, and the residue is triturated with acetone (≃5 ml); further quantities (200 mg) of *3a* are removed on crystallization to give a total yield of 60%. Both fractions of *3a* have m.p. 216° to 218°, unaltered after one recrystallization from methanol to give colorless needles; [lit.[4] m.p. 218° or 230°]; $\lambda_{max}^{H_2O}$ 249 nm, $\lambda_{min}^{H_2O}$ 216.5 nm [lit.[4] $\lambda_{max}^{H_2O}$ 250 to 253 nm; $\lambda_{min}^{H_2O}$ 218 nm].

2,5'-Anhydro-2'-deoxyuridine (3b)

The mixture containing *3b* is filtered while hot, and the filtrate is evaporated to dryness. The residue is triturated with acetone (≃5 ml), giving crystals of pure *3b* (620 mg, 59%), m.p. 195° to 199° [lit.[6] 199° to 200°], $\lambda_{max}^{H_2O}$ 236 nm, $\lambda_{min}^{H_2O}$ 212 nm [lit.[6] λ_{max} 236, λ_{min} 212.5 nm].

2,5'-Anhydro-2'-deoxy-5-fluorouridine (3c)

Colorless insoluble crystals (400 mg of analytically pure *3c*) are obtained by filtration of the reaction mixture containing *3c*. The filtrate is evaporated to dryness, and the residue is triturated with acetone (5 ml). Further quantities of pure *3c* (240 mg) are removed on crystallization to afford a combined yield of 56%, m.p. 185° to 187°, $\lambda_{max}^{H_2O}$ 249 nm ($\varepsilon_{m\underline{M}}$ 10.80), $\lambda_{min}^{H_2O}$ 216.5 nm ($\varepsilon_{m\underline{M}}$ 3.80).

REFERENCES

1. S. Shibuya, A. Kuninaka, and H. Yoshino, *Chem. Pharm. Bull.*, 22, 719 (1974). See also the preparation (60) by Watanabe, Chu, Reichman, and Fox.

2. K. A. Watanabe and J. J. Fox, unreported experiments.

3. E. J. Reist, A. Benitez, and L. Goodman, *J. Org. Chem.*, 29. 554 (1964).

4. A. M. Michelson and A. R. Todd, *J. Chem. Soc.*, 816 (1955); R. Letters and A. M. Michelson, ibid, 1410 (1961).

5. S. S. Tang and J. S. Roth, *Tetrahedron Lett.*, 2123 (1968).

6. J. A. Secrist, *Carbohydr. Res.*, 42, 379 (1975).

7. M. Logue and N. J. Leonard, *J. Am. Chem. Soc.*, 94, 2842 (1972).

8. Y. Nishizawa, J. E. Casida, S. W. Anderson, and C. Heidelberger, *Biochem. Pharmacol.*, 14, 1605 (1965).

[49] 4-AMINO-1-(2,3,5-TRI-*O*-ACETYL-β-D-ARABINOFURANOSYL)-2(1*H*)-PYRIMIDINONE MONOHYDROCHLORIDE [1-(2,3,5-TRI-*O*-ACETYL-β-D-ARABINOFURANOSYL)CYTOSINE HYDROCHLORIDE]

Selective O-Acylation of Nucleosides

Jiří Beránek and Pavel Drašar

Institute of Organic Chemistry and Biochemistry, Czechoslovak Academy of Sciences, 16610 Prague, Czechoslovakia

INTRODUCTION

Acetylation with acetic anhydride, or acetyl chloride, or both, in the absence of a base, so-called[1] "acid acetylation," has been used as a simple method for the preparation of *O*-acetyl derivatives of pyrimidine nucleosides[a] (of ribo,[1,2] arabino,[2] 3-deoxy-3-halo-arabino,[3] 2-deoxy,[4] and 5-deoxy[4] sugars). When applied to nucleosides bearing an amino group on the aglycon[5] or in the sugar moiety,[6] the reaction proceeds as a selective *O*-acylation. The method is exemplified by the preparation of 1-(tri-*O*-acetyl-β-D-arabinofuranosyl)cytosine as the hydrochloride of *2*, a water-soluble acetyl derivative of the cancerostatic[7] 1-β-D-arabinofuranosylcytosine. This procedure is more advantageous than the acetylation of *1* with trifluoroacetic anhydride—acetic acid in

---

[a]For "acid acetylation" of purine nucleosides, see Reference 10.

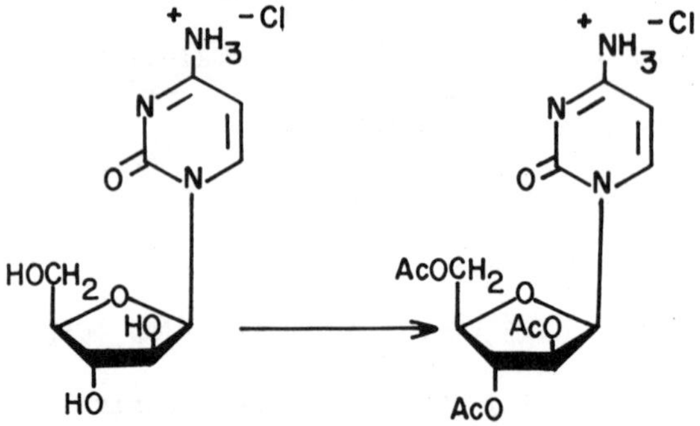

1 (279.7) → 2 (405.8)

trifluoroacetic acid,[8] or the *N*-deacetylation of *N*[4]-acetyl-1-(tri-*O*-acetyl-β-D-arabinofuranosyl)cytosine,[9] resulting in 1-(tri-*O*-acetyl-β-D-arabinofuranosyl)cytosine.

PROCEDURE

4-Amino-1-(2,3,5-tri-O-acetyl-β-D-arabinofuranosyl)-2(1H)-Pyrimidinone Monohydrochloride (2)

A suspension of 1-β-D-arabinofuranosylcytosine hydrochloride[b] (*1*, 200 mg, 0.72 mol) in acetic acid[c] (2 ml) and acetyl chloride[d] (1 ml) is stirred at room temperature. After 3 hr chloroform[e]

---

[b]Instead of the hydrochloride *1*, free 1-β-D-arabinofuranosylcytosine may be used without change in procedure or in yield of *2*. The procedure given has been performed with 5.0 g of the starting material *1* to yield 6.0 g (72%) of crystalline *2* and an additional 0.66 g (8%) of *2* by processing the mother liquors.

[c]Glacial acetic acid is distilled from phosphorus pentaoxide.

[d]Acetyl chloride is distilled prior to use.

[e]Chloroform is added, to cause dissolution of the *2* formed.

(1 ml) is added, and stirring is continued for 21 hr at room temperature.[f] The mixture is then evaporated to dryness under reduced pressure,[g] and the residue is coevaporated with ethanol (3 × 5 ml). The resulting glass (300 mg) is crystallized from 2-propanol—ethanol; yield, 270 mg of chromatographically pure *2*. On recrystallization[h] from ethanol (3 ml), 240 mg (71.9%) of the hydrochloride *2* is obtained, m.p. 171° to 174°, $[\alpha]_D^{25}$ +96.9° (*c* 0.7, ethanol); $\lambda_{max}^{H_2O}$ 271 nm (log ε 3.88), $\lambda_{min}^{H_2O}$ 243 nm (log ε 3.70); λ_{max}^{nujol} 3310, 3115, and 1677 cm^{-1} (NH), 2720 and 1545 cm^{-1} (N^+H), 1756 and 1742 cm^{-1} (C=O, acetyl), and 1716 cm^{-1} (C=O, cytosine).

REFERENCES

1. J. Beránek and J. Pithа, *Collect. Czech. Chem. Commun.*, 29, 625 (1964).
2. J. Brokeš and J. Beránek, *Collect. Czech. Chem. Commun.*, 39, 3100 (1974).
3. J. Brokeš and J. Beránek, *Collect. Czech. Chem. Commun.*, 40, 3071 (1975).
4. P. Drašar, L. Hein, and J. Beránek, *Nucleic Acids Res. Spec. Publ.* No. 1, s61 (1975).

---

[f]During the reaction time, the mixture remains heterogenous; neither the starting compound *1* nor the product *2* is completely soluble in the reaction medium. To obtain a solution, the mixture is heated to ≃50° at the end of the reaction.

[g]The solution is evaporated in a rotary evaporator at 20 Torr (water pump) and a bath temperature of 35° to 40°.

[h]The mother liquors give 30 mg of a residue containing mainly product *2* (as shown by TLC).

5. J. Beránek and P. Drašar, Czechoslovakian Patent 164,080 (1973).

6. J. Brokeš and J. Beránek, *Collect. Czech. Chem. Commun.*, 40, 3061 (1975).

7. G. P. Bodey, E. J. Freireich, R. W. Monto, and J. S. Hewlett, *Cancer Chemother. Rep.*, 53, 59 (1969).

8. A. P. Martinez, W. W. Lee, and L. Goodman, *J. Med. Chem.*, 9, 268 (1966).

9. J. A. Montgomery and H. J. Thomas, *J. Med. Chem.*, 15, 116 (1972).

10. J. Beránek and H. Hřebabecký, *Nucleic Acid Res. Spec. Publ.* No. 1, s57 (1975).

[50] ANOMERIC 1-(2-DEOXY-D-*ERYTHRO*-PENTOFURANOSYL)-2(1*H*)-PYRIMIDINONES

Synthesis of Analogs of 2'-Deoxycytidine by the Chloromercuri Procedure

Peter Kohler, Ernst Volz, Urs Séquin, and Christoph Tamm

Institut für Organische Chemie der Universität, CH-4056 Basel, Switzerland

INTRODUCTION

Several methods for the synthesis of 1-(2-deoxy-β-D-*erythro*-pentofuranosyl)-2(1*H*)-pyrimidinone have been published. Laland and Serck-Hanssen[1] described the reduction of 2'-deoxyuridine with sodium amalgam, leading to isolation of the corresponding 2'-deoxy-D-*erythro*-pentosyl-pyrimidin-2-one. Øyen and co-workers[2] prepared the compound by incubation of 2-pyrimidinone in a modified 2-deoxy-D-*erythro*-pentofuranosyl transferase system that used 2'-deoxycytidine as the donor. In 1973, Wightman and Holý[3] reported the conversion of 1-(2-deoxy-β-D-*erythro*-pentofuranosyl)-4-thiouracil into 1-(2-deoxy-β-D-*erythro*-pentofuranosyl)-2(1*H*)-pyrimidinone. A description of the preparation of anomeric deoxyglycosylpyrimidinones under mild conditions in good yield by the chloromercuri procedure is given here. The

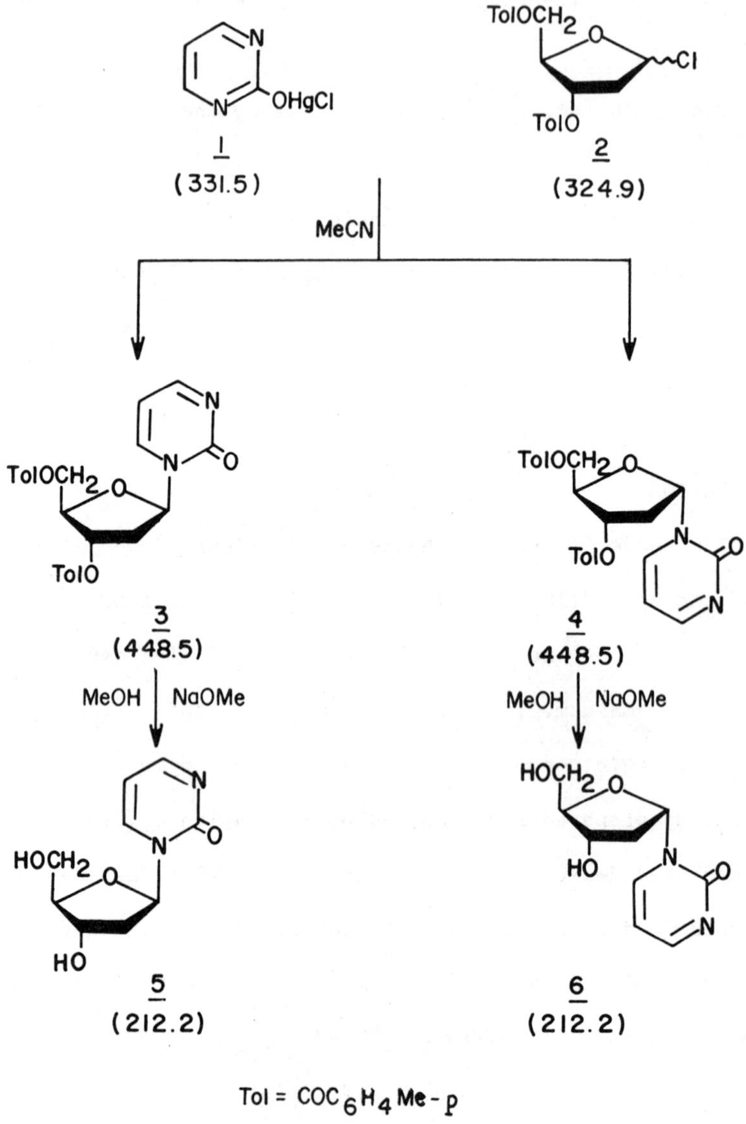

β-D-anomer has no growth-inhibitory effect on *Escherichia coli*,[2] but the corresponding 5'-nucleotide is a very potent inhibitor of thymidylate synthetase.[3]

The arrangement of the heteroaromatic nitrogen atoms in the 2-deoxy-D-*erythro*-pentofuranosylpyrimidinones corresponds to that responsible for base-pairing in naturally-occurring nucleoside components.

PROCEDURE

(Chloromercuri)-2-pyrimidinol[4] *(1)*

To a solution of 2-pyrimidinone hydrochloride (3.66 g, 0.028 mol) in water (200 ml) containing sodium hydroxide (1.6 g, 0.02 mol) is added dropwise, with vigorous stirring, a solution of mercuric chloride (5.43 g, 0.02 mol) in hot ethanol (50 ml). The resulting yellowish precipitate is removed by filtration, washed successively with water, ethanol, and ether, and dried in a vacuum desiccator for 20 hr at 40° to give a yellowish-white powder; yield 75.9%.

Analysis calculated for $C_4H_3ClHgN_2O$: C, 14.6; H, 0.9; Cl, 10.0; N, 8.6. Found: C, 14.5; H, 0.9; Cl, 10.7; N, 8.5.

*Anomeric 1-(3,5-Di-*O-p-*toluoyl-2-deoxy-*D-erythro-*pentofuranosyl)-2(1*H*)-pyrimidinones (3 and 4)*

A suspension of dry (chloromercuri)-2-pyrimidinol (*1*)[a] (9.77 g, 29.5 mmol) and 3,5-di-*O*-*p*-toluoyl-2-deoxy-D-*erythro*-pentofurano-

---

[a]Dissolved in few ml of dry benzene and evaporated to dryness; this treatment is repeated five times.

syl chloride[5] (*2*) (11.5 g, 29.6 mmol) in dry acetonitrile[b] (125 ml) is stirred for 20 min at room temperature, boiled under reflux with vigorous stirring for 5 min, and cooled. The mixture is evaporated to dryness, the residue is treated with chloroform (250 ml), and the suspension is filtered. The filtrate is concentrated to ≃70 ml under diminished pressure. The concentrate is washed successively with 30% aqueous potassium iodide (3 × 70 ml) and water (3 × 75 ml), dried, and evaporated at 30°. The residue is chromatographed on a column (5.5 × 50 cm) (1 kg of aluminum oxide[c]) with benzene and increasing proportions of ethyl acetate. A nonpolar side product is eluted in the first fractions. The β-nucleoside (*3*) is eluted next, followed by a mixture of the anomers (*3* and *4*), and finally, the α-D-anomer (*4*). The mixture of anomers is rechromatographed as before, until complete separation is achieved. The separate anomers are recrystallized from acetone—pentane; the total yield of nucleosidic material varies between 9.5 and 47% (14% to 17% of the β and 30% to 33% of the α-D-anomer).

Compound *3* (fine, colorless needles) has m.p. 136° to 138°, $[\alpha]_D^{24}$ +73° (*c* 1.13, methanol); λ_{max}^{EtOH} in nm 308 (ε_{mM} 5.559), 282 (3.908), 272.5 (3.749), and 239 (34.434); NMR data (100 MHz,

---

[b]Puriss. p.a., Fluka A.G., Buchs, Switzerland.

[c]Akt. stufe II-III, E. Merck A.G., Darmstadt, Germany.

C_5D_5N): δ 8.58 (d of d, H-4), 8.36 (d of d, H-6), 6.61 (d of d, H-1'), 6.21 (d of d, H-5), 5.82 (m, H-3'), 4.88 (m, H-5',5', H-4'), 3.24 (m, H-2'), and 2.61 (m, H-2').

Analysis calculated for $C_{25}H_{24}N_2O_6$: C, 66.95; H, 5.39; N, 6.25. Found: C, 67.05; H, 5.41; N, 6.44.

Compound *4* (fine, colorless needles) has m.p. 191° to 193°, $[\alpha]_D^{24}$ -176° (*c* 0.87, methanol); λ_{max}^{EtOH} in nm 308 (ε_{mM} 5.728), 282.5 (3.944), 273 (3.690), and 241 (32.96); NMR data (100 MHz, C_5D_5N): δ 8.66 (m, H-4), 8.33 (d of d, H-6), 6.51 (d of d, H-1'), 6.37 (d of d, H-5), 5.76 (d, H-3'), 5.19 (t, H-4'), 4.71 (d, H-5', 5'), 3.17 (m, H-2'), and 2.86 (d, H-2').

Analysis calculated for $C_{25}H_{24}N_2O_6$: C, 66.95; H, 5.39; N, 6.25. Found: C, 66.83; H, 5.44; N, 6.49.

*1-(2-Deoxy-β-D-*erythro*-pentofuranosyl)-2(1H)-pyrimidinone (5)*

A solution of di-*p*-toluate *3* (1.01 g, 2.25 mmol) in absolute methanol (80 ml) is treated with 82 m*M* sodium methoxide in absolute methanol (73 ml), and the mixture is kept for 18 hr at room temperature and made neutral with Amberlite IRC-50 (H^+)[d] ion-exchange resin (32 to 45 mesh, 60 ml), and the suspension is filtered. The resin is washed with absolute methanol (500 ml). The filtrate and washings are combined and evaporated under diminished pressure, to afford a colorless residue, which is dissolved

---

[d] Fluka A.G., Buchs, Switzerland

in a mixture of water (100 ml) and petroleum ether (150 ml). The organic phase is extracted with water (3 × 100 ml), and the extracts and water layer are combined, washed with petroleum ether (150 ml), and evaporated to dryness under diminished pressure. The residue is dissolved in a few ml of ethanol; acetone and ether are added to incipient opalescence, and the solution is cooled to give 396.2 mg (83%) of colorless crystals of 5, m.p. 136° to 138°, $[\alpha]_D^{24}$ +197° (*c* 0.74, methanol); λ_{max}^{MeOH} in nm 308 (ε_{mM} 5.70) and 213 (9.75); NMR data (100 MHz, Me_2SO-d_6): δ 8.55 (m, H-4,7), 6.48 (d of d, H-5), 6.11 (d of d, H-1'), 5.25 (d, OH-3'), 5.06 (t, OH-5'), 4.28 (m, H-3'), 3.95 (q, H-4'), 3.66 (s, H-5',5'), 2.39 (m, H-2'), and 2.04 (m, H-2').

Analysis calculated for $C_9H_{12}N_4O$: C, 50.95; H, 5.7. Found: C, 51.16; H, 5.9.

*1-(2-Deoxy-α-D-*erythro*-pentofuranosyl)-2(1H)-pyrimidinone (6)*

This compound is prepared as described for the β-D-anomer. The yield of crude product is 95%. It is recrystallized from methanol by adding a few ml of ether to give yellowish crystals, m.p. 156° to 157°, $[\alpha]_D^{24}$ -176° (*c* 0.89 methanol); λ_{max}^{EtOH} in nm 307 (ε_{mM} 5.78) and 212 (10.05); NMR data (100 MHz, Me_2SO-d_6): δ 8.52 (d of d, H-4), 8.24 (m, H-6), 6.47 (d of d, H-5), 6.02 (d of d, H-1'), 5.03 (broad signal, OH-3',5'), 4.28 (m, H-3',4'), 3.45 (d, H-5',5'), 2.60 (m, H-2'), and 2.02 (d of d, H 2').

Analysis calculated for $C_9H_{12}N_2O_4$: C, 50.95; H, 5.7; N, 13.2. Found: C, 51.4; H, 5.9, N, 13.5.

REFERENCES

1. S. G. Laland and G. Serck-Hanssen, *Biochem. J.*, 90, 76 (1974).

2. T. B. Øyen, L. Helgeland, and S. G. Laland, *Biochem. J.*, 92, 27*p* (1964).

3. R. Wightman and A. Holý, *Collect. Czech. Chem. Commun.*, 38, 1381 (1973).

4. M. Ikehara, *Chem. Pharm. Bull.*, 8, 308 (1960).

5. M. Hoffer, *Chem. Ber.*, 93, 2777 (1960); C. C. Bhat, in W. W. Zorbach and R. S. Tipson (eds.), *Synthetic Procedures in Nucleic Acid Chemistry*, Vol. 1, Interscience Publishers, New York, 1968, p. 521.

[51] 3'-*O*-(CARBOXYMETHYL)THYMIDINE

Carboxymethylation of Thymidine at O-3'

M. D. Edge, M. H. Halford, and A. S. Jones

Department of Chemistry, University of Birmingham, Birmingham B15 2TT, United Kingdom

INTRODUCTION

Thymidine (*1*) is converted into its 5'-trityl ether (*2*) by a procedure previously detailed.[1] Treatment of 5'-*O*-tritylthymidine with 2.5 molecular proportions of sodium hydride in dimethyl sulfoxide and then with sodium chloroacetate gives 3'-*O*-(carboxymethyl)-5'-*O*-tritylthymidine (*3*) in ≈90% yield. Hydrolysis of *3* with aqueous acetic acid gives the desired product in 64% yield.[2,3]

PROCEDURE

All evaporations are conducted under diminished pressure in a rotary evaporator (water pump), unless otherwise stated.

3'-O-(Carboxymethyl)-5'-O-tritylthymidine (3)

5'-*O*-Tritylthymidine (monobenzene adduct; 2.27 g, 4 mmol)[1] is dried *in vacuo* over phosphoric anhydride, and then dissolved in dry dimethyl sulfoxide (20 ml). Sodium hydride (240 mg, 10 mmol) is added, and the solution is stirred under anhydrous conditions

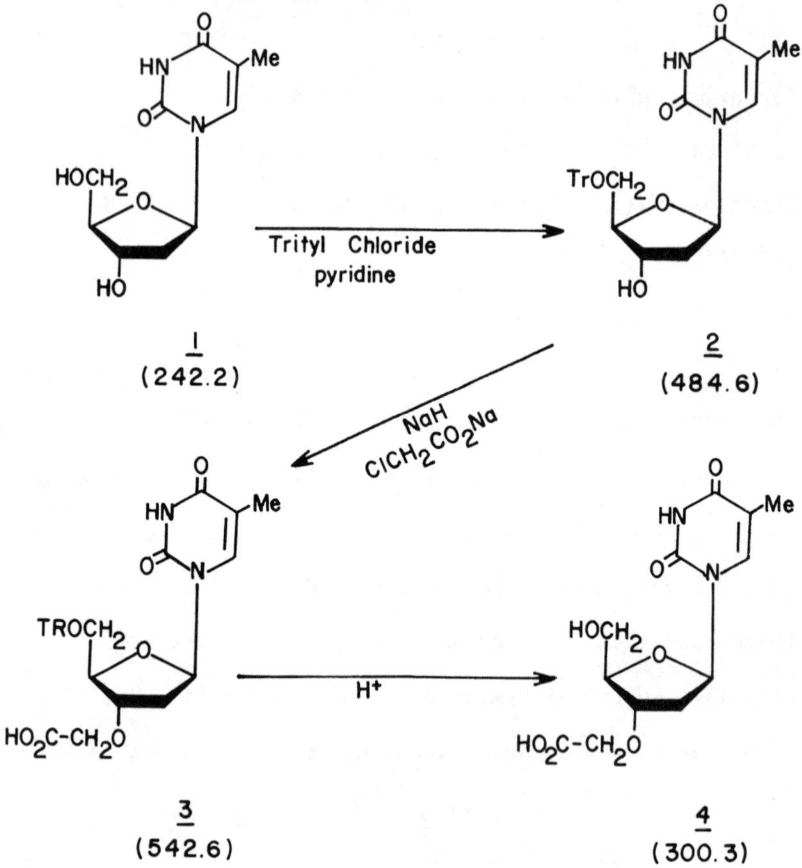

until the evolution of hydrogen is complete (≈1 hr). Sodium chloroacetate (519 mg, 4.4 mmol) is added, and the mixture is shaken at ≈20°, giving a yellow solution which is stirred at room temperature for 3 days. A white precipitate that forms is dissolved by the addition of water (40 ml) and ethanol (40 ml), and the solution is evaporated under high vacuum at 40° to remove the dimethyl sulfoxide. The residue is dried *in vacuo* over phosphoric anhydride, pulverized, and extracted with boiling, 1:3 (v/v) anhydrous chloroform—ethanol (6 × 200 ml, 30 min each). The hot extracts are combined and filtered, and the filtrate is cooled to room temperature and then kept for 18 hr at 4°. The resulting, gelatinous mass is filtered with suction, and the solid is retained. The filtrate is evaporated to dryness, and the residue is extracted with chloroform—ethanol, as before, to give a second quantity of solid. A further amount of product is obtained by repeating the process. The total yield of the combined solids, consisting of the sodium salt of 3'-*O*-(carboxymethyl)-5'-*O*-tritylthymidine monohydrate, is ≈90%; λ_{max}^{EtOH} 267 nm ($\varepsilon_{m\underline{M}}$ 9.55).

3'-O-(Carboxymethyl)thymidine (4)

A solution of 3'-*O*-(carboxymethyl)-5'-*O*-tritylthymidine, sodium salt, monohydrate (300 mg) in 4:1 acetic acid—water (50 ml) is heated for 20 min at 100°, cooled, and extracted several times with hot heptane. The aqueous layer is separated and evaporated to dryness, and traces of acetic acid are removed by repeated coevaporation with water. The colorless oil so obtained is crystallized by the addition of dry ethanol, to give the desired product as the sodium salt, monohydrate (150 mg, 64% yield). After recrystallization from ethanol, it has $\lambda_{max}^{pH\ 2}$ 267 nm ($\varepsilon_{m\underline{M}}$ 9.51); $\lambda_{min}^{pH\ 2}$ 234 nm ($\varepsilon_{m\underline{M}}$ 2.21).

REFERENCES

1. H. S. Munson, Jr., in W. W. Zorbach and R. S. Tipson (eds.), *Synthetic Procedures in Nucleic Acid Chemistry*, Wiley—Interscience, New York, Vol. I, pp. 321 to 322.

2. M. H. Halford and A. S. Jones, *J. Chem. Soc.*, *C*, 2667 (1968).

3. M. D. Edge and A. S. Jones, *J. Chem. Soc.*, *C*, 1933 (1971).

[52] 3',5'-DI-*O*-NITROTHYMIDINE

Introduction of O-Nitro Groups into the Glycosyl Group of a Nucleoside

Robert Duschinsky

Swiss Institute for Experimental Cancer Research, 1011 Lausanne, Switzerland

INTRODUCTION

When treated with fuming nitric acid, thymine and 5-halouracils add nitric acid across the 5,6 double bond[1,2]; uracil is nitrated at C-5 under more energetic conditions.[3] By contrast, the corresponding nucleosides undergo esterification of the hydroxyl groups of the glycosyl group.[a,b] The *O*-nitro group behaves similarly to a mesyl or tosyl group as a blocking and leaving group permitting a variety of changes in the sugar moiety and having the additional advantage of facile removal by hydrogenolysis.[2,7]

PROCEDURE

3',5'-Di-O-nitrothymidine[2,4] (2)

---

[a] 2-Deoxy-D-*erythro*-pentosylpyrimidines give good to excellent yields of 3',5'-dinitrates.[2,4] The esterification of fully hydroxylated glycosylpyrimidines is less satisfactory, as mixtures of partially substituted derivatives are obtained.[4,5]

[b] 5-Nitrouridine was obtained by nitration of a suitably protected uridine.[6]

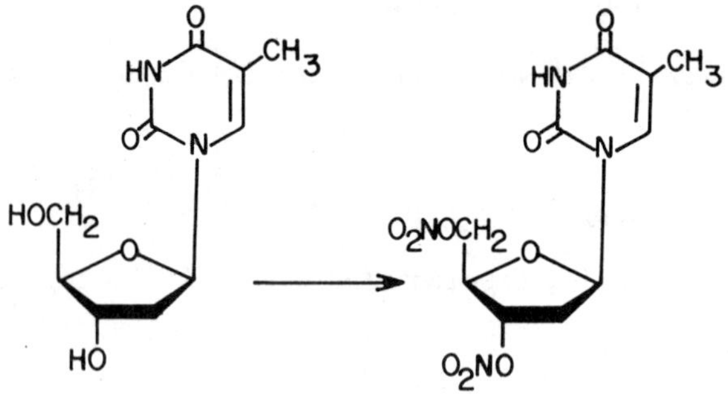

1
(242.2)

2
(332.2)

To 8 ml of fuming nitric acid (d 1.52), kept in an ice bath, thymidine (4 g, 16.5 mmol) is gradually added with continuous stirring giving a solution immediately. After reaction for 1 hr, TLC reveals[c] that formation of the ester is complete. The reaction is quenched by the addition of crushed ice (30 g) with stirring, and a white solid is formed. The mixture is kept overnight in a refrigerator and filtered; the crystalline material is washed with cold water until the washings are neutral and is dried *in vacuo* at 60°; yield 4.48 g (81%) of fairly pure compound *2*, suitable as an intermediate for some further transformations.[7]

A solution of this product in hot methanol (45 ml) is clarified by filtration. Addition of warm water (45 ml) to the filtrate, with stirring readily induces crystallization, which is

---

[c]For TLC on silica gel, developed with 1:1 ethanol—ethyl acetate samples are mixed with ice and extracted with ethyl acetate; R_F of *1*, 0.47; of *2*, 0.78.

completed by cooling in an ice bath. The crystals are removed by filtration and successively washed with a little cold methanol and ether; yield 3.9 g (70%); m.p. 131° to 132° (on slow heating a gradual change of crystal structure becomes apparent at ≃120°), $[\alpha]_D$ -11.6° (*c* 0.3, ethyl acetate); λ_{max}^{EtOH} 265 nm ($\varepsilon_{m\underline{M}}$ 9.50); ν_{max}^{KBr} disappearance of the band at 3333 cm^{-1} (OH) and appearance of a band at 1563 cm^{-1} (ONO_2); δ (Me_2SO-$\underline{d}_6$) 1.80 (s, 5-CH_3), 2.58 (m, H-2'), 4.50 (m, H-4'), 4.86 (d, *J* 5 Hz, H-5'), 5.69 (m, H-3'), 6.12 (t, *J* 7.3 Hz, H-1'), 7.53 (s, H-6), and 11.30 (s, NH).

REFERENCES

1. T. B. Johnson, *Am. Chem. J.*, 40, 19 (1908).
2. R. Duschinsky and U. Eppenberger, *Tetrahedron Lett.*, 5103 (1967).
3. T. B. Johnson and I. Matsuo, *J. Am. Chem. Soc.*, 41, 782 (1919).
4. R. Duschinsky, Swiss Patent 492,721 (1967); *C. A.*, 73, 131,272 (1970), and unpublished results.
5. F. W. Lichtenthaler and H. J. Müller, *Angew. Chem.*, 85, 765 (1973).
6. I. Wempen, I. L. Doerr, L. Kaplan, and J. J. Fox, *J. Am. Chem. Soc.*, 82, 1624 (1960).
7. J. Kàra and R. Duschinsky, *Biochim. Biophys. Acta*, 186, 223 (1969).

[53] 1-(2,3-DIDEOXY-3-FLUORO-β-D-RIBOFURANOSYL)THYMINE (3'-DEOXY-3'-FLUOROTHYMIDINE)

Cleavage of a 2,3'-Anhydro Ring by Hydrogen Fluoride

G. Kowollik and P. Langen

Department of Cell Kinetics, Institute of Molecular Biology, Academy of Sciences of GDR, GDR 1115, Berlin, Germany

INTRODUCTION

1-(2,3-Dideoxy-3-fluoro-β-D-ribofuranosyl)thymine (*4*) has proved to be a selective inhibitor of DNA synthesis *in vitro*, by incorporation in the growing DNA chain, thus preventing further chain growth.[1] Compound *4* shows cytostatic[1] and virostatic activity. In the synthesis here described for *4*, the 5'-*O*-mesyl anhydronucleoside[2] *1* is first converted into the fluoro nucleoside *2* by cleavage of the anhydro ring with anhydrous hydrogen fluoride in 1,4-dioxane; in the second step, compound *2* is treated with potassium acetate, to give the acetate *3*; finally, deacetylation of *3* yields the desired *4*. This route[3] is more expedient than the use[4] of *1* unprotected at O-5'.

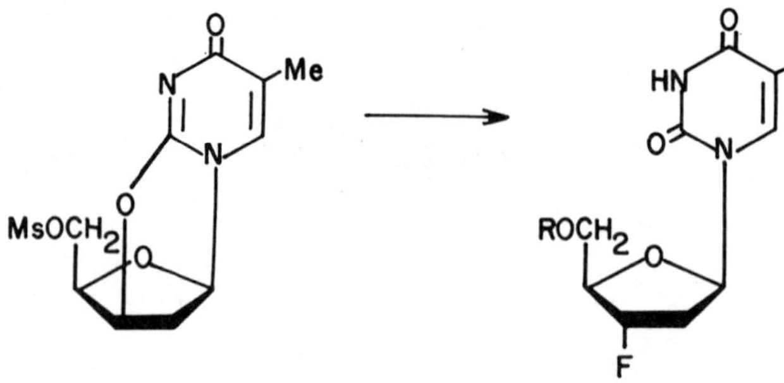

1
(302.3)

2, R = Ms
(322.3)
3, R = Ac
(286.3)
4, R = H
(244.2)

PROCEDURE

1-[2,3-Dideoxy-3-fluoro-5-O-(methylsulfonyl)-β-D-ribofuranosyl]-thymine (2)

2,3'-Anhydro-1-(2-deoxy-5-*O*-(methylsulfonyl)-β-D-xylofuranosyl)-thymine[2] (*1*) (3 g, 0.01 mol) is heated in a solution of hydrogen fluoride in 1,4-dioxane [600 ml, containing 0.1% (v/v) of hydrogen fluoride] in the presence of aluminum fluoride (1 to 3 g) as the catalyst in a stainless-steel container for 1 hr at[a] 170°. The mixture is rapidly cooled to room temperature, and water (≃150 ml) is added. Calcium carbonate powder is added to the suspension

---

[a]Before the reaction is started the bath temperature should be ≃220°; frequent shaking of the vessel is necessary. The use of a pressure bottle resistant to hydrogen fluoride, and a mechanical stirrer is more convenient.

with stirring to neutralize the acid, the suspension is filtered with suction, and the solid is well washed with warm acetone. The clear filtrate and washings are combined, and evaporated to dryness under diminished pressure. The crude product[b] is purified by chromatography on a column of "Kieselgel" (Merck) (1.0 kg) with 3:1 (v/v) chloroform—acetone as the eluant. The fractions are combined and evaporated to give crystalline *2* which is washed with chloroform; yield 2.05 g (66.5%), m.p. 162.5° to 163.5° (from isopropyl alcohol); λ_{max}^{MeOH} 263.5 nm ($\varepsilon_{m\underline{M}}$ 9.05); λ_{min}^{MeOH} 233 nm ($\varepsilon_{m\underline{M}}$ 2.47).

1-(5-O-Acetyl-2,3-dideoxy-3-fluoro-β-D-ribofuranosyl)thymine (3)

Compound *2* (2 g, 6 mmol) is heated with potassium acetate (2 g) in acetic anhydride (250 ml) for 3 hr at 135°, and the mixture is evaporated to dryness. The residue is dissolved in water (≃400 ml), and *3* is extracted from the solution with chloroform (500 ml, in portions). The extracts are combined, dried (sodium sulfate), and evaporated to dryness. The crude[b] *3* is purified by chromatography on a column of "Kieselgel" (Merck) with 19:1 (v/v) chloroform—methanol as the eluant. The combined fractions of pure *3* are evaporated to dryness; yield 1.5 g (85%), m.p. 98° to 99°

---

[b] This product can be directly converted into *3* without further purification. Furthermore, crude *3* can also be treated with methanolic ammonia to give *4*. In this case, column chromatography of *4* must be performed with 49:1 (v/v) chloroform—methanol as the eluant. The overall yield is, however, somewhat lower.

(from methanol); λ_{max}^{MeOH} 264.5 nm (ε_{mM} 9.53); λ_{min}^{MeOH} 233 nm (ε_{mM} 2.28).

1-(2,3-Dideoxy-3-fluoro-β-D-ribofuranosyl)thymine (4)

A solution of *3* (1.5 g, 5 mmol) in methanol presaturated with ammonia at 0° (100 ml) is kept overnight at 0°, and evaporated, and the residue is crystallized from ethyl acetate; yield 1.15 g (90%), m.p. 174.5° to 176° $[\alpha]_D^{20}$ −6.3° (*c* 0.96, methanol); $\lambda_{max}^{H_2O}$ 266 nm, $\lambda_{min}^{H_2O}$ 245.5 nm; PMR data [$(CD_3)_2SO$, tetramethylsilane as the standard]: δ 5.22 (m), 1 H, H-3'; $J_{2',F}$ 35 Hz, $J_{2'',F}$ 27 Hz, $J_{3',F}$ 53 Hz, and $J_{4',F}$ 35 Hz.

REFERENCES

1. P. Langen, G. Kowollik, G. Etzold, H. Venner, and H. Reinert, *Acta Biol. Med. Ger.*, 29, 483 (1972).

2. J. P. Horwitz, J. Chua, J. A. Urbanski, and M. Noel, *J. Org. Chem.*, 28, 942 (1963); J. J. Fox and N. C. Miller, ibid., 28, 936 (1963).

3. G. Kowollik, G. Etzold, M. von Janta-Lipinski, K. Gaertner, and P. Langen, *J. Prakt. Chem.*, 315, 895 (1973).

4. G. Etzold, R. Hintsche, G. Kowollik, and P. Langen, *Tetrahedron*, 27, 2463 (1971).

[54] 1-[2,3-DIHYDRO-5-(HYDROXYMETHYL)FURAN-2-YL]THYMINE (3'-DEOXY-3',4'-DIDEHYDROTHYMIDINE)

Synthesis of a 3',4'-Unsaturated 2',3'-Dideoxynucleoside

Jiří Žemlička and Jerome P. Horwitz

Michigan Cancer Foundation and Department of Oncology, Wayne State University, School of Medicine, Detroit, Michigan 48201

INTRODUCTION

The conversion of thymidine (*1*) into 3'-deoxy-3',4'-didehydrothymidine (*6*) is considered typical of the preparation of 3',4'-unsaturated 2',3'-dideoxynucleosides,[1] which are of interest as potential antimetabolites. Thus, the oxidation of *2* by oxygen in buffered aqueous solution[2] in the presence of platinum[3] provides thymidine-5'-carboxylic acid (*2*). The latter is converted, by treatment with ethanol and ethyl orthoformate in the presence of hydrogen chloride, into the corresponding ethyl ester (*3*), which is then transformed into the 3'-*O*-(methylsulfonyl) derivatives (*4*) by reaction with methanesulfonyl chloride in pyridine. Elimination of the methylsulfonyloxy group from *4* is effected with triethylamine in *N*,*N*-dimethylformamide at 100°, to give ethyl 3'-deoxy-3',4'-didehydrothymidine 5'-carboxylate (*5*). Reduction of *5* to the title compound (*6*) is accomplished with

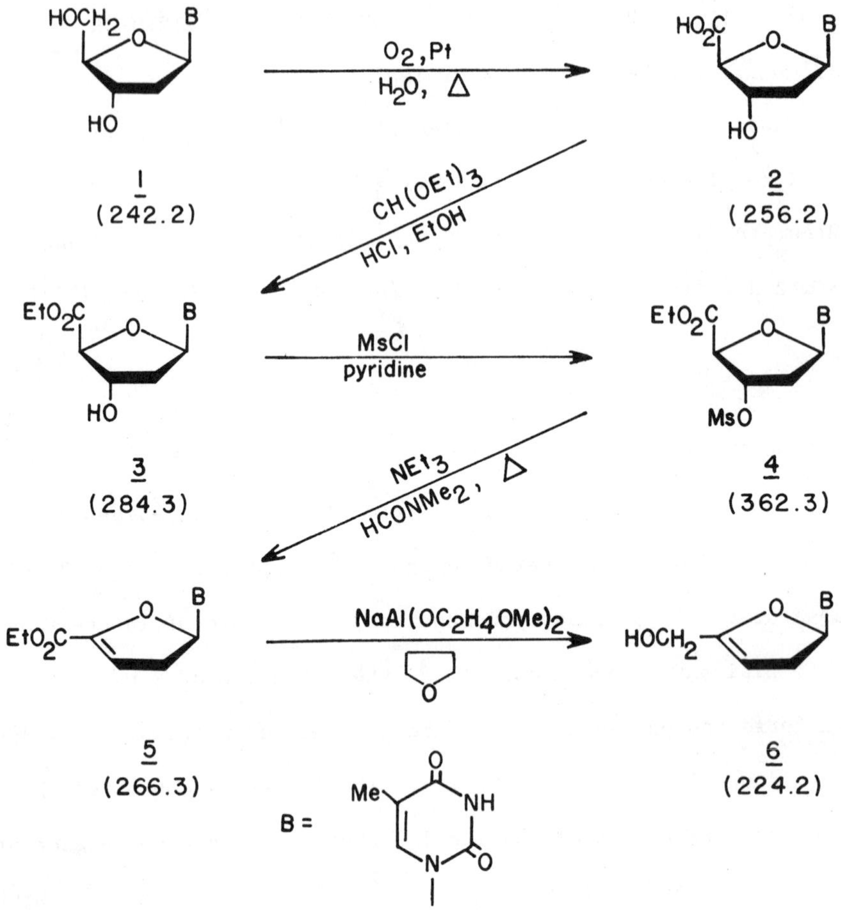

sodium bis(methoxyethoxy)aluminum hydride.

PROCEDURE

Thymidine-5'—Carboxylic Acid (2)

To a solution of thymidine (*1*) (14.31 g, 62 mmol), sodium carbonate (0.57 g, 5 mmol), and sodium hydrogen carbonate (4.95 g,

59 mmol) in water (1.900 l)[a] is added freshly prepared[b] platinum catalyst, and oxygen is bubbled through the rapidly stirred mixture for 7 hr at 90° to 100° (bath temperature). The suspension is cooled, the catalyst is removed by filtration and washed with water, and the clear filtrate[c] is stirred with a strong cation-exchange resin,[d] (H^+) (≃100 g) for 30 min at room temperature. The resin is removed by filtration and washed with water, and the filtrate is concentrated[e] to ≃50 ml. The crystalline product that has separated after standing overnight at 0° is collected, washed with acetone, and dried at 1 mtorr at room temperature to give 10.2 g (65%) of *2*, m.p. 258° to 260° (dec.) [lit.[2] m.p. 263° to 265° (dec.)]; homogeneous in TLC in 1:1 chloroform—methanol.[f]

---

[a]A 3-l, three-necked Morton flask equipped with a water-cooled, high-speed stirrer, two reflux condensers in series, and a fritted-glass, gas-inlet tube is used.

[b]Platinum oxide (83%, 7.27 g), Engelhard Industries, Newark, N. J.. is hydrogenated in water (100 ml) in a Parr apparatus for 2 hr at room temperature. The hydrogen is then removed by repeated evacuation.

[c]At this point, any cloudiness is removed by filtration through a thin bed of Celite.

[d]Dowex 50-W X-4 (200 to 400 mesh) is one of several which may be used

[e]All evaporations unless stated otherwise, are conducted *in vacuo* (aspirator) in a rotary evaporator. The bath temperature is kept below 40°.

[f]Aluminum foil (2 × 6 cm) precoated with a layer (0.25-mm thick) of silica gel F-254 (Merck, Darmstadt, Germany) is used. Ultraviolet-absorbing compounds are detected by use of a Mineralight lamp.

Ethyl Thymidine-5'-carboxylate (3)

Through a stirred suspension of compound *2* (5 g, 19.6 mmol) in ethanol[g] (100 ml) containing ethyl orthoformate (10 ml) and cooled in an ice-bath, hydrogen chloride is bubbled directly from a tank.[h] Most of the starting material dissolves within several minutes. After saturation of the mixture with hydrogen chloride, a white crystalline product precipitates, and the resultant thick suspension is kept overnight at room temperature. Dry ether[i] (200 ml) is then added, and the product is removed by filtration, washed with ether, and dried at 1 m Torr to give 4.74 g (86%) of *3*, m.p. 239° to 241° (lit.[1] m.p. 244° to 245°), homogeneous in TLC in 1:1 chloroform—methanol.[b]

Ethyl 3'-(O-Methylsulfonyl)thymidine-5'-carboxylate (4)

A stirred solution of ethyl thymidine-5'-carboxylate[j] (3)(4.74 g. 17 mmol) in pyridine[g] (100 ml) is cooled in an ice bath, and methane sulfonyl chloride[i] (2.15 ml, 34 mmol) is added.[j] After 1 hr, the solution is allowed to warm to room temperature. The progress of the reaction is monitored by TLC in 9:1 chloroform—

---

[g]Commercial reagent-grade product is dried for at least one week with Linde Molecular sieves 3A.

[h]A 500-ml three-necked flask equipped with a drying tube (calcium chloride), a magnetic stirrer, and a gas-inlet tube is used.

[i]Commercial reagent-grade product is used.

[j]A 250-ml, three-necked flask equipped with a drying tube (calcium chloride) and a magnetic stirrer is used. Methyl sulfonyl chloride is added in one portion from a pipet.

methanol.[f] After 48 hr a second portion of methanesulfonyl chloride (2.15 ml, 34 mmol) is added, and the mixture is kept for 72 hr at room temperature. The mixture, containing precipitated pyridine hydrochloride, is filtered through a thin layer of Celite. Ethanol (20 ml) is added to the filtrate, and the solution is evaporated[e] to a syrup. A solution of the residue in chloroform (250 ml) is washed with saturated aqueous, sodium hydrogen carbonate solution (50 ml), the aqueous layer is extracted with chloroform (50 ml), and the organic phases are combined, washed with water (3 × 50 ml), dried (anhydrous magnesium sulfate), and evaporated[e] to a syrup that is coevaporated with ethanol (2 × 50 ml). After addition of dry ether[i] (200 ml), a crystalline product is deposited.[k] The mixture is kept overnight at 0°, and petroleum ether (b.p. 30° to 60°; 200 ml) is added to effect precipitation. The solid is collected, washed with petroleum ether, and dried at 1 mtorr to give 5.79 g (96%) of *4*, m.p. 124° to 128° (lit.[1] m.p. 130° to 133°), homogeneous in TLC in 95:5 chloroform—methanol.

Ethyl 3'-Deoxy-3',4'-didehydrothymidine-5'-carboxylate (5)

The methyl sulfonate *4* (5.25 g, 14.5 mmol) is heated with triethylamine (6 ml) in *N*,*N*-dimethylformamide (100 ml) for 2 hr at 100° (bath temperature) in a well-stoppered flask. The solution is

---

[k] The crystallization is accelerated by scratching the inner sides of the flask.

cooled and evaporated at 0.1 torr (bath temperature below 40°), and the residue is dissolved in chloroform (100 ml). The solution is washed with saturated, aqueous sodium hydrogen carbonate solution (50 ml), and the aqueous portion is extracted with chloroform (100 ml). The extracts are combined, washed with water (50 ml), dried (anhydrous magnesium sulfate), and evaporated[e] to a solid (3.75 g); this is dissolved in chloroform (20 ml) and the solution is applied to a column[l] (51 × 2.5 cm) of silica gel (150 g). Elution is initiated with chloroform (1 l) followed by chloroform—2% methanol (2 l).[m] Compound 5 is eluted with chloroform—3% methanol (2 l) and the eluate evaporated to a white solid (2.1 g, 55%), m.p. 222° to 224° (lit.[1] m.p. 229° to 231°), homogeneous in TLC in 9:1 chloroform—methanol.[f] The NMR spectrum[n] of 5 in $CDCl_3$ shows a characteristic pseudoquartet (multiplet of four) at δ 6.82 (H-1') that is partially overlapped by a doublet (H-6) at 6.94, and a triplet at 6.03 (H-3').

---

[l]Silica gel ASTM (70 to 325 mesh), (Merck, Darmstadt, Germany) is used. The solumn is prepared by suspending the silica gel in chloroform, and pouring the suspension into a glass tube fitted with a sintered-glass septum and a stopcock.

[m]The progress of the chromatography is monitored by continuous recording of the UV absorption of the eluate at 254 nm with a UV detector coupled with a recorder (Instrumentation Specialties Co., Lincoln, Nebr., or LKB Produkter-AB, Bromma, Sweden) and also visually (nontransparent, white zones of compounds are seen in the otherwise transparent column).

[n]Tetramethylsilane is used as the internal reference.

3'-Deoxy-3',4'-didehydrothymidine (6)

A stirred solution of the unsaturated ester 5 (0.27 g, 1 mmol) in tetrahydrofuran[o] (30 ml) cooled in an ice-bath,[p] is treated with sodium bis(methoxyethoxy)aluminum hydride (2 mmol).[q] The progress of the reaction is monitored by TLC in 9:1 chloroform—methanol.[b] Additional 1-mmol portions of the reagent are added after 45 min and 90 min. After 2 hr TLC shows ≃90% reduction of 5. The reaction is stopped after 3 hr by addition of ethanol (20 ml), followed by dry Dowex 50 resin (NH_4^+)[r] (7 g, 36 mequiv.). The resin is removed by filtration and washed with ethanol, and the clear, colorless filtrate is evaporated[e] (bath temperature below 35°). A chloroform solution of the resultant syrup is applied to two 2-mm thick loose layers (10 × 25 cm) of silica gel containing 1% of fluorescent indicator.[s] The layers are

---

[o]Commercial, reagent-grade product is distilled from lithium aluminum hydride and stored in the dark over sodium wire.

[p]A 100-ml, three-necked flask equipped with a drying tube (calcium chloride), a magnetic stirrer, and a rubber cap is used. The solution of the reagent is added from a syringe whose needle is plunged through the rubber cap.

[q]Commercial product (Redal, Aldrich Chemical Co., Milwaukee, Wis.); a 70% solution in benzene is diluted with benzene to give 1 mmol of the reagent per ml.

[r]Prepared from Dowex 50 (H^+)[d] by successively washing it with 5% ammonia, water, and methanol, and drying it *in vacuo* (aspirator) at room temperature.

[s]For the silica gel used, see footnote *l*. The fluoroescent indicator (Leuchtpigment ZS Super, Riedel-DeHaën, Hannover, Germany) is first thoroughly mixed in a mortar with a small amount of the

developed with 9:1 chloroform—methanol containing 0.1% of triethylamine. The major, slower-moving[t] zone is eluted with 4:1 chloroform—methanol containing 0.1% of triethylamine, and the elute is evaporated[e] (bath temperature below 35°). The resultant syrup crystallizes on being kept overnight at -20°, to give 0.12 g (54%),[u] of *6*, m.p. 105° to 110° (transition point; remelting at 160° to 165°, dec. above 230°); $[\alpha]_D^{23}$ -134.6° (*c* 0.05, 1,4-dioxane); λ_{max}^{EtOH} 267 nm (ε_{mM} 8.70), λ_{min} 235 (2.40); the NMR spectrum[n] (acetone-d_6) shows a characteristic pseudoquartet (cf. compound *5*) at δ 6.65 (H-1') and a triplet at 5.02 (H-3').

REFERENCES

1. J. Žemlička, J. V. Freisler, R. Gasser, and J. P. Horwitz, *J. Org. Chem.*, 38, 990 (1973).

2. G. P. Moss, C. B. Reese, K. Schofield, R. Shapiro, and A. R. Todd, *J. Chem. Soc.*, 1149 (1963).

3. K. Heyns and H. Paulsen, *Adv. Carbohydr. Chem.*, 17, 169 (1962).

---

silica gel. The mixture is now added to the main batch of silica gel, which is then shaken for 1 hr. The loose layers of silica gel are prepared by spreading the material on glass plates, smoothing them with a glass rod leaving a piece of rubber (or plastic) tubing of the desired thickness at each end, and carefully placing the plates in glass tanks (the angle should not exceed ≃20°). The developing time does not exceed 4 to 6 hr.

[t]The minor, faster-moving component is the starting material *5*.

[u]According to the elemental analysis, compound *6* contains 0.25 mol of water per mol.

[55] 1-(2-DEOXY-β-D-*ERYTHRO*-PENTOPYRANOSYL)URACIL AND ITS α-D ANOMER

Acid-Catalyzed Isomerization of the Glycosyl Group in 5,5-Dibromo-2'-deoxy-5,6-dihydro-6-hydroxyuridine

Jean Cadet

Laboratoire de Radiobiologie, Departement de Recherche Fondamentale, Centre d'Etudes Nucleaires de Grenoble, France

INTRODUCTION

The preparation of 1-(2-deoxy-α-D-*erythro*-pentopyranosyl)uracil (*4*) and its β anomer *5* was initially accomplished by the condensation of 2,4-bis(trimethylsiloxy)uracil with 2-deoxy-3,4-di-*O*-*p*-toluoyl-D-*erythro*-pentopyranosyl chloride.[1] An alternative procedure involves the acid-catalyzed isomerization of the glycosyl group of sodium 2'-deoxy-5,6-dihydro-6-*O*-sulfouridine[2] or of the readily available 5,5-dibromo-2'-deoxy-5,6-dihydro-6-hydroxyuridine[3] (*3*). The warming of both diastereoisomers of the latter compound, generated *in situ*, with M hydrobromic acid gives a mixture of the anomers of the pyranoid and furanoid glycosyl derivatives of uracil, 5-bromouracil and 5,5-dibromo-5,6-dihydro-6-hydroxyuracil. The hydrogenolysis of the halogenated compounds gives *4* and *5*, with uracil and the corresponding anomeric

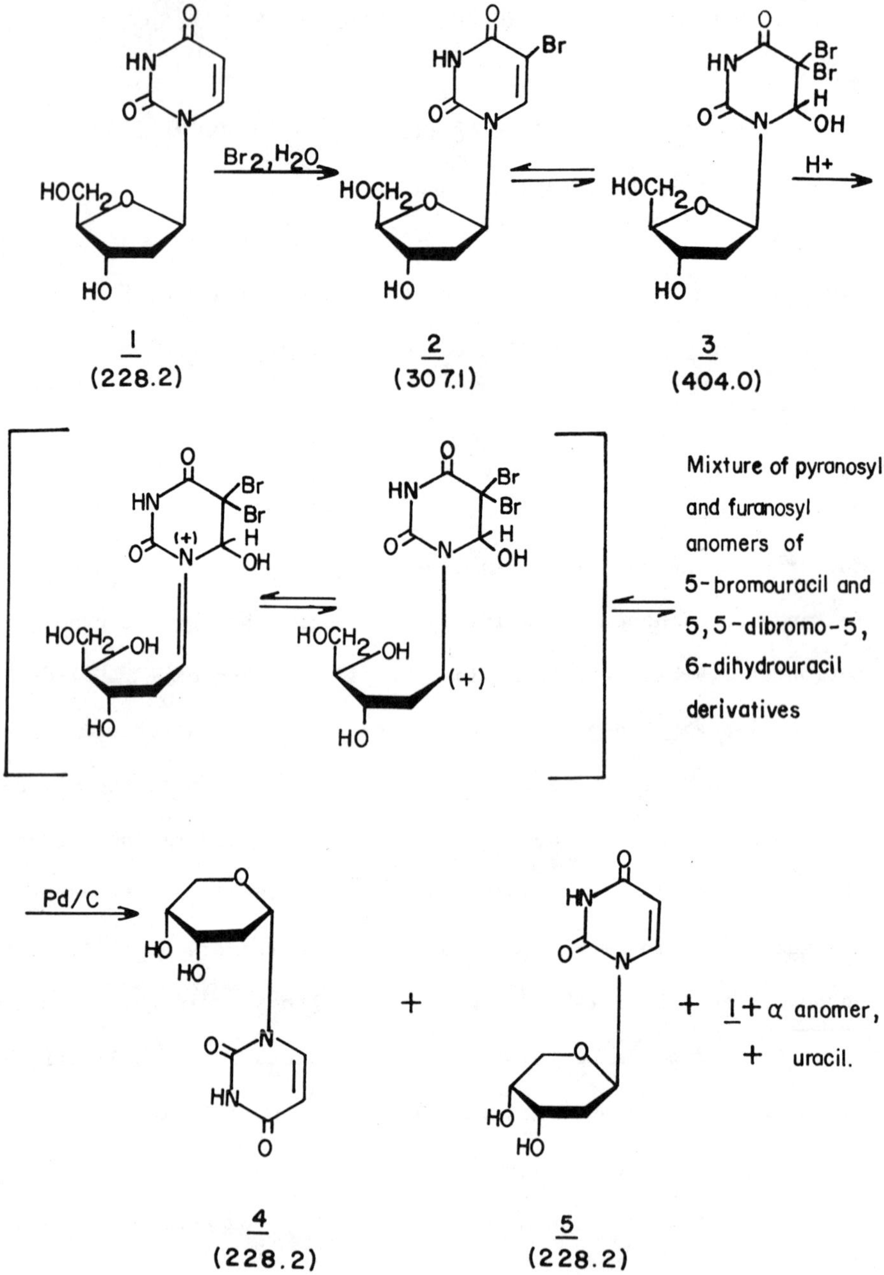

Br_2, H_2O
H+
1
(228.2)
2
(307.1)
3
(404.0)
Mixture of pyranosyl and furanosyl anomers of 5-bromouracil and 5,5-dibromo-5, 6-dihydrouracil derivatives
Pd/C
+ 1 + α anomer,
+ uracil.
4
(228.2)
5
(228.2)

furanoid isomers.

PROCEDURE

*1-(2-Deoxy-α-*D-erythro-*pentopyranosyl)uracil (4)*

Bromine (0.26 ml, 8 mmol) is added dropwise at room temperature to a well-stirred solution of 0.91 g (4 mmol) of 2'-deoxyuridine (*1*) in water (4 ml) in a 24-ml, round-bottomed flask.[a] The flask is stoppered and the mixture is shaken for 30 min in an oven at 90°. The solution is cooled and diluted with methanol (200 ml) containing 10% palladium-on-carbon (240 mg). The mixture is hydrogenated for 2 hr at room temperature in a Parr hydrogenation apparatus at an initial pressure of 15 atmospheres[4] and is then filtered through a Celite pad. The filtrate is rendered neutral by the addition, in small portions, of a saturated, aqueous solution of sodium hydrogen carbonate (10 ml). The solution is concentrated[b] to ≃10 ml under diminished pressure and stirred with AG 11 A 8 (H^+, OH^-) mixed-bed ion-exchange resin (3 g). The resin is removed by filtration and washed thoroughly with methanol (50 ml). The filtrates are combined and evaporated to dryness, giving a colorless syrup. Preparative thin-layer

---

[a]The aqueous solution is molar with respect to hydrobromic acid.

[b]Evaporations are conducted with a rotary evaporator at a bath temperature < 25°.

[c]Approximately 80 mg of mixed products may conveniently be separated on a plate (40 × 20 × 0.1 cm) of silica gel.

chromatography is conducted on Macherey Nagel MN-N-HR/UV 254 silica gel with 75:16:9 ethyl acetate—2-propanol—water[5] as the developer.[c] The slowest-moving UV-absorbing zone,[d] R_F 0.33, is carefully excised and extracted with methanol (3 × 20 ml). Evaporation of the solution yields 163 mg (18%) of homogeneous, analitically *pure*, noncrystalline *4*, $[\alpha]_D^{25}$ -43° (*c* 1.3, water) [lit.[1] $[\alpha]_D^{20}$ -43.9° (*c* 1.84, water)]; λ_{max} 261 nm; m.s.: *m/e* (relative intensity) 228 (M^+ 3.6), 117 (glycosyl group, 93.1), 113 (base + 2HH, 38.9), 112 (base + H, 11), and 99 (117-H_2O, 100).[e]

*1-(2-Deoxy-β-D-*erythro-*pentopyranosyl)uracil (5)*

Evaporation of the methanolic extract of the middle UV-absorbing zone R_F 0.42 gives a colorless syrup (415 mg), which is deposited on 6 plates (40 × 20 × 0.1 cm) of silica gel. The β-D-pyranosyl compound *5* is fractionated from the mixture of *1* with its α anomer by using, as the developing solvent, the lower phase of 4:2:1 chloroform—methanol—water with 5% of methanol added. The faster-moving, UV-absorbing zone R_F 0.39 is extracted with methanol (3 × 10 ml). Evaporation of the solution to dryness gives a colorless syrup which is recrystallized from *absolute* ethanol; yield 120 mg (12%), m.p. 200° to 202°; $[\alpha]_D^{25}$ +27.6 (*c* 2.1, water);

---

[d] Two other zones, which correspond respectively to uracil and the mixture of *1* and its α-D-anomer, are detected with a Wood lamp.

[e] High-resolution, mass spectra are obtained on a GEC AEI MS 9 mass spectrometer, with sample introduction by direct probe.

lit.[1] m.p. 200° to 202°, $[\alpha]_D^{20}$ +28.60 (*c* 2.52, water): $\lambda_{max}^{H_2O}$ 261.5 nm; mass spectrum: (II) *m/e* (relative intensity) 228 (M^+, 10.3), 117 (glycosyl group, 100), 113 (base + 2 H, 27.1), 112 (base + H, 64.5), 99 (117 $-H_2O$, 90.3).

REFERENCES

1. E. Wittenburg, *Chem. Ber.*, 93, 2132 (1968).
2. J. Cadet, *Tetrahedron Lett.*, 867 (1974).
3. H. Quelo, J. Cadet, and R. Teoule, *C. R. Acad. Sci., Ser. C*, 275, 1137 (1972).
4. An alternative procedure was reported by M. M. Ponpipom and S. Hanessian, *Can. J. Chem.*, 50, 246 (1972).
5. J. Cadet and R. Teoule, *J. Chromatogr.*, 76, 407 (1973).

[56] 1-(2-DEOXY-4-THIO-β-D-*ERYTHRO*-PENTOFURANOSYL)-5-FLUOROURACIL AND ITS α ANOMER

The Trimethylsilyl Synthesis of 2'-Deoxy-4'-thionucleosides in the Presence of Mercuric Bromide and Mercuric Oxide

Yun-Lung Fu and Miroslav Bobek

Department of Experimental Therapeutics and Grace Cancer Drug Center, Roswell Park Memorial Institute, New York State Department of Health, Buffalo, New York 14263

INTRODUCTION

The synthesis of the title nucleosides is an example of condensation[1] of a very unstable glycosyl chloride with the trimethylsilyl derivative of 5-fluorouracil catalyzed by mercuric oxide—mercuric bromide. This reaction, which proceeds under mild conditions, provides a favorable yield of the usually desired β-D-anomer (*6b*), whereas use of stannic chloride as the catalyst gives primarily the α anomer *6a*, and only a trace of *6b*. Compounds *6a* and *6b*, the first nucleosides known that contain a 2-deoxy-4-thio-D-*erythro*-pentofuranosyl group, show a marked inhibitory activity to leukemia L-1210 cells *in vitro*.[2]

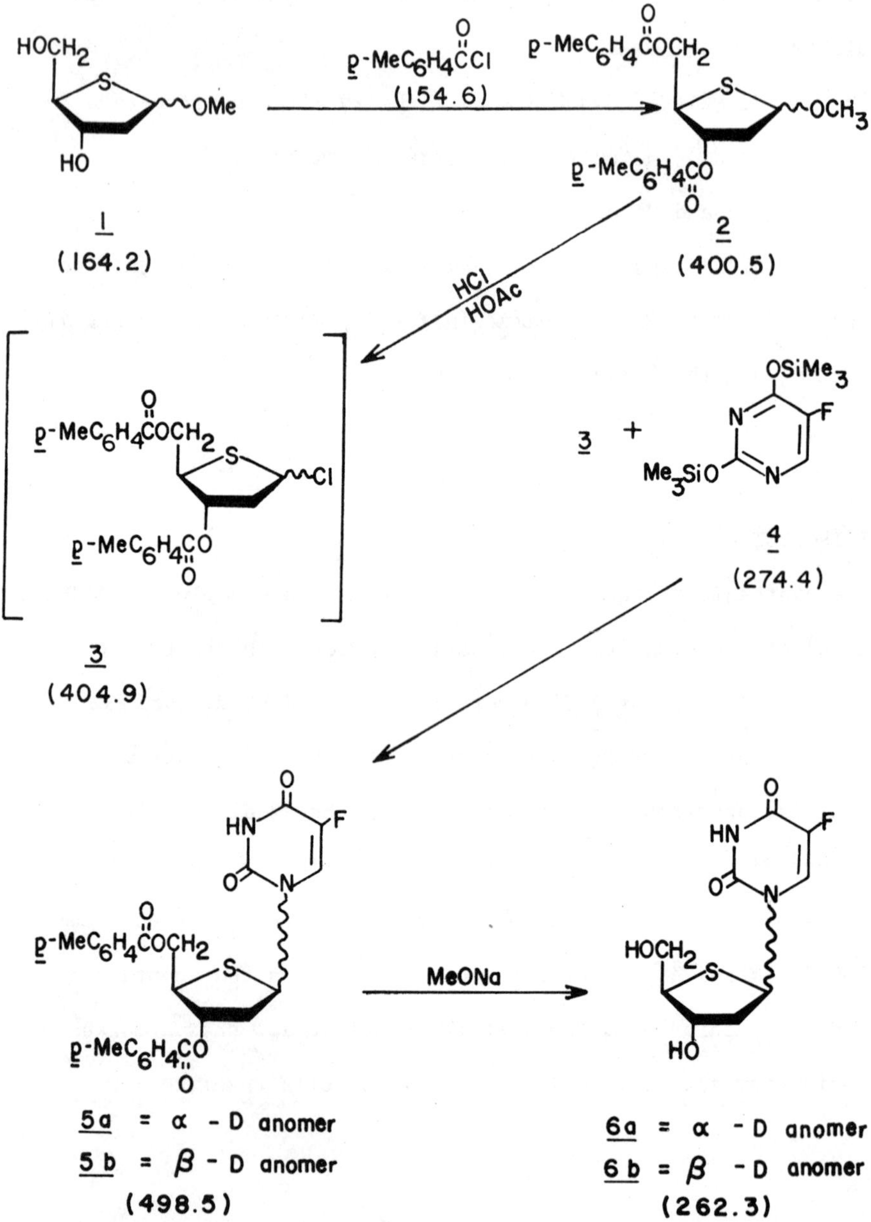

HOCH2
S
OMe
HO
1
(164.2)
p-MeC6H4CCl
O
(154.6)
p-MeC6H4COCH2
OCH3
p-MeC6H4CO
2
(400.5)
HCl
HOAc
p-MeC6H4COCH2
Cl
3
(404.9)
3 +
OSiMe3
N
F
Me3SiO
4
(274.4)
HN
O
F
MeONa
5a = α - D anomer
5b = β - D anomer
(498.5)
6a = α - D anomer
6b = β - D anomer
(262.3)

PROCEDURE

Methyl 2-Deoxy-4-thio-3,5-di-O-p-toluoyl-α,β-D-erythro-pentoside (2)

A solution of methyl 2-deoxy-4-thio-α,β-D-*erythro*-pentofuranoside (*1*, 5.0 g, 30.5 mmol) in dry pyridine (100 ml) is cooled in an ice bath and treated with *p*-toluoyl chloride (12 g, 77.6 mmol). The mixture is stirred for 1 hr at 0° and for 16 hr at room temperature and is then poured with stirring into ice water (500 ml). The mixture is stirred for 16 hr and extracted with chloroform (4 × 100 ml). The extracts are combined, washed successively with cold dilute sulfuric acid, water, saturated sodium hydrogen carbonate solution, and water (until neutral), and evaporated[a] to yield a syrup that is crystallized twice from 95° ethanol to give a mixture of the anomeric glycosides (*2*); yield 12 g (98%), m.p. 82° to 85°, $[\alpha]_D^{25}$ -62.5° (*c* 1.00, chloroform); NMR[b] data (acetone-d_6—Me_4Si): δ 2.37 and 2.40 (2 s, 6 H, 2 CH_3), 2.60 (m, 2 H, H-2), 3.32 (s, 3 H, OCH_3), 3.98 (m, 1 H, H-4), 4.46 (2 br s, 2 H, H-5), 5.25 (m, 1 H, H-3), 5.80 (m, 1 H, H-1), and 7.18 to 7.94 (m, 8 H, aromatic H).

2-Deoxy-4-thio-3,5-di-O-p-toluoyl-D-erythro-pentosyl chloride (3)

---

[a]Evaporations are performed under diminished pressure at a bath temperature of 40° in a rotary evaporator.

[b]Nuclear magnetic resonance spectra are recorded with a Varian XL-100 spectrometer.

Compound *2* (3.5 g, 8.7 mmol) is dissolved in cold (10°) glacial acetic acid (50 ml) presaturated with hydrogen chloride at 10°. After 40 min[c] the solution is evaporated to dryness in a rotary evaporator under diminished pressure (oil pump) at 0° to 5°, and traces of acetic acid are removed by coevaporation (0° to 5°) of the residue with dry toluene (20 ml). Chloride *3* is very unstable, even at 10° and is used immediately in the nucleosidation.

*1-(2-Deoxy-4-thio-3,5-di-*O-p-*toluoyl-α-*D-erythro-*pentofuranosyl)-5-fluorouracil (5a) and 1-(2-Deoxy-4-thio-3,5-di-*O-p-*toluoyl-β-*D-erythro-*pentofuranosyl)-5-fluorouracil (5b)*

A mixture of compound *3* [obtained from 3.5 g (8.7 mmol) of compound *2*] and the trimethylsilyl derivative[3] *4* [obtained from 3 g (23.1 mmol) of 5-fluorouracil] is stirred in dry dichloromethane (200 ml) in the presence of yellow mercuric oxide (2 g, 9.2 mmol) and mercuric bromide (3.2 g, 8.8 mmol) for 24 hr at 0°. Water (50 ml) is added, and the precipitate is removed by filtration and extracted thoroughly with chloroform (to remove the nucleosides). The filtrate and extracts are combined, washed once with

---

[c]The course of the reaction is monitored by TLC on silica gel N-HR/UV_{254} (Brinkman) with 9:1 (v/v) benzene—acetone. The spots are detected by UV absorption and by spraying the plates with a 1:9 (v/v) sulfuric acid—ethanol, and heating. For compound *2*, R_F 0.86; for compound *3*, 0.75. When the third spot, R_F 0.57, appears, the reaction is stopped by evaporating all of the acetic acid and hydrogen chloride.

water (50 ml), and evaporated to give a white powder, which is transferred to the top of a dry column of silica gel and eluted with 15:1 (v/v) benzene—ethyl acetate to obtain 0.4 g of unreacted *2*. The column is then eluted with chloroform to obtain 2.1 g of the protected nucleosides. The mixture is dissolved in 1,4-dioxane (50 ml), and to this solution is added boiling acetone (200 ml). Compound *5b* crystallizes as needles, and it is collected by filtration (filtrate *A*). Recrystallization from absolute ethanol gives 0.53 g of *5b*, m.p. 259.5° to 260°, $[\alpha]_{D}^{25}$ -66.3° (*c* 0.08, chloroform); NMR data (Me_2SO-d_6—Me_4Si): δ 2.41 (s, 6 H, 2 CH_3), 2.7 (m, 2 H, H-2'), 3.96 (m, 1 H, H-4'), 4.60 (m, 2 H, H-5'), 5.77 (m, 1 H, H-3'), 6.40 (m, 1 H, H-1'), 7.3 to 7.9 (m, 8 H, aromatic H), 8.28 (d, 1 H, $J_{H,F}$ 7.2 Hz, H-6), and 11.93 (s, 1 H, H-3).

Filtrate *A* is evaporated to dryness, the residue is dissolved in boiling chloroform (30 ml), and boiling, absolute ethanol (100 ml) is added. Compound *5a*, which crystallizes in needles, is collected by filtration (filtrate *B*). Recrystallization from absolute ethanol gives 0.6 g of compound *5a*, m.p. 250° to 251°, $[\alpha]_{D}^{25}$ -56.0° (*c* 0.25, chloroform); NMR data (Me_2SO-d_6—Me_4Si): δ 2.40 (s, 6 H, 2 CH_3), 2.53 to 3.0 (m, 2 H, H-2'), 4.41 (br s, 3 H, H-4',5'), 5.65 (m, 1 H, H-3'), 6.28 (m, 1 H, H-1'), 7.20 to 7.98 (m, 8 H, aromatic H), 8.42 (d, 1 H, $J_{H,F}$ 7.5 Hz, H-6), and

11.85 (s, 1 H, H-3).

Filtrate *B* is evaporated and, by repeating the fractional recrystallization process used to separate *5a* from *5b*, additional crops (0.4 g of *5a* and 0.3 g of *5b*) are obtained. The total yields of *5a* and *5b* are 25.9 and 21.5%, respectively (based on the *2* that has reacted).

*1-(2-Deoxy-4-thio-α-D-*erythro*-pentofuranosyl)-5-fluorouracil (6a) and 1-(2-Deoxy-4-thio-β-D-*erythro*-pentofuranosyl)-5-fluorouracil (6b)*

A solution of *5a* (1.33 g) in anhydrous 1,4-dioxane (50 ml) is stirred at room temperature with a freshly prepared solution of sodium methoxide (from 0.02 g of sodium in 100 ml of absolute methanol). After 5 hr the solution is rendered neutral with Dowex 50W-X8 (H^+) ion-exchange resin, the suspension is filtered, and the resin is washed with methanol. The filtrates are combined and evaporated to dryness, and methyl *p*-toluate is removed at 50°/0.01 torr. Crystallization of the residue from absolute ethanol gives 0.63 g (90%) of *6a*, m.p. 215° to 215.5°, $[\alpha]_D^{25}$ +53.7° (*c* 0.27, water); $\lambda_{max}^{H_2O}$ in nm 272 (ε_{mM} 9.83) and 206 (10.78); NMR data (D_2O—DSS): δ 2.16 to 2.82 (m, 2 H, H-2'), 3.52 to 3.88 (m, 3 H, H-4',5'), 4.52 (m, 1 H, H-3'), 6.24 (q of d, 1 H, $J_{1',2',2''}$ 7.4 and 3.6 Hz, $J_{1',F}$ 1.8 Hz, H-1'), 8.60 (d, 1 H, $J_{6,F}$ 7.4 Hz, H-6).

By the use of the same method as described for the preparation of the α-D anomer (*6a*), 300 mg of *5b* yielded 139 mg (88%) of *6b*, m.p. 202° to 204°, $[\alpha]_D^{25}$ -27.5° (*c* 0.32, water); $\lambda_{max}^{H_2O}$ in nm 273 (ε_{mM} 8.40) and 204 (10.10); NMR data (D_2O—DSS): δ 2.24 (m, 2 H, H-2'), 3.57 (m, 1 H, H-4'), 3.86 (2 br s, 2 H, H-5'), 4.58 (m, 1 H, H-3'), 6.34 (t, 1 H, $J_{1',2'}$ 6.9 Hz, H-1'), and 8.37 (d, 1 H, $J_{6,F}$ 6.8 Hz, H-6).

REFERENCES

1. E. Wittenburg, *Chem. Ber.*, 101, 1095 (1968).

2. Y.-L. Fu, M. Bobek, and A. Bloch, 172nd ACS National Meeting, San Francisco, August 30, 1976; Med. Chem. Abstract #69.

3. Hoffman-La Roche and Co., A. G., British Patent 1,080,491 (Cl C07d,f) (August 23, 1967).

[57] (+)-(R)-5-FLUORO-1-(TETRAHYDROFURAN-2-YL)URACIL[a]

Decarboxylative Elimination of 2'-Deoxynucleoside-5'-carboxylic Acids

Jiří Žemlička, Robert Gasser, and Jerome P. Horwitz

Michigan Cancer Foundation and Department of Oncology, Wayne State University, School of Medicine, Detroit, Michigan 48201

INTRODUCTION

Tetrahydrofuryl derivatives of nucleic acid bases are lipophilic analogs of nucleosides, and some of them exhibit antitumor activity.[1,2] Their racemic (R,S) mixtures are prepared by an acid-catalyzed addition of purine bases to 2,3-dihydrofuran.[1,3] The corresponding pyrimidine derivatives are obtained by reaction of 2-chlorotetrahydrofuran with mercury salts[4] or with trimethylsilyl derivatives[5] of bases, or by a modified Hilbert-Johnson procedure from 2,4-dimethoxypyrimidines, 2,3-dihydrofuran, and hydrogen chloride.[6] The racemic form of the title compound is an anticancer drug used in the U.S.S.R. under the name Ftorafur (Fluorofur).[2] The synthesis of Ftorafur followed the procedures just outlined.[4,5] Ftorafur was also obtained by direct fluorination of 1-(tetrahydrofuran-2-yl)uracil.[7] Both the R and S

---

[a]Alternative name: (+)-(R)-2-(5-fluorouracil-1-yl)tetrahydrofuran.

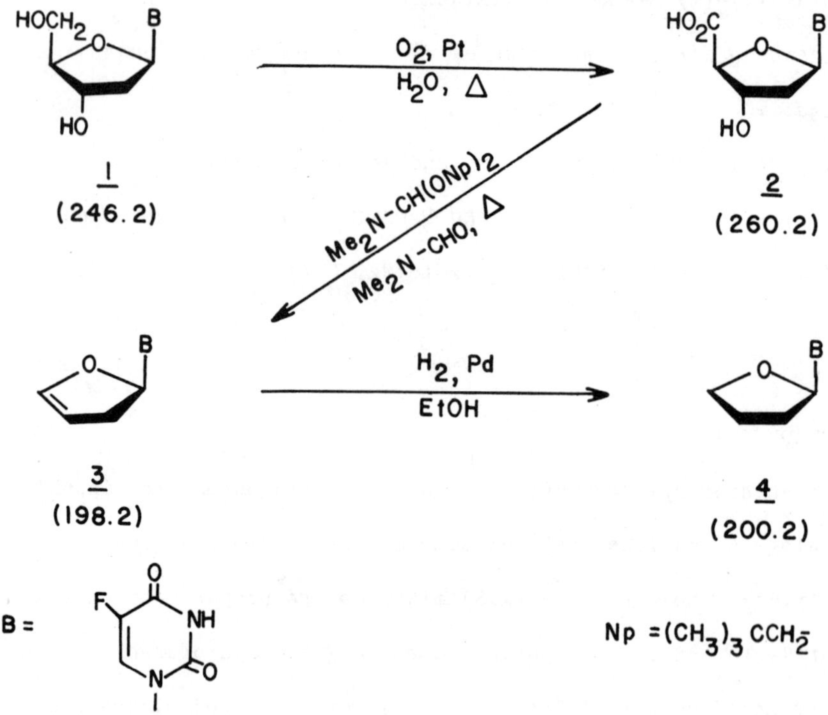

isomers of 5-fluoro-1-(tetrahydrofuran-2-yl)uracil were prepared by decarboxylative elimination of the anomers of 2'-deoxy-5-fluorouridine-5'-carboxylic acid.[8,9] The procedure is presented for the *R* isomer *4*. 2'-Deoxy-5-fluorouridine (*1*) is oxidized catalytically to the corresponding 5'-carboxylic acid *2*. On heating with *N*,*N*-dimethylformamide dineopentyl acetal in *N*,*N*-dimethylformamide, compound *2* is converted into the 2,3-dihydrofuryl derivative *3* which, by hydrogenation, gives the title compound *4*. The *S* isomer of *4* is obtained similarly, starting from 2'-deoxy-

5-fluoro-α-uridine.[9]

Decarboxylative elimination constitutes a potentially useful procedure for selective removal of the 4'-(hydroxymethyl) group from a 2'-deoxynucleoside in two simple steps.

PROCEDURE

2'-Deoxy-5-fluorouridine-5'-carboxylic Acid (2)

A mixture of 2'-deoxy-5-fluorouridine[b] (*1*) (5 g, 0.02 mol), the platinum catalyst[c] (2.9 g), and sodium hydrogen carbonate (0.64 g, 7.6 mmol) in water (500 ml) is rapidly stirred in a 1-l Morton flask[d] at 60° to 65° (bath temperature) while oxygen is bubbled in. After 25 hr, TLC with 3:1 chloroform—2-propanol[e] shows that the oxidation is complete. The catalyst is filtered off, Norit A is added to the turbid filtrate, and the mixture is filtered through a thin bed of Celite. The filtrate is evaporated[f] to a viscous mass, which is dissolved in water (50 ml). Dowex 50-W X-4 (H^+) ion-exchange resin (100 to 200 mesh; 20 g) is added, and the suspension is magnetically stirred for 1 hr at room

---

[b] Product of Hoffmann-LaRoche, Inc., Nutley, N. J.

[c] For preparation of the catalyst see footnote *b* of Reference 10.

[d] For a detailed description of the apparatus, see footnote *a* of Reference 10 (contribution #54 of the present volume).

[e] Precoated, 0.25-mm thick plates (6 × 2 cm) of silica gel F-254 (Merck, Darmstadt, Germany) are used. Ultraviolet-absorbing compounds are detected with a Mineralight lamp.

[f] All evaporations are conducted *in vacuo* (aspirator) in a rotary evaporator at a bath temperature below 40°.

temperature. The resin is removed by filtration and washed with water; the filtrate and washings are combined, treated with Norit A, and filtered through a thin layer of Celite. Evaporation of the filtrate[f] affords a syrupy residue that crystallizes after trituration with water (10 ml); yield of *2*, 1.84 g. Evaporation of the mother liquors gives an additional crop of *2*, 3.02 g (total yield, 4.86 g; 92%), m.p. 224° to 225°, lit.[11] m.p. 218° to 221°.

(-)-(R)-1-(2,3-Dihydrofuran-2-yl)-5-fluorouracil (3)

The carboxylic acid *2* (0.26 g, 1 mmol) in *N*,*N*-dimethylformamide (10 ml) is heated with *N*,*N*-dimethylformamide dineopentyl acetal[g] (0.69 g, 3 mmol) for 24 hr at 80° to 90° (bath temperature). The progress of the reaction is monitored by TLC in 9:1 chloroform—methanol.[e] The mixture is heated with additional acetal (0.69 g, 3 mmol) and *N*,*N*-dimethylformamide (5 ml) for 5 hr, bringing the reaction to completion. The solution is then evaporated at 0.1 torr (bath temperature below 45°), and the residual syrup is partitioned between chloroform (2 × 20 ml) and saturated aqueous sodium carbonate (5 ml). The chloroform layer is dried over anhydrous magnesium sulfate, and the filtered solution is evaporated.[f] The syrupy residue crystallizes on addition of ether followed by petroleum ether (b.p. 30° to 60°); the product is

---

[g]Product of Aldrich Chemical Co., Milwaukee, Wisconsin.

removed by filtration and dried at 1 mtorr to give 50 mg (25%) of *3*, m.p. 169° to 170°.

The aqueous layer is stirred with Dowex 50-W X-4 (ammonium form),[h] the resin is removed by filtration and washed with ethanol, and the filtrate and washings are combined and evaporated,[f] to give an additional crop of *3* (0.1 g, 50%), m.p. 170° to 173°. Both fractions are homogeneous by TLC in 9:1 chloroform—methanol.[e] Recrystallization of the second crop of crystals from benzene gives *3* (60 mg), m.p. 178° to 180°, $[\alpha]_{D}^{24}$ -175.2° (*c* 0.5, chloroform); λ_{max}^{EtOH} 268 nm (ε_{mM} 9.60), λ_{min} 235 (1.70). The PMR spectrum (deuteriochloroform)[i] shows a typical doublet of a pseudoquartet at δ 6.70 (H-2) partially overlapped with a signal at 6.50 (H-5), and a quartet at 5.18 (H-4).[j]

(+)-(R)-5-Fluoro-1-(tetrahydrofuran-2-yl)uracil (4)

A magnetically stirred solution of compound *3* (0.25 g, 1.25 mmol) in ethanol[k] (20 ml) containing 10% palladium-on-charcoal catalyst (0.25 g) is hydrogenated at room temperature and atmospheric pressure in a Brown apparatus[l] for 4 hr. The catalyst is removed

---

[h]This is conveniently obtained by a procedure described in footnote *r* of reference 10 (contribution #54 in the present volume).

[i]Tetramethylsilane is used as the internal reference.

[j]Signals are numbered according to the nomenclature given in the title.

[k]Rossville Gold Shield Alcohol, 200-proof, Commercial Solvents Corp., Terre Haute, Indiana.

[l]The apparatus is described in detail in Reference 12.

by filtration, washed with ethanol, and the clear, colorless filtrate is evaporated to dryness.[6] Petroleum ether (b.p. 30° to 60°; 10 ml) is added to the solid residue, and the product (*4*) is removed by filtration; yield 0.2 g (80%), m.p. 165° to 167°. Crystallization from benzene, after removal of a small insoluble portion, gives 0.11 g (44%) of *4*, m.p. 174° to 176°, $[\alpha]_{\underline{\underline{D}}}^{23}$ +56° (*c* 0.5, chloroform); λ_{max}^{EtOH} 270 nm ($\varepsilon_{\underline{mM}}$ 7.20), λ_{min}^{EtOH} 235 nm (1.40). The PMR spectrum (deuteriochloroform)[i] shows a poorly resolved multiplet at δ 6.00 (H-2) and a multiplet at 2.12 that includes H-4 and H-5.

REFERENCES

1. W. A. Bowles, F. H. Schneider, L. R. Lewis, and R. K. Robins, *J. Med. Chem.*, 6, 471 (1963).

2. D. V. Meiren and A. K. Belousova, *Vopr. Med. Khim.*, 18, 288 (1972) and references cited therein.

3. L. R. Lewis, F. H. Schneider, and R. K. Robins, *J. Org. Chem.*, 26, 3837 (1961).

4. S. A. Hiller, R. A. Zhuk, and M. Yu. Lidak, *Dokl. Akad. Nauk. SSSR*, 176, 332 (1967).

5. S. A. Hiller, M. Yu. Lidak, R. A. Zhuk, A. E. Berzinya, K. Ya. Pets, I. N. Getsova. and E. I. Bruk, *Khim. Geterotsikl. Soedin.*, 375 (1969).

6. C. W. Noell and C. C. Cheng, *J. Heterocycl. Chem.*, 5, 25 (1968).

7. R. A. Earl and L. B. Townsend, *J. Heterocycl. Chem.*, 9, 1141 (1972).

8. J. Žemlička, R. Gasser, J. V. Freisler, and J. P. Horwitz, *J. Am. Chem. Soc.*, 94, 3212 (1972).

9. J. P. Horwitz, J. J. McCormick, K. D. Philips, V. M. Maher, J. R. Otto, D. Kessel, and J. Zemlicka, *Cancer Res.*, 35, 1301 (1975).

10. J. Žemlička and J. P. Horwitz, *Nucleic Acid Chemistry, New or Improved Syntheses, Methods, and Techniques*, Vol. I, Wiley, New York, 1978, contribution #54 of the present volume.

11. K. C. Tsou, N. J. Santora, and E. E. Miller, *J. Med. Chem.*, 12, 173 (1969).

12. C. A. Brown and H. C. Brown, *J. Am. Chem. Soc.*, 84, 2829 (1962).

[58] 6-METHYL-3-β-D-RIBOFURANOSYLURACIL

Selective Glycosylation by the Mercuri Procedure

Miroslav Prystasz

Institute of Organic Chemistry and Biochemistry, Czechoslovak Academy of Sciences, 166 10 Prague 6, Czechoslovakia

INTRODUCTION

Glycosylation of substituted uracils by the mercuri procedure proceeds almost exclusively at the N-1. 6-Methyluracil, however, provides[1] a rare example of a reverse regioselectivity of the glycosylation reaction; D-ribofuranosylation of its mercuri salt (*1*) gives rise to the isomeric nucleosides *3* and *4* with a preponderance of the former. 6-Methyluracil reacts analogously with the glycosyl halide *2* in nitromethane, in the presence of mercuric cyanide.[2] Esters *3* and *4* have been methanolyzed to afford crystalline nucleosides 5 and *6*.

PROCEDURE

6-Methyl-3-(2,3,5-tri-O-benzoyl-β-D-ribosyl)uracil (3) and 6-Methyl-1-(2,3,5-tri-O-benzoyl-β-D-ribosyl)uracil (4)

A mixture of the mercuric salt *1* (7.89 g, 35 mequiv; predried by codistillation with 100 ml of toluene) and 0.15M solution of the glycosyl halide *2* in acetonitrile (220 ml, 33 mmol) is boiled

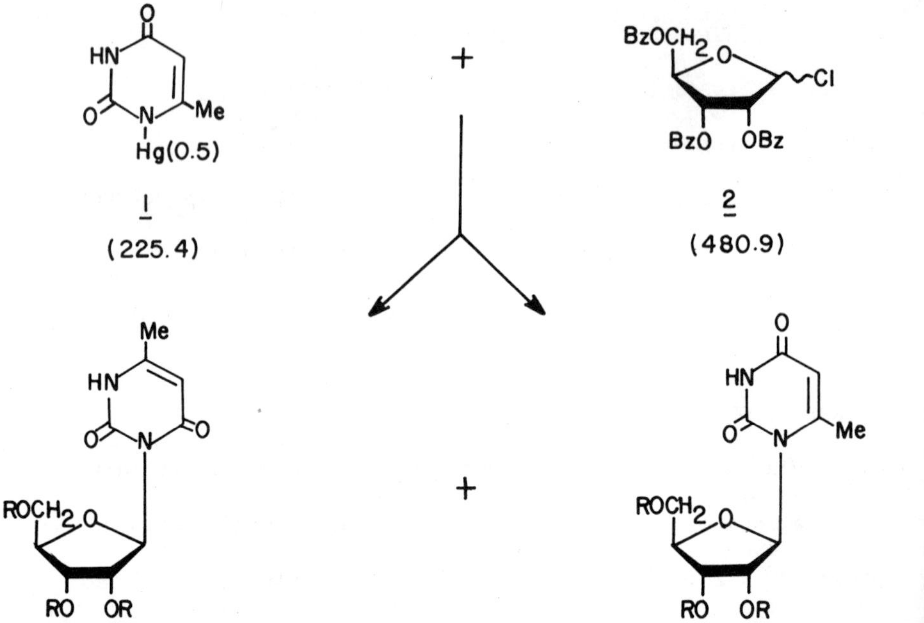

<u>3</u>, R = Bz (570.5) <u>4</u>, R = Bz (570.5)

<u>5</u>, R = H (258.2) <u>6</u>, R = H (258.2)

under reflux with mechanical stirring for 1.5 hr, cooled, and evaporated *in vacuo*. The residue is dissolved in a mixture of benzene (300 ml) and 25% aqueous potassium iodide (120 ml), and the organic layer is separated, washed with saturated aqueous sodium hydrogen carbonate (50 ml), dried (anhydrous sodium sulfate), and applied to a column of 500 g of neutral alumina (Brockmann activity II and III). The column is successively eluted with the following eluants: benzene (1.5 l), 5:1 benzene—ethyl acetate (1.5 l, giving fractions 1 to 6), 3:1 benzene—ethyl acetate (2 l, giving fractions 7 to 14), and 1:2 benzene—

ethyl acetate (1.5 l, giving fractions 15 to 20). The homogenous fractions 5 to 10 (monitored by TLC on a loose layer of alumina in 5:3 benzene—ethyl acetate; R_F 0.65) are combined, and evaporated, and the residue is triturated with light petroleum (150 ml), to give 13.4 g (71%) of compound *3* as an amorphous powder (dried for 1 hr at 75°/0.1 torr); NMR data ($CDCl_3$): δ 2.07 (s, 6-CH_3), 4.5 to 4.8 (m, H-4',5'), 5.45 (s, H-5), 5.95 to 6.25 (m, H-2',3'), 6.63 (s, H-1', $J_{1',2'}$ 0.5 Hz), 7.0 to 8.1 (m, $3 \times C_6H_5$), and 10.5 to 10.6 (broad s, N-1-H). The combined fractions 16 to 18 (R_F 0.1) are evaporated *in vacuo*, and the residue is crystallized from ether (20 ml) to afford 1.95 g (10%) of the protected nucleoside *4*, m.p. 125° to 128°; NMR data ($CDCl_3$): δ 2.21 (s, 6-CH_3), 4.5 to 4.85 (m, H-4',5'), 5.53 (s, H-5), 5.66 (d, H-1', $J_{1',2'}$ <1 Hz), 5.95 to 6.20 (m, H-2',3'), 7.0 to 8.1 (m, $3 \times C_6H_5$), and 9.3 to 9.5 (broad s, N-3-H).

6-Methyl-3-β-D-ribofuranosyluracil (5)

A mixture of the tribenzoate *3* (5.70 g, 10 mmol) and 0.05M methanolic sodium methoxide (200 ml) is kept overnight at room temperature, the resulting solution is neutralized by addition of Dowex-50 (H^+) ion-exchange resin, and the suspension is filtered. The filtrate is evaporated to dryness *in vacuo*, the residue is dissolved in water (50 ml) and chloroform (100 ml), and the aqueous solution is washed with chloroform (100 ml), filtered with charcoal, and evaporated under diminished pressure. Crystallization of the syrupy residue from isopropyl alcohol, and the usual processing of the mother liquors gives 2.3 g (89%) of the free nucleoside *5*, m.p. 145° to 146° (lit.[3] m.p. 144° to 146°); $\lambda_{max}^{pH\ 7}$ 208 nm ($\varepsilon_{\underline{mM}}$ 8.50) and 265 (8.70), $\lambda_{max}^{pH\ 14}$ 289 nm ($\varepsilon_{\underline{mM}}$ 11.20) [lit.[3] $\lambda_{max}^{pH\ 1\ to\ 4}$ 265 nm ($\varepsilon_{\underline{mM}}$ 8.50), $\lambda_{max}^{pH\ 14}$ 289 (11.20)]; ORD (*c* 0.03, water): 282 nm ([Φ] -2,400°), 255 nm (+2,400°), 226 nm (-4,100°).

6-Methyluridine (6)

A solution of the tribenzoate *4* (570 mg, 1.0 mmol) in 0.1M methanolic sodium methoxide (10 ml) is kept for 10 hr at room temperature and processed as in the preceding paragraph, to give 220 mg (85%) of *6*, m.p. 179° to 181° (from 85% aqueous ethanol) [lit.[3] m.p. 177° to 178°]; $\lambda_{max}^{pH\ 7}$ 207 nm ($\varepsilon_{\underline{mM}}$ 8.10) and 261 (11.50), $\lambda_{max}^{pH\ 11}$ 262 nm ($\varepsilon_{\underline{mM}}$ 9.95) [lit.[3] $\lambda_{max}^{pH\ 1\ to\ 4}$ 261 nm ($\varepsilon_{\underline{mM}}$ 11.20), $\lambda_{max}^{pH\ 11}$ 262 (6.15)]; ORD (*c* 0.03, water): 285 nm ([Φ] -1,500°), 257 nm (-700°), 224 nm (-6,000°); NMR data [$(CD_3)_2SO$]: δ 2.23 (s, 6-CH_3), 3.48 (q, H-5', $J_{4',5'}$ 6 Hz, J_{gem} 12 Hz), 3.71 (t, H-4', $J_{3',4'}$ 6 Hz), 4.09 (t, H-3', $J_{2',3'}$ 6 Hz), 4.56 (q, H-2', $J_{1',2'}$ 4.5 Hz), 5.46 (d, H-1'), and 5.51 (s, H-5).

REFERENCES

1. M. Prystasz and F. Sorm, *Collect. Czech. Chem. Commun.*, 34, 2316 (1969).

2. R. S. Klein, I. Wempen, K. A. Watanabe, and J. J. Fox, *J. Org. Chem.*, 35, 2330 (1970).

3. M. W. Winkley and R. K. Robins, *J. Org. Chem.*, 33, 2822 (1968).

[59] 5'-AMINO-2',5'-DIDEOXY-5-IODOURIDINE

Synthesis of a 5'-Amino-5-iodo Pyrimidine Nucleoside

Tai-Shun Lin,[a] William H. Prusoff,[a] and David C. Ward[b]

Department of Pharmacology[a] and Departments of Human Genetics and Molecular Biophysics and Biochemistry,[b] Yale University, School of Medicine, New Haven, Connecticut 06510

INTRODUCTION

5'-Amino-2',5'-dideoxy-5-iodouridine (AIU), an analog of thymidine, has been found to be not only a potent inhibitor of *Herpes simplex* virus in cell culture,[1] but also an effective agent in the therapy of experimental *Herpes keratitis* in rabbits. The original synthesis of AIU involved iodination of the 5-mercuriacetate of 5'-amino-2',5'-dideoxyuridine.[2] For the preparation of AIU an alternative procedure more readily suited to large-scale production is described.

PROCEDURE

2'-Deoxy-5'-O-p-tolylsulfonyl-5-iodouridine (2)

To a solution of 2'-deoxy-5-iodouridine[3] (17.7 g, 50.0 mmol) in dry pyridine (100 ml) at 0° (ice-bath) is added *p*-toluenesulfonyl chloride (12.4 g, 65.0 mmol), and the mixture is refrigerated for 48 hr (3°). The solvent is evaporated at room temperature/0.1

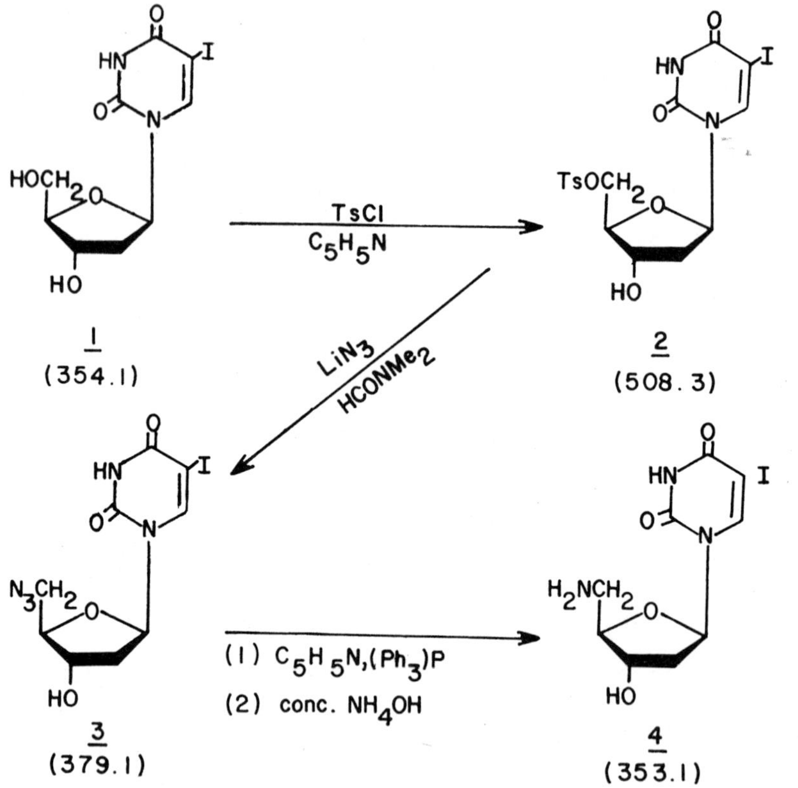

torr to give a gum which is extracted with ether (2 × 100 ml) and then triturated with ice-water. The product solidifies as a white crystalline mass, which is collected by filtration and successively washed with water, a small amount of ice-cooled ethanol, ether, and dried to afford 24.3 g (95%) of *2*. One recrystallization from ethanol gives an analytically pure sample, m.p. 164° to 165° (dec.).

Analysis calculated for $C_{16}H_{17}IN_2O_7$: C, 37.80; H, 3.37; N, 5.51. Found: C, 37.97; H, 3.50; N, 5.63.

5'-Azido-2',5'-dideoxy-5-iodouridine (*3*)

A mixture of *2* (21.0 g, 41.3 mmol) and lithium azide (6.07 g, 124 mmol) in *N*,*N*-dimethylformamide (120 ml) is heated for 2 hr at 80° (oil bath) and then evaporated at 60°/0.1 torr to a syrup, which is triturated with ice water. The product crystallizes as a white solid which is collected by filtration, washed thoroughly with ice-cooled water and ether, and dried to afford 9.62 g (61%) of the product. One recrystallization from ethanol yields analytically pure material, m.p. 184° to 185° (dec.).

Analysis calculated for $C_9H_{10}IN_5O_4$: C, 28.51; H, 2.66; N, 18.47. Found: C, 28.78; H, 2.94; N, 18.20.

5'-Amino-2',5'-dideoxy-5-iodouridine[1,2] (*4*)

Compound *3* (5.10 g, 13.5 mmol) and triphenylphosphine[4] (5.79 g, 22.1 mmol) are dissolved in pyridine (35 ml), and the solution is kept for 1 hr at room temperature. Concentrated ammonium hydroxide (5 ml) is then added, and the solution is kept for 2 hr and evaporated to dryness at room temperature/0.1 torr to a syrup that is washed with ether (2 × 100 ml). The residue is then extracted with M ammonium hydroxide solution (2 × 150 ml), and the excess of triphenylphosphine and triphenylphosphine oxide is removed by filtration. The filtrate is successively washed with benzene (1 × 100 ml) and ether (2 × 150 ml) to remove residual triphenylphosphine, and then evaporated to dryness *in vacuo* at 50°

to yield 4.43 g (93%) of a solid that is washed with boiling ethanol (2 × 150 ml). The insoluble material is isolated by filtration, washed successively with ethanol and ether, and dried to afford 2.51 g (53%) of an analytically pure product; m.p. 201° to 202° (dec.); $\lambda_{max}^{0.01\underline{M}\ HCl}$ 285 nm ($\varepsilon_{m\underline{M}}$ 7.42), $\lambda_{min}^{0.01\underline{M}\ HCl}$ 246 nm, $\lambda_{max}^{0.01\underline{M}\ NaOH}$ 279 nm ($\varepsilon_{m\underline{M}}$ 5.53), $\lambda_{min}^{0.01\underline{M}\ NaOH}$ 252 nm.

Analysis calculated for $C_9H_{12}IN_3O_4$: C, 30.61; H, 3.43; N, 11.90. Found: C, 30.49, H, 3.59; N, 11.98.

REFERENCES

1. Y. C. Cheng, J. P. Neenan, B. Goz, D. C. Ward, and W. H. Prusoff, *J. Virol.*, 1284 (1975).

2. T. S. Lin, J. P. Neenan, Y. C. Cheng, W. H. Prusoff, and D. C. Ward, *J. Med. Chem.*, 19, 495 (1976).

3. W. H. Prusoff, *Biochim. Biophys. Acta*, 32, 295 (1959).

4. W. S. Mungall, G. L. Greene, G. A. Heavner, and R. L. Letsinger, *J. Org. Chem.*, 40, 1659 (1975).

[60] 2,5'-ANHYDROURIDINE AND 2,5'-ANHYDRO-5-FLUOROURIDINE

*One-Step Conversion of Uridine and 5-Fluorouridine into Their Corresponding 2,5'-Anhydronucleosides**

K. A. Watanabe, C. K. Chu, U. Reichman, and Jack J. Fox

Laboratory of Organic Chemistry, Memorial Sloan-Kettering Cancer Center, Sloan-Kettering Institute, Sloan-Kettering Division of Graduate School of Medical Sciences, Cornell University, New York, New York 10021

INTRODUCTION

Although the synthesis of 2,5'-anhydro derivatives of 2',3'-*O*-isopropylidene- and 2',3'-di-*O*-acetyluridine by reaction of the corresponding 5'-deoxy-5'-iodouridines with silver acetate has been described,[1] difficulties were encountered[1] in the preparation of the 2',3'-unsubstituted 2,5'-anhydronucleoside (*2a*) by this procedure, but the preparation of *2a* has been achieved[2] by this approach. Wada and Mitsunobu[3] described the direct synthesis of the 2',3'-isopropylidene acetal of *2a* by reaction of 2',3'-*O*-isopropylideneuridine with triphenylphosphine, followed by the addition of diethyl azodicarboxylate. Later Shibuya and colleagues[4] synthesized *2a* in one step from uridine (*1a*) by

*The authors acknowledge the support of the U.S. Public Health Service, Grant Nos. CA-08748 and CA-18601.

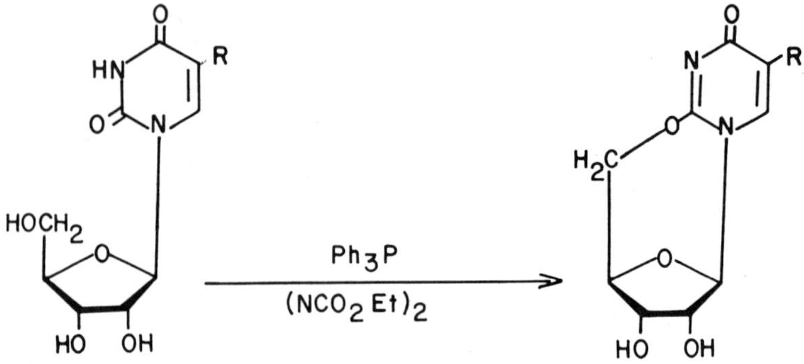

1a	R=H	(244.2)	2a	(226.2)
1b	R=F	(262.2)	2b	(244.2)

modification of the Wada—Mitsunobu procedure. The synthesis of *2a* and its hitherto unreported 5-fluoro analog (*2b*), reported here, utilize the latter approaches.[3,4] The yields of *2a* and *2b* are ≃60%.

PROCEDURE

General Procedure

Diethyl azodicarboxylate (6.26 g, 0.036 mol) is added dropwise to a stirred suspension of *1* (0.02 mol) and triphenyl phosphine (9.44 g, 0.023 mol) in 1,4-dioxane (50 ml). The red color of the azodicarboxylate disappears as the exothermic reaction proceeds. The mixture is stirred for 2.5 hr, diluted with water (5 ml), boiled under reflux for 35 min, and evaporated to dryness *in vacuo*. The residue is dried by azeotropic distillation with ethanol (50 ml) and toluene (50 ml); it is then mixed with benzene (50 ml) and the mixture is boiled under reflux for 10 min. Colorless crystals are collected by filtration of the hot mixture, washed with hot benzene (50 ml), and recrystallized from a minimal volume of water to afford the anhydronucleoside *2*.

2,5'-Anhydrouridine (2a)

2,5'-Anhydrouridine (*2a*) is obtained as colorless needles (2.57 g, 57%) m.p. 213° to 215° (dec.), $\lambda_{max}^{H_2O}$ in nm 240 ($\varepsilon_{\underline{mM}}$ 12.00), $\lambda_{min}^{H_2O}$ 214 (2.90) [lit.[4] m.p. 213° to 214° (dec.); $\lambda_{max}^{H_2O}$ 240 nm (12.00)]

2,5'-Anhdyro-5-fluorouridine (2b)

2,5'-Anhydro-5-fluorouridine (*2b*) crystallizes from water as colorless needles that change into cubic crystals within a few minutes; yield, 2.79 g (57%), m.p. 215° to 219° (dec.); $\lambda_{max}^{H_2O}$ 252 nm ($\varepsilon_{\underline{mM}}$ 4.00). This product is analytically pure.

The mother liquor from the crystallization contains the starting material *1* and the product *2* in various proportions; without purification the mixture can be recycled with triphenylphosphine and diethyl azodicarboxylate.

REFRENCES

1. D. M. Brown, A. R. Todd, and S. Varadarajan, *J. Chem. Soc.*, 868 (1957).

2. I. L. Doerr and J. J. Fox, unpublished work. Their sample of *2a* was used by E. G. Lovett and D. Lipkin [*J. Am. Chem. Soc.*, 95, 2312 (1973)] for mass-spectral studies.

3. M. Wada and O. Mitsunobu, *Tetrahedron Lett.*, 1279 (1972).

4. S. Shibuya, A. Kuninaka, and H. Yoshino, *Chem. Pharm. Bull.*, 22, 719 (1974).

[61] 2'-DEOXY-L-URIDINE

Total Synthesis of a Uracil 2'-Deoxynucleoside from a Sugar 2-Aminooxazoline Through a 2,2'-Anhydronucleoside Intermediate

Antonín Holý

Institute of Organic Chemistry and Biochemistry, Czechoslovak Academy of Sciences, 166 10 Praha 6, Czechoslovakia

INTRODUCTION

The preparation of 2'-deoxy-L-uridine has been accomplished by a total synthesis starting from L-arabinose, which on treatment with cyanamide affords 2-amino-β-L-arabinofurano[1',2':4,5]oxazoline. This intermediate affords 2,2'-anhydro-L-uridine by reaction with ethyl 1-propynoate.[1,2] On benzoylation with benzoyl cyanide[3] the 3',5'-di-*O*-benzoyl derivative of the anhydronucleoside is obtained; this is transformed into the 2'-chloro-2'-deoxyuridine derivative by heating with hydrogen chloride in *N,N*-dimethylformamide. The reductive dehalogenation of the 2'-chloro derivative affords 3',5'-di-*O*-benzoyl-2'-deoxy-L-uridine, which gives the free nucleoside on methanolysis.[4] 2'-Deoxy-L-uridine serves as a starting compound for other pyrimidine 2'-deoxy-L-*erythro*-pentonucleosides[4] of potential biological interest. This method can be conveniently used for both the preparation of

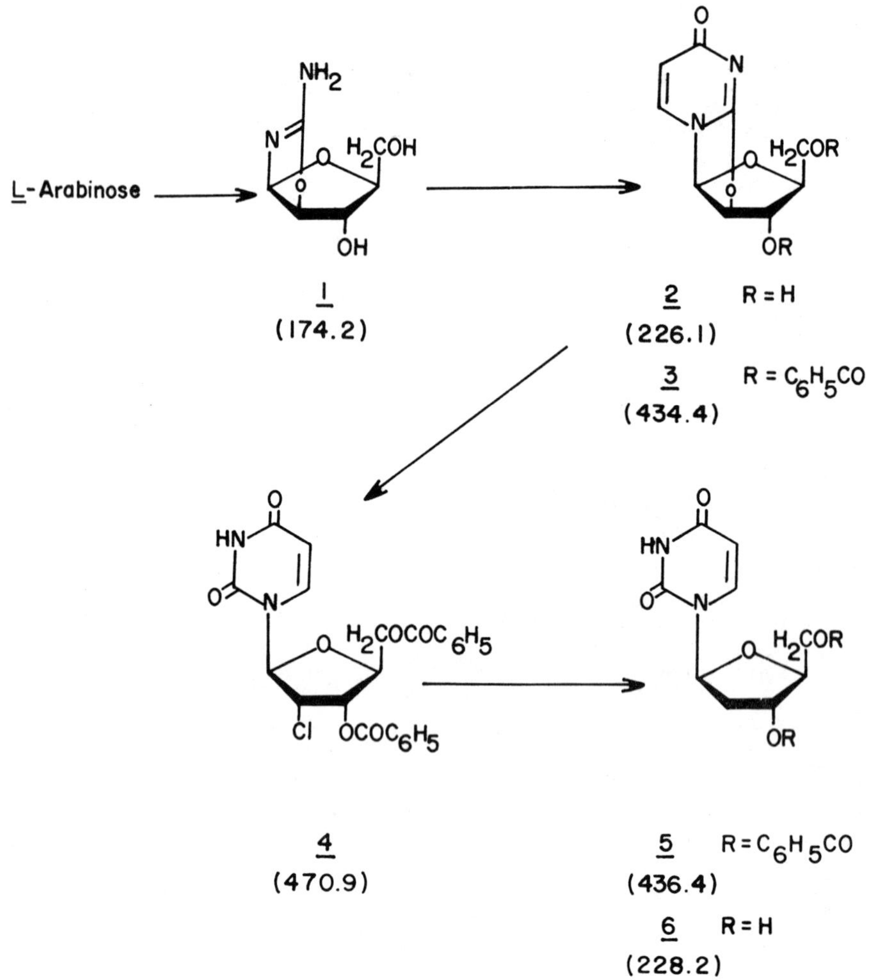

2'-deoxyuridine from D-arabinose and for the syntheses of 2-deoxypentofuranosyluracils modified either in the heterocyclic base[5] or the sugar moiety.[6-8]

PROCEDURE

2-Amino-β-L-arabinofurano[1',2':4,5]oxazoline (1)

To a suspension of L-arabinose (170 g, 1.1 mol) and cyanamide (100 g) in methanol (300 ml) in a 1-l Erlenmeyer flask is added 6M aqueous ammonia (50 ml), the flask is stoppered, and the mixture is magnetically stirred overnight at room temperature.[a] The mixture is then kept overnight at -10°, and the solid is collected by filtration, washed with methanol (3 × 200 ml), and dried over phosphorus pentaoxide *in vacuo*. The pure *1* weighs 135 to 150 g (70% to 77%), m.p. 180° $[\alpha]_D^{25}$ -20.6° (*c* 1, water).

2,2'-Anhydro-L-uridine (2)

A mixture of compound *1* (174 g, 1 mol) and ethyl 1-propynoate (150 ml) in 50% aqueous ethanol (2 l) is boiled under reflux, with stirring, in a 3-l round-bottomed flask equipped with a reflux condenser and a mechanical stirrer. After 5 hr the mixture is cooled to ≈40° and concentrated to half its volume *in vacuo* (water pump). The resulting suspension is cooled in ice, the product is collected by filtration and washed with ethanol (500 ml). On concentration *in vacuo* to ≈300 ml, and diluting with acetone (1 l), the mother liquor affords an additional crop of the product, which is collected by filtration and washed with acetone (500 ml). After being dried *in vacuo* over phosphorus pentaoxide the combined product weighs 150 to 160 g (66% to 71%). The analytically pure compound *2* has m.p. 239°.

---

[a]During the reaction the starting material gradually dissolves, and the product crystallizes from the clear warm solution.

3',5'-Di-O-benzoyl-2'-deoxy-L-uridine (3)[b]

In a 2-l, 3-necked, round-bottomed flask equipped with a stirrer and a drying tube (calcium chloride) are placed dry *2* (100 g, 0.43 mol), benzoyl cyanide (125 g, 0.95 mol), and dry *N,N*-dimethylformamide (800 ml). To the well-stirred suspension is added triethylamine in 5-ml portions until an exothermic reaction sets in (15 to 20 ml).[c] During the reaction the temperature rises to 50° to 60° (no external cooling is necessary), and nearly all of the solid dissolves. The product crystallizes spontaneously from the mixture.[d] The mixture is stirred for 2 hr, ethanol (100 ml) is added, and the whole is poured into ether (2 l). The product is removed by filtration with suction, successively washed with ether (1 l), ethanol (200 ml, to remove colored impurities), and ether (500 ml). The product is air dried, and then dried *in vacuo* over phosphorus pentaoxide to afford 170 to 180 g (91% to 97%) of analytically pure compound *3* as a white microcrystalline material, m.p. 257° to 258°.

---

[b]The procedure must be conducted in a good hood due to the evolution of hydrogen cyanide during the reaction.

[c]The amount of catalyst depends on the purity of the sample of benzoyl cyanide. In the procedure a technical-grade (95%) commercial preparation may be used. Generally, triethylamine is added until the spontaneous, exothermic reaction sets in.

[d]In some cases, the product crystallizes before complete dissolution of the starting material has occurred. Therefore, efficient stirring is necessary (to ensure quantitative conversion).

3',5'-Di-O-benzoyl-2'-chloro-2'-deoxy-L-uridine (4)

In a 2-l, 3-necked, round-bottomed flask equipped with a stirrer and a drying tube (calcium chloride) are placed compound *3* (87 g, 0.2 mol) and *N,N*-dimethylformamide (800 ml). The flask is placed in an oil bath and the mixture is heated, with stirring, to 100°. A 6M solution of hydrogen chloride in *N,N*-dimethylformamide (160 ml) is added, and the mixture is stirred for 90 min at 100°. The greenish solution is then slowly poured into well-stirred water (3 l) in a 5-l beaker, and the precipitated product is removed by filtration with suction and washed with water until the final filtrate is neutral. The moist product is dissolved in chloroform (1 l), and the solution is dried (magnesium sulfate), and evaporated *in vacuo*. The foamy residue is rapidly dissolved in boiling ethanol (1.2 l), and the solution is kept overnight at 4° for crystallization.[e] The product is collected by filtration, washed with ethanol, and dried *in vacuo* over phosphorus pentaoxide. The yield amounts to 38 to 41 g (81% to 87%) of *4* m.p. 166° to 167°.

3',5'-Di-O-benzoyl-2'-deoxy-L-uridine (5)

In a 2-l, three-necked, round-bottomed flask equipped with a stirrer and a reflux condenser closed by a drying tube (calcium chloride) and placed in an oil bath, a mixture

---

[e]On prolonged heating, the crystalline material is formed, and this is very insoluble in boiling ethanol.

of compound *4* (47 g, 0.1 mol), tributyltin hydride[6] (95 g) and azabisisobutyronitrile (0.4 g) in benzene (600 ml) is boiled under reflux, with stirring, for 90 min. The reaction is monitored by TLC in 2:3 ethyl acetate—benzene on plates of silica. After compound *4* has disappeared the mixture is cooled, and the solvent is evaporated under vacuum (water pump) at 40°. To the oily residue is added petroleum ether (b.p. 40° to 60°, 1 1), and the product is removed by filtration, washed with petroleum ether (200 ml), and dried *in vacuo*. The yield of chromatographically and analytically pure compound *5* amounts to 40 to 43 g (92% to 99%), m.p. 218° to 219°. The product may be recrystallized from boiling ethanol (5 g of compound per liter).

2'-Deoxy-L-uridine (6)

A suspension of compound *5* (40 g, 89 mmol) in 0.1*M* sodium methoxide in methanol (400 ml) is stirred magnetically in a 1-1 stoppered Erlenmeyer flask overnight at room temperature. The clear solution is then rendered neutral by stepwise addition of Dowex 50 X-8 (H^+) ion exchange resin prewashed with methanol; the resin is collected by filtration and washed with methanol (100 ml), and the filtrate and washings are combined and evaporated to dryness at 40° under vacuum (water pump). To the residue are added water (300 ml) and ether (100 ml), and the aqueous layer is washed with

---

[6]For the preparation of this reagent, see contribution #172.

ether (2 × 100 ml), evaporated to dryness, and the residue coevaporated with ethanol (3 × 50 ml). The crystalline residue is recrystallized from 99% ethanol (≃300 ml) by keeping the solution overnight at 4°; the crystalline material is removed by filtration, washed successively with ethanol (50 ml) and ether (100 ml), and dried *in vacuo*. Analytically pure compound *6* is obtained; 16 to 17 g (79% to 84%); m.p. 158°. An additional crop of compound *6* may be obtained from the mother liquor by evaporation and crystallization of the residue from ethanol.

REFERENCES

1. A. Holý, *Collect. Czech. Chem. Commun.*, 37, 4072 (1972).
2. R. A. Sanchez and L. E. Orgel, *J. Molec. Biol.*, 47, 531 (1970).
3. A. Holý and M. Souček, *Tetrahedron Lett.*, 185 (1971).
4. A. Holý, *Tetrahedron Lett.*, 189 (1971).
5. A. Holý and D. Cech, *Collect. Czech. Chem. Commun.*, 40, 3157 (1975).
6. A. Holý, *Collect. Czech. Chem. Commun.*, 38, 100 (1973).
7. A. Holý, *Collect. Czech. Chem. Commun.*, 38, 428 (1973).
8. A. Holý, *Nucleic Acid Res.*, 1, 289 (1974).

[62] 2'-DEOXY-5-FORMYLURIDINE

Arieh Kampf and Mathias P. Mertes

Department of Medicinal Chemistry, University of Kansas, Lawrence, Kansas 66045

INTRODUCTION

The synthesis of 2'-deoxy-5-formyluridine (*3*) has been approached by direct condensation of bistrimethylsilylated 5-formyluracil[1] and by a tedious route via the 5-benzyloxymethyl derivative.[2] However, the easiest preparation is by way of the dimethyl acetal derivative *1*, as described.

PROCEDURE

1-(2-Deoxy-3,5-di-O-p-toluoyl-β-D-erythro-pentofuranosyl)-5-formyluracil (2) and its α Anomer

A suspension of 5-formyluracil dimethyl acetal[a] (*1*, 6.0 g, 32 mmol) in hexamethyldisilazane (70 ml) containing ammonium sulfate (50 mg) is boiled under reflux with stirring and exclusion of moisture for 4 hr. The pale-yellow solution is evaporated under diminished pressure (bath temperature, 80°), and the oily residue is distilled at 105° to 110°/0.2 torr. The bistrimethylsilyl

---

[a]Prepared from 5-formyluracil by refluxing in methanol with *p*-toluenesulfonic acid; m.p. 199° to 200°.

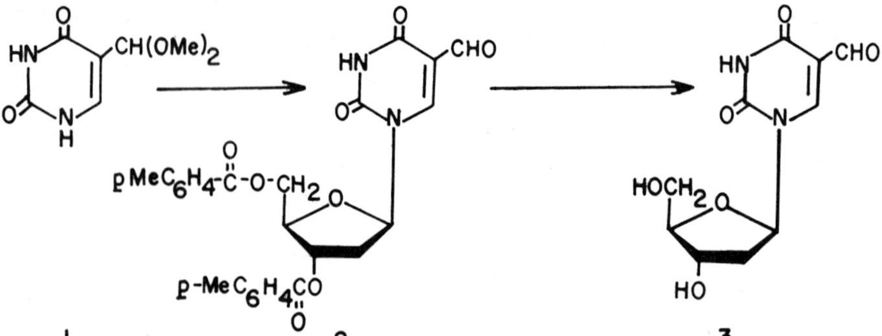

1 (174.2) 2 (492.5) 3 (256.2)

derivative of *1* is obtained as a colorless oil, 7.9 g (74%).

To a solution of this compound (7.9 g, 24 mmol) and 2-deoxy-3,5-di-*O*-*p*-toluoyl-D-*erythro*-pentofuranosyl chloride (7.5 g, 19 mmol) in dry 1,2-dichloroethane[3] (190 ml) at 2° is added dropwise, with stirring and exclusion of moisture, a solution of stannic chloride (0.41 ml) in 1,2-dichloroethane (95 ml). A clear solution is obtained after a short time; it is stirred for 2 hr at 2°, diluted with 1,2-dichloroethane (100 ml), and shaken with saturated sodium hydrogen carbonate solution (500 ml). The layers are filtered through moist Celite, and the organic layer is washed with water, dried, and evaporated to yield a mixture of the α and β (*2*) anomers as a white solid (12.0 g). The anomers are separated by fractional recrystallization from hot acetone. The yield of pure β anomer (*2*), which separates first, ranges from 25 to 45%; m.p. 195° to 196°. The yield of the pure α anomer is only 15%; m.p. 188° to 189°.

2'-Deoxy 5-formyluridine (3)

Compound *2* (0.73 g, 1.5 mmol) is suspended in a mixture of dry benzene (15 ml) and dry methanol (15 ml) and transesterified by addition of a freshly prepared methanolic solution of sodium methoxide until the solution is distinctly alkaline. The mixture is kept overnight at room temperature in a stoppered flask, with occasional stirring. The mixture is diluted with dry benzene (20 ml), and the precipitate of the sodium salt of the nucleoside is collected by filtration and washed well with dry benzene. The precipitate is dissolved in water (10 ml), Dowex 50 (H^+) is added until the solution is neutral to slightly acidic, and the solution is concentrated and cooled, depositing 0.27 g (71%) of crystalline *3*. Repeated recrystallization from ethanol gives analytically pure *3* as tan-yellow crystals, m.p. 175° to 175.5°.

REFERENCES

1. M. P. Mertes and M. T. Shipchandler, *J. Heterocycl. Chem.*, 7, 751 (1970).

2. M. P. Mertes and M. T. Shipchandler, *J. Heterocycl. Chem.*, 8, 133 (1971).

3. U. Niedbolla and H. Vorbruggen, *J. Org. Chem.*, 39, 3654 (1974).

[63] 2'-DEOXY-5-(HYDROXYMETHYL)URIDINE AND 2'-DEOXY-5-FORMYLURIDINE

Selective Bromination of the 5-Methyl Group of 5-Methylpyrimidine Nucleosides

D. Bärwolff and P. Langen

Department of Cell Kinetics, Institute of Molecular Biology, Academy of Sciences of GDR, GDR - 1115 Berlin, Germany

INTRODUCTION

5-Methylpyrimidine nucleosides brominated on the 5-methyl group, such as 5-(bromomethyl)-1-(3,5-di-*O*-acetyl-2-deoxy-β-D-*erythro*-pentofuranosyl)uracil (*2*) and 1-(3,5-di-*O*-acetyl-2-deoxy-β-D-*erythro*-pentofuranosyl)-5-(dibromomethyl)uracil (*3*) are new, important, and versatile intermediates[1] for the synthesis of various 5-substituted 2'-deoxyuridine derivatives[2] by a facile, nucleophilic displacement of bromine therefrom. The procedure described here is the most direct and effective of those reported for the synthesis of 2'-deoxy-5-(hydroxymethyl)uridine[3] (*4*) and 2'-deoxy-5-formyluridine[4] (*5*).

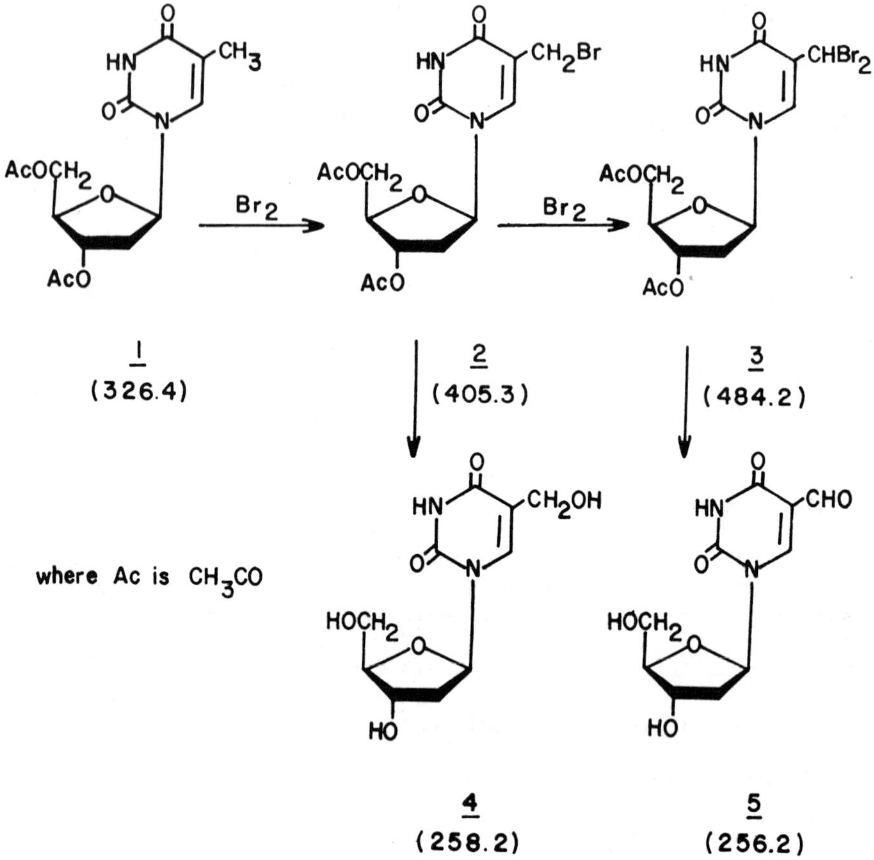

PROCEDURE

5-(Bromomethyl)-1-(3,5-di-O-acetyl-2-deoxy-β-D-erythro-pentofuranosyl)uracil (2)

3',5'-Di-*O*-acetylthymidine[5] (*1*) (5 g, 15.3 mmol) is dissolved in anhydrous carbon tetrachloride[a] (1 l) in a 2-l three-necked flask

---

[a]Only *1*, bromine, and carbon tetrachloride of the highest purity and dryness are to be used.

(A; see Figure 1) fitted with an effective reflux condenser (B), a drying tube (C), a mechanical stirrer[b] (E), and a special device (D) for introduction of bromine under reflux with use of a 500-W, reflexing photo-lamp. When dissolution is complete a stream of dry nitrogen blows bromine (0.88 ml, 0.17 mmol) in carbon tetrachloride (20 ml) from the introduction device (D) into the stirred, refluxing solution, with rigorous exclusion of moisture. Gaseous hydrogen bromide leaves the condenser, fitted with a drying tube (calcium chloride). After 2 to 3 hr, when all of the bromine has been added,[c] nitrogen is bubbled through the solution for a further 30 min to remove the hydrogen bromide. The hot solution is quickly filtered through ordinary filter paper, and the filtrate is kept in a refrigerator for 24 hr. The amorphous nucleoside is collected by filtration and washed successively with carbon tetrachloride (10 ml) and petroleum ether (20 ml); yield 3.03 g of *2* (first crop). The filtrate is evaporated to dryness under diminished pressure at 40°; it affords an additional 1.8 g of *2* (second crop), contaminated with *1* and *3*; total yield 4.83 g (78%) of *2*[d]; R_F 0.56 (ethyl acetate) $\lambda_{max}^{CH_2Cl_2}$

---

[b]The stirrer is oiled with carbon tetrachloride.

[c]Only the progress of bromination can be readily monitored by TLC plates of Merck silica gel HF (0.25-mm layer) on glass are used.

[d]Compounds *2* and *3* are extremely sensitive to moisture, but are stable for weeks when stored in the cold over sodium hydroxide and phosphorus pentaoxide.

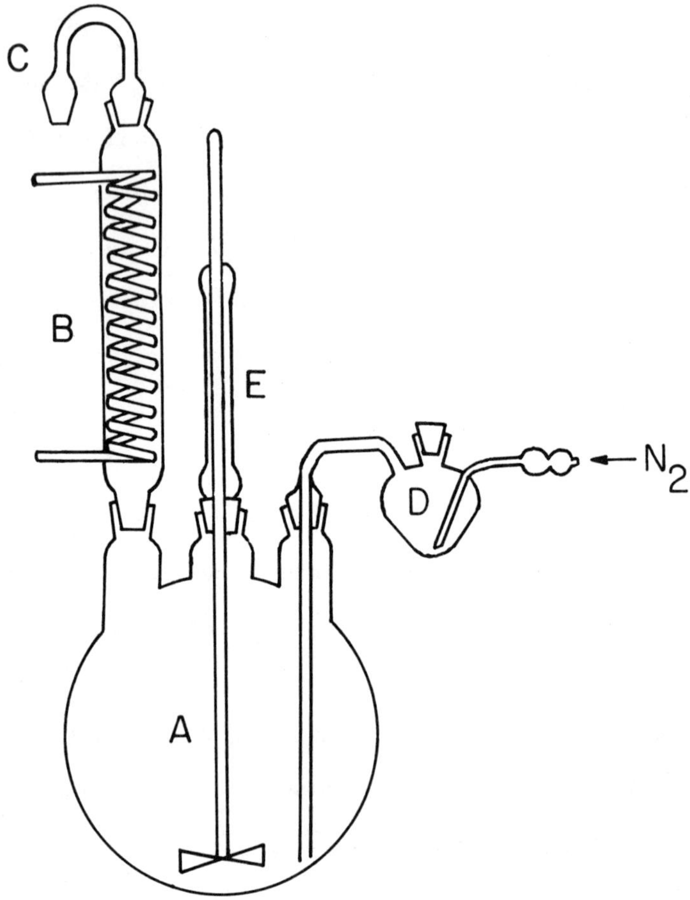

Fig. 1, —

Apparatus for the Bromination ⊙

270 nm ($\varepsilon_{\underline{mM}}$ 10.58), λ_{min} 242 nm; PMR data[e]: δ 3.9 (2 H,d, 5-CH_2-Br with ring).

2'-Deoxy-5-(hydroxymethyl)uridine (4)

To a stirred solution of sodium hydrogen carbonate (1 g) in water (20 ml) is added 3.03 g (first batch) of *2* in 1,4-dioxane (20 ml). The mixture is stirred for 60 min at room temperature, heated for 5 min at 60°, and evaporated to dryness under diminished pressure. The residue is extracted in a Soxhlet apparatus with dichloromethane (200 ml), the extract is evaporated, and methanolic ammonia (100 ml; presaturated at 0°) is added to the residue; the mixture is kept overnight and evaporated to dryness. The residue is dissolved in ethanol (10 ml), filtered (activated carbon), and the filtrate concentrated to ≃5 ml and kept for crystallization. Recrystallization from ethanol—ethyl acetate yields colorless crystals of *4*; yield 1.86 g, m.p. 178° to 179° (lit.[3] 176° to 182°); $\lambda_{max}^{H_2O\ (pH\ 7)}$ 264 nm ($\varepsilon_{\underline{mM}}$ 10.60), $\lambda_{min}^{H_2O\ (pH\ 7)}$ 232 nm ($\varepsilon_{\underline{mM}}$ 2.60); PMR data: δ 4.86 (1 H, t; the proton of the hydroxyl group of 5-CH_2OH of the pyrimidine is overlapped with that of the 5'-CH_2OH of the ribofuranosyl group).

A second batch of *2* (1.8 g) is hydrolyzed in the same way, and the residue is chromatographed on a column of silica gel (200 g; Merck, 0.2 to 0.5 mm) with ethyl acetate. Small

---

[e]Proton magnetic resonance spectra are recorded for solutions in Me_2SO-d_6, with tetramethylsilane as the internal standard.

quantities of *1* (R_F 0.41, TLC in ethyl acetate) and the 3',5'-diacetate of *4* (R_F 0.14, TLC in ethyl acetate) of m.p. 116° are separated. The chromatographically homogeneous fractions are combined, evaporated, the residue deacetylated, and the product treated as already described; yield 0.79 g of *4*; overall yield 2.65 g (67%).

*1-(3,5-Di-O-acetyl-2-deoxy-β-D-*erythro*-pentofuranosyl)-5-(dibromomethyl)uracil (3)*

3'-5'-Di-*O*-acetylthymidine (*1*) (5 g, 15.3 mmol) is dissolved as described for the preparation of *2*, and bromine (1.8 ml, 0.35 mmol) in carbon tetrachloride (20 ml) is blown by nitrogen into the refluxing, stirred solution for 3 to 4 hr.[b] When all of the bromine has been added the solution is bubbled in nitrogen for a further 30 min, the mixture is quickly filtered through ordinary filter-paper, and the filtrate is refrigerated for 24 hr. The resulting yellow, amorphous deposit is collected by filtration and successively washed with carbon tetrachloride (10 ml) and petroleum ether (20 ml); yield 3.15 g (first crop). The filtrate is evaporated to dryness under diminished pressure at 40°, affording an additional 2.45 g (second crop), which is contaminated by traces of *2* and the 3',5'-diacetate of 2'-deoxy-5-(tribromomethyl)uridine (R_F 0.70, TLC in ethyl acetate); total yield 5.6 g (75%) of *3*[c]; R_F 0.63 (TLC in ethyl acetate); $\lambda_{max}^{CH_2Cl_2}$ 285 nm (ε_{mM} 12.02), $\lambda_{min}^{CH_2Cl_2}$ 254 nm; PMR data[d]: δ 6.3 (1 H, s, 5-$CHBr_2$).

2'-Deoxy-5-formyluridine (5)

The first batch of *3* (3.15 g) is dissolved in an ice-cold mixture of pyridine (180 ml) and water (20 ml). The brown solution obtained is kept for 24 hr at 0° and then evaporated at 30°/1 torr. The residue is dissolved in ethyl acetate (20 ml), and the solution is passed through a column of silica gel (60 g; Merck, 0.2 to 0.5 mm), which is then eluted with ethyl acetate until UV absorption at 280 nm is no longer detectable, or until a spot of the eluate fails to give a positive (yellow color) (2,4-dinitrophenyl)hydrazone test. The eluate is evaporated to dryness under diminished pressure, the residue is deacetylated with methanolic ammonia (100 ml), and the solution is repeatedly evaporated with ethanol. The residue is taken up in ethanol (10 ml), and the suspension is filtered (activated carbon) and concentrated to ≃5 ml; compound *5* crystallizes on cooling, and one recrystallization from ethanol yields 1.53 g.

A second batch of *3* (2.45 g) is hydrolyzed in an ice-cold mixture of pyridine (135 ml) and water (15 ml) and then treated by the method just described. After deacetylation with methanolic ammonia (75 ml), and evaporation, the dark residue is purified by column chromatography. The chromatographic column (3 × 100 cm) is packed[6] with Celite 535 (200 g), and developed with the upper phase of 12:1:6 (v/v) ethyl acetate—isopropyl alcohol—water. The content of nucleoside *5* in the eluate may be determined by measuring the UV absorption at 280 nm or by TLC of 5 as its (2,4-dinitrophenyl)hydrazone (R_F 0.31, upper phase of 12:1:6 (v/v) ethyl acetate—isopropyl alcohol—water. The chromatographically homogeneous fractions are combined, and evaporated to dryness under diminished pressure; recrystallization of the residue yields a further 0.9 g, giving an overall yield of 2.43 g (62%); yellow crystals, m.p. 175° (lit.[4] 175° to 175.5°); $\lambda_{max}^{H_2O\ (pH\ 7)}$ 281 nm ($\varepsilon_{m\underline{M}}$ 13.20), 231 nm ($\varepsilon_{m\underline{M}}$ 10.30), $\lambda_{min}^{H_2O\ (pH\ 7)}$ 250 nm ($\varepsilon_{m\underline{M}}$ 3.05); PMR data[d]: δ 9.81 (1 H, s, 5-CHO).

REFERENCES

1. D. Bärwolff and D. Murawski, DDR Patent WP 113,361 (1973).

2. P. Langen and D. Bärwolff, *Acta Biol. Med. Ger.*, 34, K 7 - K 10 (1975); *Biochem. Pharmacol.*, 24, 1907 (1975); S. Waschke, J. Reefschläger, P. Langen, and D. Bärwolff, *Nature*, 255, 629 (1975).

3. M. Green, H. D. Barner, and S. S. Cohen, *J. Biol. Chem.*, 288, 621 (1957); R. E. Cline, R. M. Fink, and K. Fink, *J. Am. Chem. Soc.*, 81, 2521 (1959); S. Bricteux-Gregoire and W. G. Verly, *Bull. Soc. Chim. Belges*, 74, 232 (1965); R. Brossmer and E. Röhm, *Hoppe-Seyler's Z. Physiol. Chem.*, 348, 1431 (1967); V. S. Gupta and G. L. Bubbar, *Can. J. Chem.*, 49, 719 (1971).

4. M. P. Mertes and M. T. Shipchandler, *J. Heterocycl. Chem.*, 8, 133 (1971).

5. A. M. Michelson and A. R. Todd, *J. Chem. Soc.*, 816 (1955).

6. S. Zadrazil, in W. W. Zorbach and R. S. Tipson (eds.), *Synthetic Procedures in Nucleic Acid Chemistry*, Wiley—Interscience, New York, Vol. II, 1973, p. 588.

[64] 2'-DEOXY-5-(THIOCYANATO)URIDINE [1-(2-DEOXY-β-D-*ERYTHRO*-PENTOFURANOSYL)-5-(THIOCYANATO)URACIL]

Direct Thiocyanation of a Pyrimidine Nucleoside

Paul F. Torrence, James A. Waters, and Bernhard Witkop

Laboratory of Chemistry, National Institute of Arthritis, Metabolism, and Digestive Diseases, National Institutes of Health, Bethesda, Maryland 20014

INTRODUCTION

Although chlorothiocyanogen (*2*, ClSCN) has been known and used for some time in organic chemistry,[1] its application to the modification of nucleosides[2] resulted in a new class of "fraudulent" nucleosides,[3] and a facile route to 5-pyrimidinethiol nucleosides,[2,3] of interest as nucleic acid antagonists.[4] One member of this class of (thiocyanato)pyrimidine nucleosides, namely, 2'-deoxy-5-(thiocyanato)uridine (*3*), has significant antiviral activity.[3,5] It shows activity against both *Herpes simplex* and *Vaccinia* virus in cell culture[3,5] and against *Vaccinia* virus in mice[6]; in the last, it shows an efficiency greater than that of ribavirin (virazole; 1-β-D-ribofuranosyl-1,2,4-triazole-3-carboxamide) and comparable to that of 2'-deoxy-5-iodouridine.

PROCEDURE

Chlorothiocyanogen (2)

Potassium thiocyanate is dried *in vacuo* for 16 hr at 100°. Acetic acid (glacial) is usually dry enough as it comes from an unopened bottle. The critical factor in the synthesis is the accurate determination of the stoichiometric ratio of chlorine to potassium thiocyanate. Chlorine (dried by passage through a calcium chloride tube) is slowly bubbled into dry, glacial acetic acid under anhydrous conditions. After some time the flow of chlorine is stopped, and an aliquot of the solution[a] is placed in a flask containing an excess of potassium iodide in water. The iodine released is titrated with standard thiosulfate to the starch endpoint, and the chlorine content of the solution is then calculated. An aliquot of the solution is used to prepare *2* from potassium thiocyanate. After the required amount of potassium thiocyanate has been added to the solution a precipitate of potassium chloride forms. There is no need to remove this precipitate before addition of the nucleoside (it is, indeed, inadvisable, as filtration would have to be done under anhydrous conditions). The chlorine is allowed to react with the potassium thiocyanate for 15 to 30 min before the nucleoside is added.

---

[a]A solution of chlorine in acetic acid may be kept, and used, for at least one week if kept tightly stoppered.

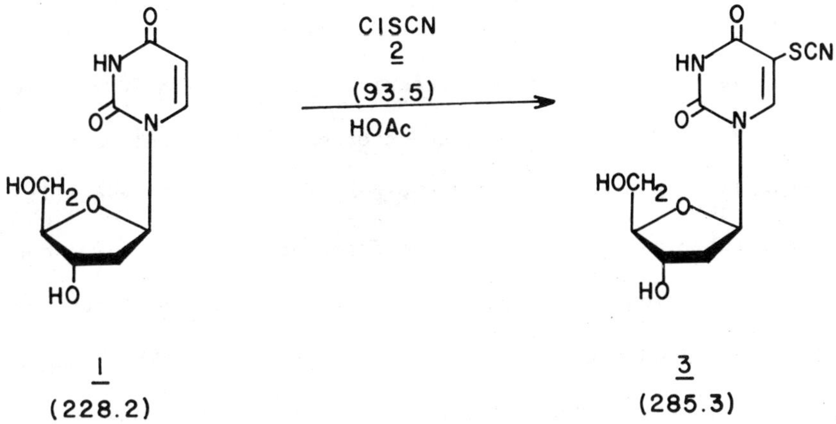

1-(2-*Deoxy*-β-D-erythro-*pentofuranosyl*)-5-(*thiocyanato*)*uracil* (3)

2'-Deoxyuridine dried for 1 hr at 100° *in vacuo* (*1*, 1.0 g, 4.2 mmol) is added to a solution of chlorothiocyanogen (*2*, from 3.55 g of chlorine and 5.35 g of potassium thiocyanate) in acetic acid (200 ml), and the resulting suspension (protected from moisture) is stirred at room temperature. The reaction is monitored by TLC on silica gel GF with chloroform—methanol. The product is clearly distinguishable from 2'-deoxyuridine as it is faster-moving. As soon as TLC shows complete disappearance of *1*, usually 20 to 30 min, the reaction is terminated by adding cyclohexene (10 ml) and stirring for 15 min; it should not be allowed to proceed longer, as additional products are then formed. Insoluble material is removed by filtration (sintered-glass funnel), and the filtrate is evaporated to dryness *in vacuo* at < 40°. Toluene is added to, and evaporated from, the residue to remove residual acetic acid, and the residue is applied to a column of silica gel (50 g) which is successively eluted with ethyl acetate (500 ml) and 20:1 ethyl acetate—acetone (or 20:1 chloroform—methanol). In many instances the product begins to crystallize in the tubes as they stand in the fraction collector. The content of fractions is analyzed by TLC and appropriate fractions are combined and evaporated; yield, 680 mg (54%). The residue is recrystallized from ethanol to give analytically pure material, m.p. 183° to 184.5°; λ_{max} 2160 cm^{-1}, UV $\lambda_{max}^{CH_3OH}$ 274 nm. When I is dissolved in 0.1<u>M</u> EDTA, pH 7.8 and dithiothreitol (DTT, Cleland's reagent) (10^{-3}<u>M</u>) is added, the UV λ_{max} shifts to 335 nm characteristic of the thiolate anion.

REFERENCES

1. A. B. Angus and R. G. R. Bacon, *J. Chem. Soc.*, 774 to 778 (1958).

2. T. Nagamachi, P. F. Torrence, J. A. Waters, and B. Witkop, *J. Chem. Soc. Chem. Commun.*, 1025 to 1026 (1972).

3. T. Nagamachi, J.-L. Fourrey, P. F. Torrence, J. A. Waters, and B. Witkop, *J. Med. Chem.*, <u>17</u>, 403 to 406 (1974).

4. T. I. Kalman and T. J. Bardos, *Mol. Pharmacol.*, <u>6</u>, 621 to 630 (1970).

5. E. De Clercq, P. F. Torrence, J. A. Waters, and B. Witkop, *Biochem. Pharmacol.*, <u>24</u>, 2171—2175 (1975).

6. E. De Clercq, M. Luczak, D. Shugar, P. F. Torrence, J. A. Waters, and B. Witkop, *Proc. Soc. Exptl. Biol. Med.*, <u>151</u>, 487—490 (1975).

[65] 3',5'-DI-*O*-ACETYLURIDINE

Monoacylation of the cis-Diol Grouping of D-Ribonucleosides via Orthoester Exchange

Geoffrey R. Owen and Colin B. Reese

University Chemical Laboratory, Lensfield Road, Cambridge, England

INTRODUCTION

3',5'-Di-*O*-acyl derivatives of D-ribonucleosides (e.g., *3*) are important intermediates in the preparation of one of the principal types of building block needed for the synthesis of oligoribonucleotides[1]; they are also potentially useful intermediates in other specific transformations involving ribonucleosides. Direct acylation of ribonucleoside 5'-esters leads to low yields of diesters,[2] and direct acylation of ribonucleosides themselves would be expected to lead to even more complex mixtures of products. However, acid-catalyzed exchange between a ribonucleoside and a small excess of trimethyl orthoacetate leads[3] to a 2',3'-*O*-methoxyethylidene derivative (e.g., *2*) in high yield. The latter may be converted into a mixture of 2'- and 3'-acetates by mild hydrolysis with acid,[3] or into a mixture of 2',5'- and 3',5'-diacetates by acetylation followed by mild hydrolysis with acid.

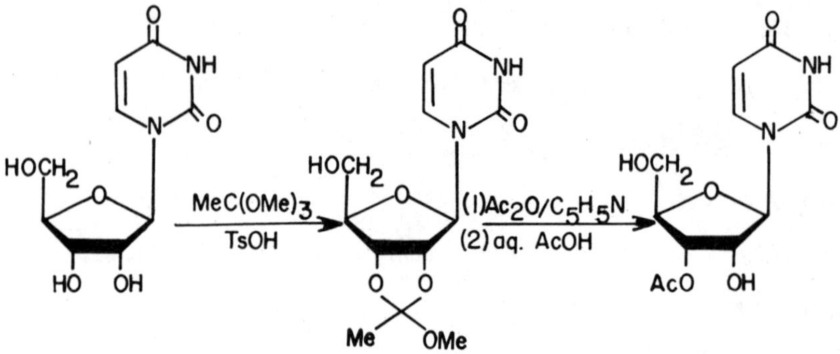

1 (244.2) 2 (300.3) 3 (328.3)

Sometimes (as with *3* and its adenosine analog), the mixture of the diacetates may be crystallized to give the pure 3',5'-diacetate in high yield.[3] The structures of the latter compounds may be established by ^{1}H NMR spectroscopy.[4]

PROCEDURE

3',5'-Di-O-acetyluridine (3)

Uridine (*1*; 50.0 g, 205 mmol), trimethyl orthoacetate[a] (34 ml, 267 mmol), *p*-toluenesulfonic acid, monohydrate[b] (1.55 g, 8.15 mmol) and anhydrous 1,4-dioxane[a] (300 ml) are stirred together[c] for 2 hr[d] at 20°. Anhydrous pyridine[a] (300 ml)

---

[a]Dried by heating under reflux with calcium hydride, and then distilling.

[b]99% pure material, purchased from the Aldrich Chemical Company, is suitable for this purpose.

[c]The reaction is conducted in a stoppered, 1-l, round-bottomed flask. The reactants are stirred magnetically.

followed by acetic anhydride[e] (100 ml, 1.06 mol) is then added. After 2 hr ethanol (100 ml) is added dropwise with stirring to the cooled mixture (ice-water bath), which is then kept for 2 hr and evaporated[f] under diminished pressure to a gum; this is dissolved in ethanol (100 ml), and the solution is reevaporated.[f] This process is repeated once more, the residual gum is dissolved in 9:1 (v/v) acetic acid—water (300 ml), and the solution is evaporated.[f] The residue is dissolved in ethanol (100 ml), and the solution reevaporated[f] under diminished pressure. The process is repeated twice more, the residue is dissolved in chloroform (400 ml), and the solution is extracted with 10% aqueous sodium hydrogen carbonate (300 ml). The layers are separated, and the aqueous layer is reextracted with chloroform (12 × 250 ml). The extracts are combined, dried (magnesium sulfate), and evaporated[f] under diminished pressure to a glass. The glass is dissolved in hot absolute ethanol[g] (400 ml) and cooled to afford crystals of *3*, yield (two crops), 60.5 g (90%), m.p. 152° to 154°.

---

[d]At this stage TLC [9:1 (v/v) chloroform—methanol] on plates coated with Merck Kieselgel GF_{254} reveals no uridine (*1*), and compound *2* as virtually the sole product.

[e]Laboratory-grade reagent may be used without redistillation.

[f]A rotary evaporator is used at water-pump pressure.

[g]The alcohol solution contains a mixture of 2',5'- and 3',5'-di-*O*-acetyluridines, from which the pure 3',5'-isomer (*3*) crystallizes. The mixed isomers in the mother liquors then reequilibrate, and more 3',5'-isomer is obtained. A period of 48 hr or more may therefore be needed to obtain a high yield of crystals of compound *3*.

REFERENCES

1. See C. B. Reese, *Colloq. Int. C.N.R.S.*, No. 182, 319 (1970).

2. D. M. Brown, G. D. Fasman, D. I. Magrath, and A. R. Todd, *J. Chem. Soc.*, 1448 (1954); D. M. Brown, Sir. A. Todd, and S. Varadarajan, ibid., 2388 (1956).

3. C. B. Reese and J. E. Sulston, *Proc. Chem. Soc.*, 214 (1964); H. P. M. Fromageot, B. E. Griffin, C. B. Reese, and J. E. Sulston, *Tetrahedron*, 23, 2315 (1967).

4. H. P. M. Fromageot, B. E. Griffin, C. B. Reese, J. E. Sulston, and D. R. Trentham, *Tetrahedron*, 22, 705 (1966).

[66] 5-FLUOROURIDINE

"Transnucleosidation" of Purine Nucleosides to Afford Pyrimidine Nucleosides

Jiří Beránek and Hubert Hřebabecký

Institute of Organic Chemistry and Biochemistry, Czechoslovak Academy of Sciences, 16610 Prague, Czechoslovakia

INTRODUCTION

The reaction of purine nucleosides with acyl halides results in the quantitative cleavage of the nucleoside bond[1] with the formation of the purine base and the acetylated glycosyl halide under relatively mild conditions, in comparison to other established methods.[2,3] The halide resulting from D-ribofuranosylpurines retains the *ribo* configuration of the glycosyl group of the starting nucleoside, and by subsequent reaction of this halide with a (trimethylsilyl)ated base affords[1] the tri-*O*-acetyl derivatives of 5-fluorouridine (*5*), 6-azauridine, and 5-methyluridine in yields of over 70%, referred to the starting material [guanosine (*3*)]. This "transnucleosidation" procedure is exemplified by the preparation of the biologically active[4] 5-fluorouridine (*6*). The present method seems to be more convenient than the classic D-ribosylation[5] of 5-fluorouracil, or the fluorination[6] of uridine.

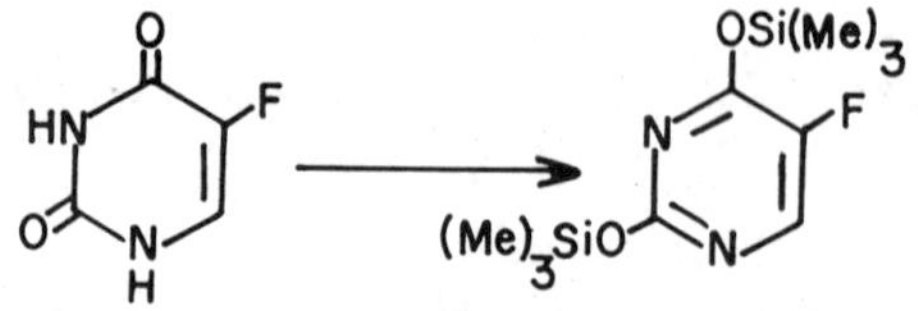

1 (130.1) 2 (274.4)

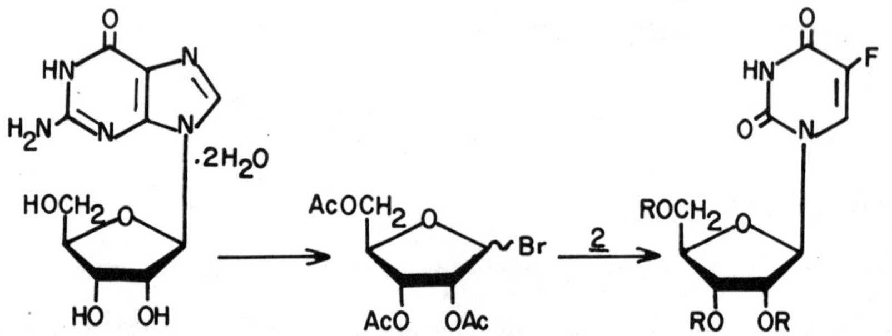

3 (319.3) 4 (338.1) 5, R = Ac (388.3) 6, R = H (262.6)

PROCEDURE

*5-Fluoro-2,4-*bis*(trimethylsilyloxy)pyrimidine* (2)

A mixture of 5-fluorouracil (*1*)[a] (2.60 g, 20 mmol), hexamethyldisilazane (60 ml), and ammonium sulfate (50 mg) is heated for 5 hr in an oil bath at 150°. The hexamethyldisilazane is then evaporated *in vacuo*[b] and the residue is coevaporated with toluene (50 ml). The residue, containing trimethylsilylated 5-fluorouracil (*2*),[c] is used without purification for the nucleosidation

[a]A product of F. Hoffmann-La Roche and Co.

[b]Solutions are evaporated in a rotary evaporator at the vacuum of a water pump (12 to 20 Torr).

[c]Bis-trimethylsilylated 5-fluorouracil is not stable under long storage.

reaction.

2',3',5'-Tri-O-acetyl-5-fluorouridine (5)

A mixture of guanosine dihydrate (*3*)[d] (6.38 g, 20 mmol), acetic acid (50 ml), and acetyl bromide (50 ml) is shaken until all of the guanosine dissolves (≈20 min); the solution is kept for 5 hr at room temperature and concentrated *in vacuo*[b] to ≈30 ml. Then toluene (70 ml) is added, and the precipitate of guanine hydrobromide is collected by filtration (sintered-glass funnel[e]) and washed with toluene (3 × 20 ml).[f] The filtrates are combined and evaporated *in vacuo*[b] and the residue is co-evaporated with toluene (3 × 40 ml).[g] The residue (containing compound *4*) is then dissolved[h] in acetonitrile (50 ml), and the solution is added to the bis-trimethylsilylated 5-fluorouracil (*2*), followed by 5.0 g

---

[d]Guanosine dihydrate is used as the cheapest commercially available purine nucleoside. The other purine nucleosides (adenosine, inosine) have been used in the reaction with the same result.

[e]The reaction mixture contains moisture-sensitive D-ribofuranosyl bromide. The filtration must be performed in dry apparatus (including the sintered-glass funnel), with exclusion of atmospheric moisture.

[f]The precipitate is guanine hydrobromide, and the solution contains the acetylated glycosyl halide. To avoid loss of the halide the precipitate should be thoroughly washed. Toluene is dried by distillation from sodium.

[g]The reaction mixture should be processed quickly. The residue (containing tri-*O*-acetyl-D-ribofuranosyl bromide) is used immediately in the nucleosidation reaction.

[h]Equimolar amounts of 5-fluorouracil (*1*) and guanosine dihydrate (*3*) are used in the preparation of the trimethylsilylated base *2* and tri-*O*-acetyl-D-ribofuranosyl bromide *4* as intermediates in the nucleosidation reaction.

of Molecular Sieve (3Å) and mercuric bromide (2 g). The mixture is stirred for 4 hr in a stoppered flask at room temperature, the molecular sieve is removed by filtration and washed with acetonitrile, and the filtrates are combined and evaporated to dryness *in vacuo*.[b] The residue is dissolved in chloroform (200 ml), and the solution is successively washed with 10% aqueous potassium iodide (3 × 30 ml) and water (2 × 30 ml), dried (anhydrous magnesium sulfate), and evaporated *in vacuo*.[b] The resiude is chromatographed on a column of silica gel (500 g, 30 to 60 μm) with 5:2 (v/v) benzene—acetone. Evaporation of fractions containing[i] compound 5 affords 5.86 g (75.5%) of 5 as a solid foam[j,k]; $[\alpha]_{\underline{D}}^{25}$ +20.7° (*c* 0.53, ethyl acetate); $\nu_{max}^{CHCl_3}$ 3384 (NH), 1752 (C=O, acetate), 1727 (sh) and 1711 (C=O, uracil), and 1676 cm^{-1} (C=C).

5-Fluorouridine (6)

A solution of compound 5 (3.88 g, 10 mmol) in 0.1$\underline{M}$ methanolic solution of sodium methoxide (100 ml) is kept for 2 hr at room temperature and rendered neutral with Dowex 50-W X-4 (H^+)[ℓ] resin. The resin is collected by filtration and washed with methanol,

---

[i] Checked by TLC on plates of silica gel.

[j] The yield from the nucleosidation reaction is readily reproducible, as also with other nucleosides.[1]

[k] The product is suitable for further synthetic use.

[ℓ] Neutralization is checked by spotting a sample of the solution on a moistened, universal pH-paper. The cation-exchange resin is washed with methanol (3 × 30 ml) prior to use; ≃10 ml of the resin is necessary.

and the filtrates are combined and evaporated to dryness *in vacuo*[b] to afford[k] 2.55 g (97.5%) of crystalline *6*, m.p. 179° to 183°. After recrystallization from ethanol, 1.60 g of 5-fluorouridine (*6*) is obtained, m.p.[m] 184° to 185°. The mother liquors afford an additional amount (0.51 g) of *6*, m.p. 183° to 185° (from ethanol); total yield, 80.4% of *6*.

REFERENCES

1. J. Beránek and H. Hřebabecký, *Nucleic Acids Res. Spec. Publ.* No. 1, s57 (1975).

2. H. K. Maurer and H. Witzel, German Patent 1,593,736 (1966).

3. M. J. Robins and R. K. Robins, *J. Am. Chem. Soc.*, 87, 4934 (1965).

4. A. Goldin, H. B. Wood, Jr., and R. R. Engle, *Cancer Chemother. Rep. Part 2*, 1, 1 (1968).

5. N. C. Yung, J. H. Burchenal, R. Fischer, R. Duschinsky, and J. J. Fox, *J. Am. Chem. Soc.*, 83, 4060 (1961).

6. D. Cech, H. Meinert, G. Etzold, and P. Langen, *J. Prakt. Chem.*, 315, 149 (1973).

---

[m]Lit.[5] m.p. 184° to 185°; lit.[6] m.p. 182°.

[67] 5-(METHOXYMETHYL), 5-(AZIDOMETHYL), AND 5-(2-CHLOROACETAMIDOMETHYL) DERIVATIVES OF 2'-DEOXYURIDINE

Ready Conversion of the Di-p-toluic Ester of 5-(Benzyloxymethyl)-2'-deoxyuridine into the 5-(Chloromethyl) Derivative

Arieh Kampf, Mohammed T. Shipchandler, Conrad Pillar, Raymond S. Wilson, and Mathias P. Mertes

Department of Medicinal Chemistry, University of Kansas, Lawrence Kansas 66045

INTRODUCTION

The synthesis of thymidine derivatives substituted on the methyl group is conveniently approached through the reactive intermediate 5-(chloromethyl)-2'-deoxy-3',5'-di-*O*-*p*-toluoyluridine[1] (*2*), prepared from 5-(benzyloxymethyl)-2'-deoxy-3',5'-di-*O*-*p*-toluoyluridine (*1*).

PROCEDURE

5-(Chloromethyl)-2'-deoxy-3',5'-di-O-p-toluoyluridine (2)

A mixture of 5-(benzyloxymethyl)-2'-deoxy-3',5'-di-*O*-*p*-toluoyluridine[2] (*1*) (0.350 g, 0.6 mmol) and dry 1,4-dioxane (15 ml) is stirred at room temperature, and hydrogen chloride is passed (through a sulfuric acid trap) into the solution for 2 hr. The flask is then fitted with a drying tube (calcium chloride), and

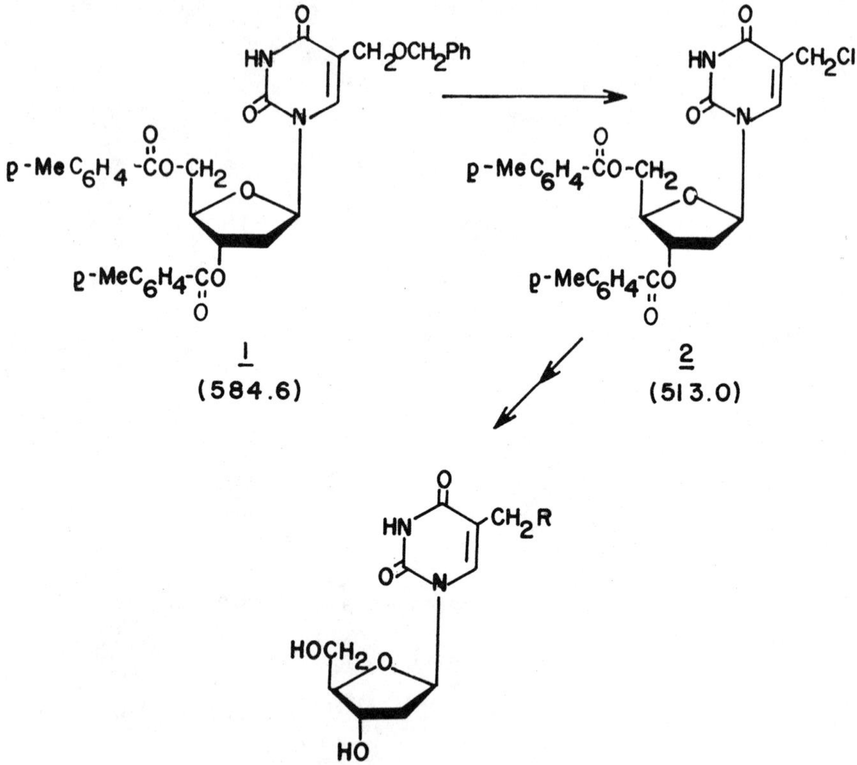

1 (584.6)

2 (513.0)

3a R = OMe (272.0)

3b R = N_3 (283.0)

3c R = $NHCOCH_2Cl$ (333.0)

the mixture is stirred overnight at room temperature, frozen in dry ice, and lyophilized. The residue is washed with hot petroleum ether (b.p. 30° to 60°) (50 ml), giving 0.298 g (97%) of *2* as a white solid, m.p. 202° to 204° (lit.[1] m.p. 206° to 207°).

2'-Deoxy-5-(methoxymethyl)uridine (3a)

A mixture of the chloromethyl derivative *2* (1.0 g, 2 mmol) with dry methanol (75 ml) is refluxed for 2 hr and cooled overnight, giving 0.8 g (81%) of crystalline 2'-deoxy-5-(methoxymethyl)-3',5'-di-*O*-*p*-toluoyluridine, m.p. 175° to 176° (lit.[3] m.p. 170° to 172°).

A mixture of this diester (520 mg, 1 mmol) with potassium carbonate (400 mg) in absolute methanol (75 ml) is stirred overnight. The resulting clear solution is rendered neutral with Amberlite IR-120 (H^+) ion-exchange resin and evaporated, and the residue is partitioned between benzene and water. The aqueous layer is evaporated, and the residue is crystallized from butanone—Skellysolve B, to give 210 mg (75%) of *3a*, m.p. 124° to 126° after two recrystallizations (lit.[3] m.p. 120° to 125°); λ_{max} in nm ($\varepsilon_{\underline{mM}}$), in ($H_2O$), 261 (11.0), λ_{min} 230 (2.5) and NMR data (D_2O): 3.5 (s, 3, OMe) and 4.3 (s, 2, CH_2-OMe).

5-(Azidomethyl)-2'-deoxyuridine (3b)

A solution of compound *2* (3.5 g, 8.7 mmol) and sodium azide (4 g, 60 mmol) in dry acetone (200 ml) is boiled under reflux for 2 hr and evaporated, and the residue is triturated with water and the suspension filtered. The solid product is 5-(azidomethyl)-2'-deoxy-3',5'-di-*O*-*p*-toluoyluridine; it is purified by recrystallization from ethanol to give 3 g (84%) of material, m.p. 155° to 157° (lit.[1] m.p. 156° to 157°).

This azidomethyl diester (2.94 g, 5.6 mmol) and potassium

carbonate (2.0 g) are stirred in absolute methanol (200 ml) for 1 hr and the mixture is processed as described for compound *2a*. The product is crystallized from butanone to give 870 mg of *3a*, m.p. 130° to 132°. After two recrystallizations from acetone—ethyl acetate, m.p. 133° to 135°; λ_{max} in nm (ε_{mM}), in H_2O, 262 (13.0), λ_{min} 231 (4.0); in <u>M</u> hydrochloric acid, λ_{max} 262 (13.0), λ_{min} 231 (4.2); in <u>M</u> sodium hydroxide, λ_{max} 261 (10.0), λ_{min} 242 (7.8); λ_{max} 2100 cm^{-1} (N_3).

5-(*Chloroacetamidomethyl*)-*2'-deoxyuridine* (*3c*)

A solution of *2* (2 g, 3.9 mmol) in liquid ammonia (125 ml) is stirred in a dry ice—acetone bath for 3 hr and then allowed to evaporate at room temperature. Thin-layer chromatography with 94:5:1 ethyl acetate—methanol—concentrated ammonium hydroxide shows one major product (low R_F) and two minor products (high R_F). Separation is effected on a column of silica by using the eluant used for TLC. Fractions containing the product (low R_F) are combined and evaporated to give 1.2 g of 5-(aminomethyl)-1-(2-deoxy-3,5-di-*O*-*p*-toluoyl-β-<u>D</u>-*erythro*-pentofuranosyl)uracil (62%), m.p. 163° to 168°. After recrystallization from absolute ethanol, m.p. 176° to 178°; NMR data ($CDCl_3$): δ 3.36 (s, 2, -CH_2NH_2) and 5.20 (broad s, 2, -CH_2-NH_2, exchangeable with D).

To an ice-cold solution of this amine (1.98 g, 4 mmol) and triethylamine (1.2 ml, 8 mmol) in chloroform (20 ml) is added dropwise a solution of chloroacetyl chloride (0.7 ml, 8 mmol) in chloroform (10 ml) during 5 to 10 min. The mixture is stirred for 1 hr at room temperature, successively washed with water and saturated sodium hydrogen carbonate solution, dried, evaporated, and the residue crystallized from ethanol to give 1.8 g of 5-(chloroacetamidomethyl)-1-(2-deoxy-3,5-di-*O*-*p*-toluoyl-β-D-*erythro*-pentofuranosyl)uracil (79%), m.p. 112° to 117°. After two recrystallizations the m.p. is 115° to 118°.

A suspension of this diester of *3c* (570 mg, 1 mmol) and potassium carbonate (280 mg, 2 mmol) in absolute methanol is stirred for 80 min at room temperature, and the resulting, clear solution is made neutral with Amberlite IR-120 (H^+) ion-exchange resin, and evaporated. The residue is triturated with ether to give a quantitative yield (310 mg) of *3c*, m.p. 183° to 185°. The product can be recrystallized from ethanol, m.p. 194° to 195°; λ_{max} in nm (ε_{mM}), in H_2O, 263 (7.20); in M hydrochloric acid, 264 (8.0).

REFERENCES

1. R. Brossmer and E. Rohm, *Hoppe-Seyler's Z. Physiol. Chem.*, 348, 1431 (1967).

2. M. P. Mertes and M. T. Shipchandler, *J. Heterocycl. Chem.*, 8, 133 (1971).

3. G. L. Bubbar and V. S. Gupta, *Can. J. Chem.*, 48, 3147 (1970).

[68] 2',3'-*O*-(METHOXYMETHYLENE)URIDINE

Protection of the cis-Diol Grouping of D-Ribonucleosides as the Cyclic Orthoformate Ester

Laval Yau and Colin B. Reese

Department of Chemistry, King's College, Strand, London WC2R 2LS, England

INTRODUCTION

In oligoribonucleotide synthesis[1] an acid-labile group which can be removed under mild conditions is needed for the protection of the 2',3'-*cis*-diol grouping of the D-ribonucleosides. The conditions of acid hydrolysis required for the removal of the isopropylidene and benzylidene protecting groups commonly used[2] are sufficiently drastic to lead to some migration and cleavage of the internucleotide linkages.[3] Consequently, the methoxymethylene protecting group, which is more labile than the isopropylidene group by some two orders of magnitude, was introduced[4] for this purpose. The acid-catalyzed exchange between ribonucleosides and trimethyl orthoformate is a general reaction which usually leads[4] to 2',3'-*O*-methoxymethylene derivatives (e.g, *2*) in good yields.

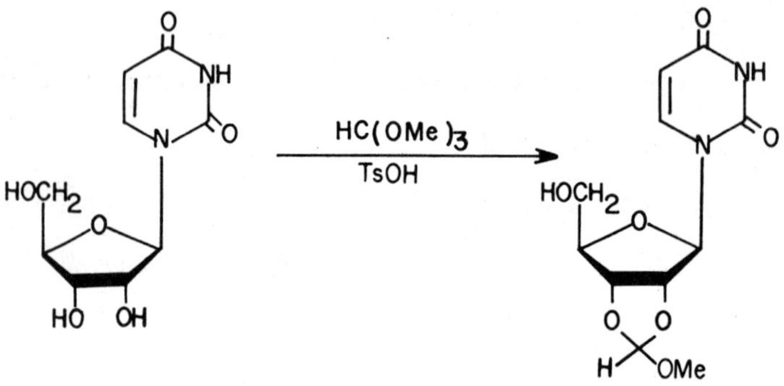

1
(244.2)

2
(286.2)

PROCEDURE

2',3'-O-(Methoxymethylene)uridine (2)

A suspension of uridine (*1*; 25.0 g, 102 mmol) and *p*-toluenesulfonic acid, monohydrate[a] (1.76 g, 9.25 mmol) in trimethyl orthoformate[b] (114 ml, 1.04 mol) is stirred at room temperature.[c] A homogeneous solution is obtained within 1 hr, but it then rapidly becomes turbid, and precipitation soon commences.[d] After a total reaction time of 16 hr, the solid is collected by filtration and washed with ice-cold, dilute, methanolic ammonia[e] (200 ml).

---

[a]Purchased from the Aldrich Chemical Company; 99% pure.

[b]Dried by heating under reflux with calcium hydride and then distilling.

[c]The reaction is conducted in a 250-ml, stoppered round-bottomed flask. The reactants are stirred magnetically.

[d]Precipitation may be facilitated by the addition of a nucleating crystal of 2',3'-*O*-(methoxymethylene)uridine (*2*).

[e]Prepared by adding one drop of concentrated aqueous ammonia (*d* 0.88) to methanol (200 ml).

Recrystallization of the residue from absolute ethanol (200 ml) gives compound[b] *2*, 25.1 g (86%), m.p. 189° to 190° (dec.).

REFERENCES

1. See C. B. Reese, *Colloq. Int. C.N.R.S.*, No. 182, 319 (1970).

2. C. B. Reese, in J. F. W. McOmie (ed.), *Prctective Groups in Organic Chemistry*, Plenum, London and New York, 1973, p. 124, ff.

3. B. E. Griffin, M. Jarman, and C. B. Reese, *Tetrahedron*, 24, 639 (1968).

4. M. Jarman and C. B. Reese, *Chem. Ind.* (Lond.), 1493 (1964); B. E. Griffin, M. Jarman, C. B. Reese, and J. E. Sulston, *Tetrahedron*, 23, 2301 (1967).

---

[b]From its ^{1}H NMR spectrum [$(Me_2SO\text{-}d_6)$: τ 3.90 (1 H, s, orthoformate H) and 6.80 (3 H, s, OMe)], this appears to be a pure diastereoisomer.

[69] *O*'-NITRO DERIVATIVES OF URIDINE

Preparation of Nitric Esters of Nucleosides

F. W. Lichtenthaler

Institut für Organische Chemie, Technische Hochschule Darmstadt, 61 Darmstadt, Germany

INTRODUCTION

Owing to the propitious fact that they possess a leaving-group capacity comparable to that of sulfonyloxy groups and yet are readily cleaved by catalytic hydrogenation, nitric ester groups on the glycosyl group of nucleosides are of considerable synthetic utility for transformations in the glycosyl group. Thus, a 5'-*O*-nitro group readily undergoes displacement reactions, affording, in the pyrimidine series, 5'-halo or 5'-azido derivatives on treatment with the respective nucleophiles,[1,2] while *N*^{3},5'-anhydronucleosides as their nitrates (salts) are obtained from adenosine and inosine 5'-nitrates.[3] Nitric esters of secondary hydroxyl groups are less readily displaced, but subjection of per-*O*-nitrated uridine or cytidine to mild alkaline conditions readily yields the 2,2'-anhydronucleosides, which on cleavage and subsequent catalytic *O*-denitration afford the corresponding arabinosylpyrimidines.[3,4] As a group for *O*-protection, the nitric ester combines susceptibility to facile removal by hydrogenation with

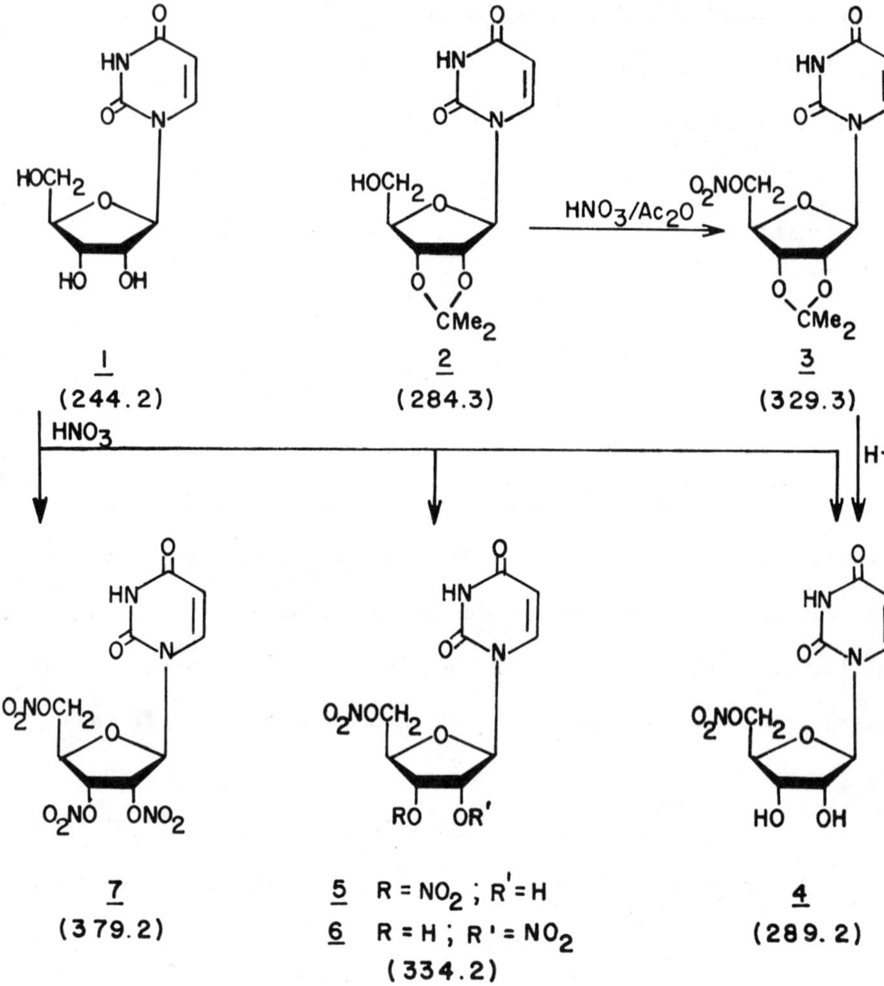

stability under a wide variety of conditions, including stability to mild alkali and dilute acid at room temperature. Thus, the standard procedures for the introduction and removal of acyl and acetal groups can be employed without affecting an *O*-nitro group.[3,4]

Regarding the preparation of nucleoside nitrates, suitable procedures for selective *O*-nitration and for per-*O*-nitration have been elaborated. Treatment of the unprotected ribonucleosides with fuming nitric acid (90% to 100%) at 0° invariably results in a mixture of di- and tri-nitrates, from which the latter may be obtained by repeated recrystallization in yields of up to 40%, or by chromatographic techniques.[1,3,5-7] At lower temperatures (-30° to -70°) the *O*-nitration can be directed toward partially nitrated products; however, their distribution is very sensitive to the concentration of the nitric acid employed. With 100% nitric acid at -70° the ribonucleosides afford mixtures of di- and tri-nitrates, from which the major products (i.e., the 3',5'-di-*O*-nitro derivatives) can be isolated in useful yields by fractional recrystallization, while separation of the minor components, the 2',5'-di-*O*-nitro- and tri-*O*-nitro-nucleosides, requires chromatography.[3,5] This procedure is exemplified here for uridine.

On treatment with 90% to 95% nitric acid at -70°, ribo-, deoxyribo-, and arabino-nucleosides afford 2:1 to 5:1 mixtures of the 5'-*O*-nitro and di-*O*-nitro compounds, from which the former may readily be isolated by preparative chromatography or, in some fortunate cases (e.g., *4*, illustrated), by fractional recrystallization, in yields of up to 50%.

For the exclusive preparation of 5'-*O*-nitroribonucleosides an alternative procedure is available, involving *O*-nitration of the respective 2',3'-*O*-isopropylidene derivatives by acetic nitric anhydride, as generated in acetic anhydride—nitric acid mixtures. The most propitious conditions were found to be 10:1 acetic anhydride—60% nitric acid during 15 min at -30°, affording the 2',3'-*O*-isopropylidene-5'-*O*-nitro derivatives of uridine (*3*), cytidine, adenosine, and inosine in good yield,[2] the only other product detectable (by TLC) in the reaction mixture being a trace of the respective, 5'-*O*-acetylated compound. This procedure is here illustrated through the conversion *2*→*3*, from which 5'-*O*-nitrouridine is obtained by deacetalation (*3*→*4*).

PROCEDURE[a]

2',3'-O-Isopropylidene-5'-O-nitrouridine[2] (*3*)

To acetic anhydride (20 ml) is gradually added, with stirring, 60% nitric acid (2.5 ml, sp. gr. 1.37), while the temperature is kept at 15° to 20° by external cooling. With continued vigorous stirring the mixture is cooled to 0°, and 2',3'-*O*-isopropylideneuridine[8] (*2*) (0.57 g, 2 mmol) is added. After 15 min at -5° to

---

[a]Thin-layer chromatography is recommended for ascertaining the purity of the products. Approximate R_F values on layers (0.2 mm) of Merck silica gel F_{254} with 10:5:1 chloroform—ethyl acetate—methanol; 0.04 (*1*), 0.12 (*4*), 0.40 (*5*), 0.48 (*6*), and 0.71 (*7*); with 6:1 chloroform-methanol; 0.55 (*2*), 0.78 (*3*), and 0.28 (*4*).

0°, the slurry formed is poured into ice-water (50 ml), ethyl acetate (100 ml) is added, and the mixture is rendered neutral by portionwise addition of solid sodium hydrogen carbonate until carbon dioxide formation ceases. The aqueous phase is reextracted with ethyl acetate (50 ml), and the extracts are combined, dried (magnesium sulfate), and evaporated *in vacuo* to a solid residue (0.61 g), which is recrystallized from methanol—water; yield 0.56 g (85%); m.p. 182° to 183°, $[\alpha]_D^{25}$ +24° (*c* 1, *N,N*-dimethylformamide).

5-O-Nitrouridine (4)

By Deacetonation[3] *of 3.* A suspension of *3* (330 mg, 1 mmol) in 5:1 trifluoroacetic acid—water (4 ml) is stirred for 1 hr at room temperature, resulting in a clear solution that is evaporated to dryness *in vacuo* (1 torr; bath temp. < 30°). Several additions and reevaporations of ethanol afford a solid residue that is recrystallized from ethanol; yield 235 mg (81%) as colorless needles; m.p. 139°, $[\alpha]_D^{25}$ +16° (*c* 1, *N,N*-dimethylformamide).

By O-Nitration[5] *of Uridine (1).* Nitric acid (90%, 7.5 ml, sp. gr. 1.48) is cooled in a dry ice—isopropyl alcohol bath to -70° with vigorous stirring; then uridine (*1*; 1.2 g, 4.9 mmol) is added in several portions, and the mixture, which soon becomes a clear solution, is stirred for 1 hr at -70°; 90% nitric acid (7.5 ml) is added, and stirring[b] continued for a further 30 min at -70°.

The mixture[c] is then stirred into ice-water (80 ml), covered with a layer of ethyl acetate (50 ml), and rendered neutral by portionwise addition of solid sodium hydrogen carbonate. The aqueous phase[d] is extracted with ethyl acetate (6 × 50 ml), and the extracts are combined, dried (magnesium sulfate), and evaporated. When the residual foam is taken up in hot ethanol, the suspension treated with charcoal and filtered, and the filtrate slowly cooled, *4* crystallizes as needles of m.p. 139°; yield 440 mg (31%).

3',5'-*Di*-O-*nitrouridine*[3] (*5*)

Uridine (*1*) (2.44 g, 10 mmol) is added portionwise with vigorous stirring to 100% nitric acid (25 ml; sp. gr. 1.52) precooled to -70°. The mixture is stirred for 30 min at -70°, poured into ice water (120 ml) covered with a layer of ethyl acetate (70 ml), and the mixture neutralized by portionwise addition of solid sodium hydrogen carbonate until formation of carbon dioxide ceases. The aqueous phase is separated, and extracted with ethyl acetate (50 ml), and the extracts are combined, dried (magnesium sulfate), and evaporated to dryness under reduced pressure. The solid

---

[b]Vigorous stirring is essential to prevent the mixture from solidifying.

[c]The reaction mixture contains *4* and the dinitrates 5 plus *6* in the ratio of ≃5:1 as well as 40% of unreacted *1*. Longer reaction times or higher temperatures (e.g., -30°), or both, increasingly shift the ratio in favor of the more highly nitrated products.

[d]Virtually all of the dinitrates and some of the mononitrates are contained in the respective ethyl acetate layer; by PLC separation,[e] a further 100 mg (7%) of *4* may be obtained.

residue, comprising an ≃3:1 mixture of di- and tri-nitrates, is recrystallized twice from methanol, to afford 700 mg (21%) of chromatographically[a] homogeneous *5* as crystals of m.p. 179° to 182°, $[\alpha]_{\mathrm{D}}^{25}$ +12° (*c* 1, *N,N*-dimethylformamide).

The combined methanolic mother-liquors are concentrated to ≃5 ml; the concentrate is depostied on two PLC plates[e] and developed with 10:5:1 chloroform—ethyl acetate—methanol. The zone corresponding to *5* ($R_F \simeq 0.41$) is removed, and thoroughly eluted with 1:1 acetone—methanol, and the eluate is evaporated to dryness. Recrystallization from methanol yields an additional 340 mg (10%) of *5*.

2',5'-Di-O-nitrouridine[3] *(6)*

Collection of the zone ($R_F \simeq 0.48$) that moves slightly faster than *5*, elution with acetone, and evaporation of the eluate to dryness, affords a syrup that crystallizes on trituration with little methanol; yield 390 mg (12%) of *6* as colorless crystals of m.p. 164° to 166°; $[\alpha]_{\mathrm{D}}^{25}$ +1.2° (*c* 0.5, *N,N*-dimethylformamide).

2',3',5'-Tri-O-nitrouridine (7)

The fastest-moving zone ($R_F \simeq 0.71$) from the PLC separation just described is collected, eluted with acetone, and evaporated to dryness, yielding *7* as a chromatographically[a] and analytically

---

[e]Preparative-layer chromatography (PLC) was performed on plates (20 × 40 cm) coated with 2-mm layers of Merck silica gel $PF_{254+366}$.

pure syrup (710 mg, 19%), not amenable to crystallization from the solvents tested; $[\alpha]_D^{25}$ 0° (*c* 1, *N*,*N*-dimethylformamide), +8° (*c* 1, methanol).

Treatment of uridine (*1*) with 100% nitric acid at -20° to -15° for 30 min yields a mixture containing *7* and the dinitrates *5* plus *6* in the approximate ratio of 2:1. Preparative-layer chromatography separation, as already described, affords syrupy *7* in 44% yield.

Denitration of O-Nitronucleosides

A solution of nitric ester in water (for 5'-nitrates) or 50% aqueous methanol is hydrogenated in the presence of 10% palladium-on-carbon (100 mg per 500 mg of educt) until uptake of hydrogen ceases (30 min to 2 hr). The solution is filtered to remove the catalyst, and the filtrate is evaporated to yield the parent nucleoside in essentially pure form.

REFERENCES

1. R. Duschinsky and U. Eppenberger, *Tetrahedron Lett.*, 5103 (1967).

2. F. W. Lichtenthaler and H. J. Müller, *Synthesis*, 199 (1974).

3. H. J. Müller, Dissertation, Technische Hochschule Darmstadt, Dec. 1975.

4. T. Kanai, C. Yamashita, and M. Ichino, Japan. Pat. 7,127,463 (1968); *C.A.*, 75, 130,077 (1971).

5. F. W. Lichtenthaler and H. J. Müller, *Angew. Chem.*, 85, 765 (1973); *Angew. Chem. Int. Ed. Engl.*, 12, 752 (1973).

6. W. Kampe, E. Fauland, M. Thiel, K. Dietmann and W. Juhran, Ger. Offen. 2,105,560 (1972); *C. A.*, 78, 4490 (1973).

7. R. Duschinsky, H. Walker, W. Wojnasowski, W. Arnold, G. Englert, and B. Pellmont, *J. Carbohydr. Nucleos. Nucleot.*, 1, 411 (1974).

8. R. S. Tipson, in W. W. Zorbach and R. S. Tipson (eds.), *Synthetic Procedures in Nucleic Acid Chemistry*, Wiley—Interscience, New York, Vol. I, 1968, p. 431.

[70] 2'-*O*-(*o*-NITROBENZYL)URIDINE

o-Nitrobenzylation via 2',3'-O-(Dibutylstannylene)uridine, and Deprotection by Photoirradiation

Eiko Ohtsuka, Shoji Tanaka, and Morio Ikehara

Faculty of Pharmaceutical Sciences, Osaka University, Suita, Osaka 565, Japan

INTRODUCTION

o-Nitrobenzyl derivatives have been used as photo-labile, protecting groups for carboxyl groups of amino acids,[1] and for hydroxyl groups[2] of glycosyl compounds, and phosphoric monoesters.[3] 2'-*O*-(*o*-Nitrobenzyl)uridine (*2*) was synthsized[4] by using 2',3'-*O*-(dibutylstannylene)uridine[5] (*1*) and *o*-nitrobenzyl bromide. Compound *2* was converted into the 5'-(monomethoxytrityl) ether, and this was used for the synthesis of dinucleoside monophosphates by condensing it with protected nucleoside 5'-monophosphates. The removal of the 2'-*O*-(*o*-nitrobenzyl) group, without damage to the pyrimidine ring,[4] was effected by irradiation with UV light of wavelength >280 nm. The 2'-(*o*-nitrobenzyl) ethers of adenosine[6] cytidine,[6] and N-isobutytylguanosine[7] can be prepared with *o*-nitrobenzyl bromide in the presence of sodium hydride.[6] Reactions using diazo-(*o*-nitrophenyl)methane to give the 2'-(*o*-nitrobenzyl) ethers

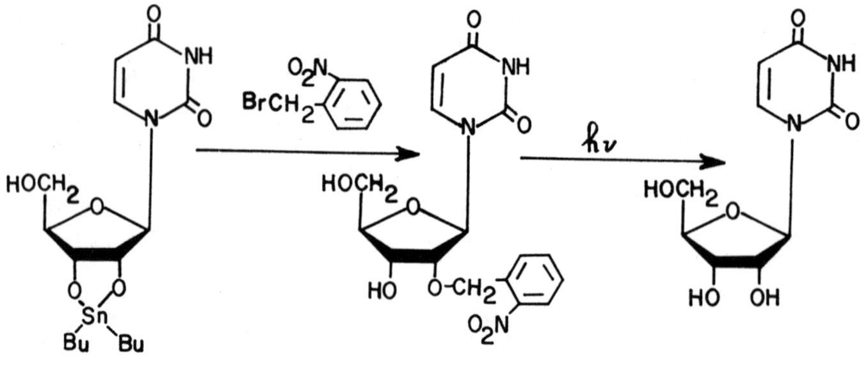

1 (475.1) 2 (379.3) 3 (244.2)

of adenosine, cytidine, and uridine have been reported.[8]

PROCEDURE

2',3'-O-(Dibutylstannylene)uridine (1)

A suspension of dry uridine[a] (2.44 g, 10 mmol) and dibutyltin oxide (2.5 g, 10 mmol) in methanol (500 ml) is heated until a clear solution is obtained (≃50 min). The product is crystallized by evaporating the methanol. The yield is 4.18 g (90%); m.p. 232° to 233° (lit.[5] m.p. 232° to 234°).

2'-O-(o-Nitrobenzyl)uridine (2)

A solution of compound *1* (4.75 g, 10 mmol) and *o*-nitrobenzyl bromide (4.32 g, 20 mmol) in *N,N*-dimethylformamide (60 ml) is heated for 3 to 4 hr at 110° until no increase in *2* is observed,[b] the

---

[a]Uridine is dried for 8 hr at 110° *in vacuo*.

[b]The increase of the product is checked by TLC on silica gel with 5:1 chloroform—ethanol. The R_F of *2* is 0.49.

solvent is removed *in vacuo*, the residue is vigorously stirred with water (50 ml) and ether (30 ml), and the ether layer is removed. The aqueous suspension is washed with ether (20 ml), mixed with water (100 ml), and heated. The aqueous solution is kept for 4 hr at room temperature and then overnight in a refrigerator. The precipitate (923 mg) is collected by filtration and the filtrate is evaporated to obtain a second crop. Recrystallization of the combined precipitates from ethanol (65 ml) gives pure[c] *2*; yield 892 mg (24%), m.p. 204° to 205°; $\lambda_{max}^{pH\ 1}$ 263 nm ($\varepsilon_{\underline{mM}}$ 13.50), $\lambda_{max}^{H_2O}$ 263 nm ($\varepsilon_{\underline{mM}}$ 13.70), $\lambda_{max}^{pH\ 14}$ 263 nm ($\varepsilon_{\underline{mM}}$ 11.30), λ_{max}^{EtOH} 263 ($\varepsilon_{\underline{mM}}$ 15.10); mass spectral data (70 eV): *m/e* 379 (M^+); NMR[d] (Me_2SO-d_6): δ 3.65 (m, 2 H, H-5',5'), 3.85 to 4.30 (m, 3 H, H-2', 3',4'), 4.93 and 5.02 (AB q, 2 H, J_{gem} 6 Hz), 5.08 (t, 1 H, 5'-OH), 5.27 (d, 1 H, 3'-OH, $J_{3'\text{-OH-H}}$ 6 Hz), 5.58 (q, 1 H, H-5, $J_{5,6}$ 8 Hz, $J_{NH\text{-}H\text{-}5}$ 1.5 Hz), 5.93 (d, 1 H, $J_{1',2'}$ 5 Hz, H-1'), 7.4 to 8.1 (m, 5 H, Ar-4H and H-6).

Photodeprotection of the 2'-O-(o-nitrobenzyl)uridine

A solution of compound *2* (10 mg) in 40% ethanol (3.5 ml) in a Pyrex tube (1.5 mm thick, 10 mm diameter) is irradiated (through a

---

[c]The location of the benzyl group on *O*-2' was proved by condensation of 5'-monomethoxytritylated *2* with *N,O*-protected pA and complete hydrolysis of the resulting dinucleoside phosphate with[4] ribonuclease A.

[d]Further evidence of the 2'-*O*-substitution is obtained by acetylation of *2* and decoupling[4] of the C-1'-H.

Pyrex filter inserted in a water jacket[e]) for 2 hr.

REFERENCES

1. J. A. Barltrop, P. J. Plant, and P. Shofield, *Chem. Commun.*, 822 (1966); A. Pachornik, B. Amit, and R. B. Woodward, *J. Am. Chem. Soc.*, 92, 633 (1970).

2. U. Zahevi, B. Amit, and P. Pachornik, *J. Org. Chem.*, 37, 2281 (1972); U. Zehavi and A. Pachornik, ibid., 37, 2285 (1972); B. Amit, U. Zehavi, and A. Pachornik, ibid., 39, 192 (1974); U. Zehavi and A. Pachornik, *J. Am. Chem. Sco.*, 95, 5673 (1973).

3. M. Rubinstein, B. Amit, and A. Pachornik, *Tetrahedron Lett.*, 1445 (1975).

4. E. Ohtsuka, S. Tanaka, and M. Ikehara, *Nucleic Acid Res.*, 1, 1351 (1974).

5. D. Wagner, J. P. H. Verheyden, and J. G. Moffatt, *J. Org. Chem.*, 39, 24 (1974).

6. E. Ohtsuka, S. Tanaka, and M. Ikehara, *Chem. Pharm. Bull.*, 25, 949 (1977).

7. E. Ohtsuka, S. Tanaka and M. Ikehara, *Synthesis*, 453 (1977).

8. D. G. Bartholomew and A. D. Broom, *Chem. Commun.*, 38 (1975).

---

[e]The photolysis apparatus has a 300-W, high-pressure mercury lamp (Eikosha Co. Model PIH 300) with a water-circulating quartz jacket.

[71] SYNTHESIS OF 3'-DEOXYURIDINE

Takayoshi Uematsu and Robert J. Suhadolnik

Department of Biochemistry, Temple University School of Medicine, Philadelphia, Pennsylvania 19140

The synthesis of 3'-deoxyuridine (*2*) has been accomplished by: (a) the condensation of 2,4-*bis*(trimethylsilyl)uracil methyl 2,5-*O*-dibenzoyl-3-deoxy-β-D-ribofuranoside (*1*) or (b) a transglycosylation reaction using N^6,N^6-dibenzoyl-2',3'-di-*O*-benzoyl-3'-deoxyadenosine (*3*). The syntheses are conducted in 1,2-dichloroethane in the presence of stannic chloride. The yields by methods (a) and (b) are 90% and 60%, respectively. Although chemical syntheses of 3'-deoxyuridine have been reported,[1-3] method (b) is especially useful, as cordycepin[a] is readily available and can serve as a convenient starting compound for the preparation of other important 3'-deoxynucleosides.

---

[a]Cordycepin (3'-deoxyadenosine) is isolated from the culture filtrates of *Cordyceps militaris* by adsorbing the filtrate onto a column of Amberlite XAD-4 ion-exchange resin. The column is washed with water until A_{260} is 0.2. The cordycepin is eluted with 50% aqueous methanol, the eluate is evaporated to dryness, the residue is dissolved in water, and the solution is added to a column of Dowex 1 (OH^-). The column is washed with water until A_{260} is 0.1. The cordycepin is eluted with 30% methanol, the eluate is evaporated to dryness, and the cordycepin is crystallized from water.

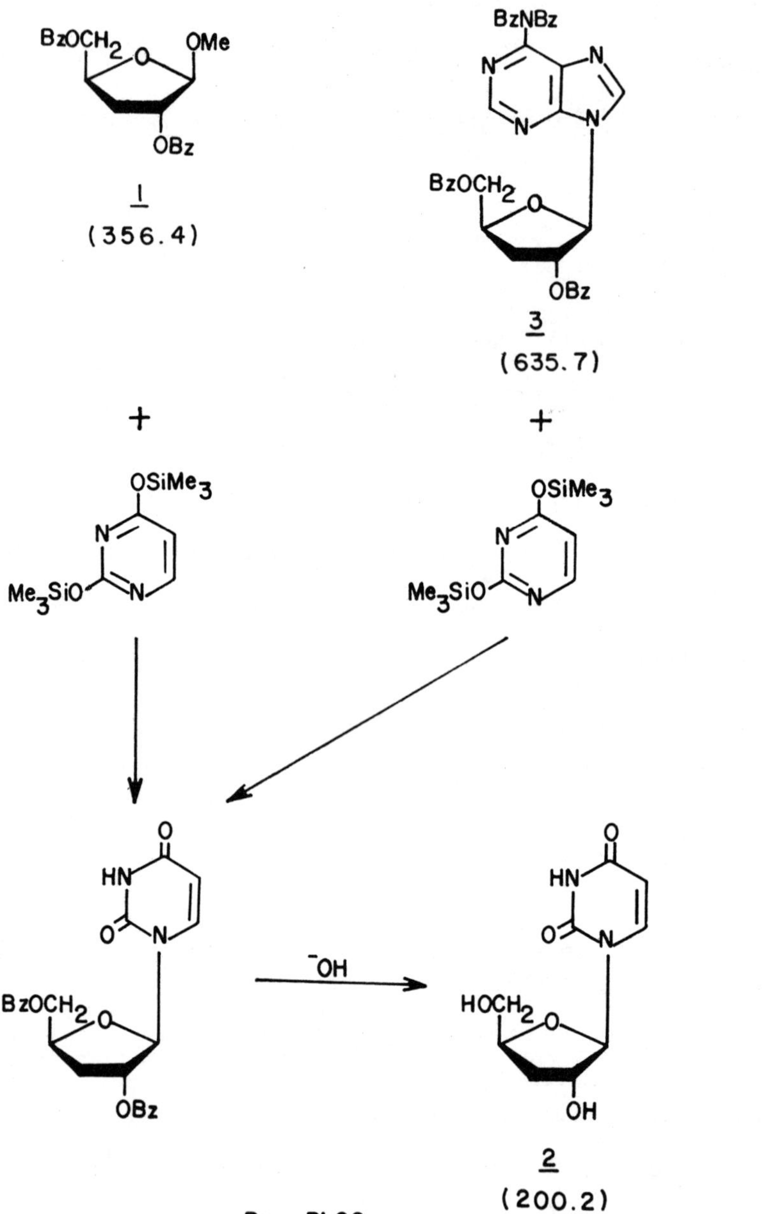
BzOCH2
OMe
OBz
1
(356.4)
BzNBz
N
BzOCH2
OBz
3
(635.7)
+
OSiMe3
Me3SiO
HN
O
-OH
HOCH2
OH
2
(200.2)

Bz = PhCO

PROCEDURE

Synthesis of 3'-Deoxyuridine (2)

Method A. From Methyl 2,5-O-Dibenzoyl-3-deoxy-β-D-ribofuranoside (1). To a solution of *1* (200 mg, 562 μmol) in dry 1,2-dichloroethane (3 ml) in a 10-ml, stoppered, round-bottomed flask (to exclude moisture) are added 2,4-(trimethylsilyl)-uracil (0.5 ml, 1.400 mmol) and stannic chloride (0.2 ml) with stirring. The slightly yellow solution is stirred overnight at room temperature. Thin-layer chromatography with 4:1 (v/v) benzene—methanol indicates the completion of the reaction by the absence of *1* (R_F 0.90), and the reaction is stopped by addition of methanol (0.5 ml). The suspension is stirred for 10 min at 0° and filtered through a coarse, sintered-glass filter, the solid being washed with 1,2-dichloroethane (30 ml). The filtrate and washings are combined and successively washed with cold, saturated, aqueous sodium hydrogen carbonate solution (2 × 15 ml) and water (2 × 15 ml), dried (anhydrous sodium sulfate), and evaporated to dryness *in vacuo*. A solution of the syrup in methanol (12 ml) and 2M sodium hydroxide (1 ml) in a 50-ml, round-bottomed flask is boiled under reflux for 3 hr, cooled, and evaporated to dryness *in vacuo*. The residue is dissolved in water (2 ml), and the solution is put on a column of AG-1 X-8 (formate) as the bottom layer (4 ml), and AG-50 X-8 (H^+) (4 ml, top layer) ion-exchange resins. The column is developed with water, 10-ml fractions being collected. Compound *2* is eluted in tubes 1 through 6; yield 0.51 mmol (91%, based on A_{260}). These fractions are combined and evaporated to dryness under reduced pressure. The residue is dissolved in 95% ethanol (1 ml) and crystallized to give *2*, m.p. 178° to 180° (lit.[1-3] m.p. 175.5° to 179°). Paper[b] (Whatman 3MM) and thin-layer (Avicel F[b] and Silica gel G[c]) chromatography show no contamination by UV absorbing material; Avicel F and paper chromatography used solvent. The UV spectra of *2* in acid, alkali, and water at pH 7 are identical with those of uridine, indicating that the product is an N^1-ribosyl compound; the NMR spectrum shows $J_{1,2}$ 1.6, so it is the β anomer (lit.[3] 1.8).

Method B. From N^6,N^6*-Dibenzoyl-9-(2,5-di-O-benzoyl-3-deoxy-β-D-ribofuranosyl)adenine (3).* Cordycepin (251 mg, 1.0 mmol) is suspended in dry pyridine (4 ml; dried over calcium hydride), and benzoyl chloride (0.8 ml) is added dropwise, with stirring, at 0°. The mixture is then heated with stirring for 2 hr at 55°, cooled to 0°, and poured onto crushed ice. The mixture is

---

[b]With 5:1:4 (v/v/v) 1-butanol—acetic acid—water.
[c]With 4:1 (v/v) benzene—methanol.

extracted with two 30-ml portions of chloroform, and the extracts are combined, washed successively with cold, saturated, aqueous sodium hydrogen carbonate solution (2 × 30 ml) and water (2 × 30 ml), dried (anhydrous sodium sulfate), and evaporated to dryness *in vacuo*. The syrup is dissolved in methanol (12 ml) and the solution cooled to give crystalline *3*; yield 546 mg (0.82 mmol; 82%), m.p. 179.5° to 180.5°. Crystalline *3* (500 mg) is recrystallized from 5:1 (v/v) ethanol—1,2-dichloroethane; yield 470 mg, m.p. 181°.

A solution of *3* (280 mg, 0.42 mmol) in dry 1,2-dichloroethane (3 ml) in a 10-ml round-bottomed flask is cooled in a dry ice—acetone bath. 2,4-Bis(trimethylsilyl)uracil (0.6 ml, 1.68 mmol) is added, followed by stannic chloride (250 µl). The mixture is allowed to warm to room temperature and is stirred for 20 hr. It is then cooled in a dry ice—acetone bath, and methanol (0.5 ml) is added. The mixture is allowed to thaw and is then filtered through a coarse, sintered-glass filter; the solid is washed with 1,2-dichloroethane (5 × 10 ml), and the filtrate and washings are combined, washed successively with cold, saturated, aqueous sodium hydrogen carbonate solution (2 × 25 ml) and water (2 × 25 ml), dried (anhydrous sodium sulfate), and evaporated *in vacuo*. The syrup is dissolved in methanol (17 ml) containg 2<u>M</u> sodium hydroxide (2 ml), and the mixture is boiled under reflux for 2.5 hr and evaporated to dryness *in vacuo*. The residue is dissolved in water (3 ml), and the solution is poured onto a column of anion—cation exchangers (see Method a). The column is washed with water, and 10-ml fractions are collected. The contents of tubes 2 through 4 are combined and evaporated to dryness; yield 0.28 mmol (67%, based on the value of A_{260}). Because a small amount of uracil is detected as a contaminant (by high-pressure liquid-chromatography), the residue is dissolved in methanol (5 ml), the solution is mixed with silicic acid (500 mg), and the methanol is removed by evaporation *in vacuo*. Chloroform is added, and the slurry is added to the top of a column (50 ml) of silicic acid packed with chloroform. The column is successively eluted with 200-ml portions of 3:17 (v/v) acetone—chloroform, 3:7 (v/v) acetone—chloroform, 1:1 (v/v) acetone—chloroform, and 3:7 (v/v) ethanol—chloroform, 20-ml fractions being collected. 3'-Deoxyuridine is eluted by the ethanol—chloroform; these eluates are combined and evaporated *in vacuo*. The syrup is dissolved in ethanol (1 ml), and then crystallizes; yield 0.25 mmol (60%). This preparation shows no contamination when examined by high-pressure liquid-chromatography.

REFERENCES

1. D. M. Brown, D. B. Parihar, A. R. Todd, and S. Varaharajan, *J. Chem. Soc.*, 3028 to 3035 (1958).

2. G. Kowollik and P. Langen, *Chem. Ber.*, 101, 235 to 239 (1968).

3. E. Walton, F. W. Holly, G. E. Boxer, and R. F. Nutt, *J. Org. Chem.*, 31, 1163 to 1169 (1966).

[72] 2- AND 4-SELENOURIDINE

Replacement of the 2- and 4-Oxo Groups of Uridine by a Seleno Group

Dean S. Wise and Leroy B. Townsend

Department of Biopharmaceutical Sciences and Department of Chemistry, University of Utah, Salt Lake City, Utah 84112

INTRODUCTION

Thiopyrimidine nucleosides are of considerable interest, primarily owing to their occurrence as constituents of certain transfer ribonucleic acids (t-RNA).[1] Two general procedures are described that illustrate the synthesis of selenopyrimidine nucleosides[2]: (a) direct D-ribosylation of preformed, selenopyrimidine heterocycles and (b) nucleophilic displacement on a reactive site of a preformed, pyrimidine nucleoside.

PROCEDURE

1-(2,3,5-Tri-O-benzoyl-β-D-ribofuranosyl)-4-pyrimidinone-2-selone (2',3',5'-Tri-O-benzoyl-2-selenouridine) (1)

2-Selenouracil[3] (1.75 g, 10 mmol), ammonium sulfate (5 mg), and 3 drops of chlorotrimethylsilane are added to hexamethyldisilazane (50 ml), and the mixture is boiled under reflux with the exclusion of moisture for 12 hr. The excess of hexamethyldisila-

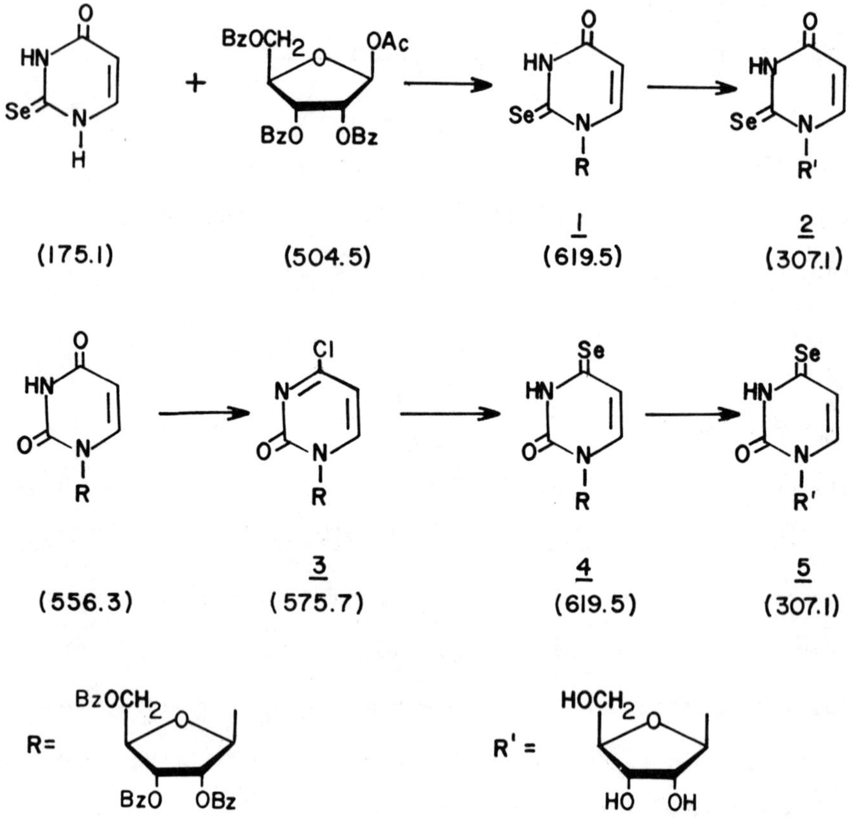

zane is removed from the clear, yellow solution by distillation under diminished pressure (water aspirator). Dry xylene (30 ml) is added to and evaporated from the residue to remove any residual hexamethyldisilazane. Dry 1,2-dichloroethane (50 ml) is added to the residue, followed by 1-*O*-acetyl-2,3,5-tri-*O*-benzoyl-β-D-ribofuranose (5 g, 9.9 mmol) and stannic chloride (3.72 g, 14.3 mmol). The reaction vessel is stoppered, the mixture is stirred for 18 hr, pyridine (3 ml) is added, and stirring is continued for 1 hr. The heavy precipitate is removed by suction filtration (Celite pad) and washed with chloroform (3 × 30 ml). The filtrate and washing are combined, successively washed with saturated sodium hydrogen carbonate solution (3 × 20 ml) and water (2 × 10 ml), dried (anhydrous magnesium sulfate) for 18 hr, and evaporated under diminished pressure to yield a yellow glass, which is triturated with methanol (10 ml) to afford 3.9 g of crystalline *1*. The solid is recrystallized from ethanol to yield 3.3 g of pure *1*, m.p. 94° to 96° (54%); λ_{max} in nm ($\varepsilon_{\underline{mM}}$), in MeOH 229 (52.7), and 311 (14.3); at pH 1, 248 (29.7) and 330 (16.1); at pH 11, 233 (52.0) and 286 (15.5).

2-Selenouridine (2)

To a suspension of *1* (200 mg, 0.3 mmol) in absolute methanol (5 ml) is added sufficient methanolic solution of sodium methoxide to bring the apparent pH to ≃9. The mixture is stirred at room temperature, the reaction being monitored by TLC on SilicAR 7GF with 1:10 methanol—chloroform until no starting material is observed.[a] Dowex 50W-X4 cation-exchange resin is added, the mixture is stirred for 30 min, and the resin is removed by filtration. The resin is washed with hot methanol (2 × 30 ml), SilicAR CC-7 (2 g) is added to the combined filtrate and washings, and

---

[a]The reaction time is generally 30 to 40 min. (Prolonged reaction-times increase the decomposition of 2-selenouridine to uridine).

the mixture is evaporated to dryness. This impregnated powder is applied to a column (2.5 × 8 cm)[b] of dry-packed SilicAR CC-7, which is first eluted with 20:1 chloroform—methanol (to remove methyl benzoate) and then with 10:1 chloroform—methanol. The fractions containing the product are slowly evaporated at room temperature to produce crystals (long needles) of *3*, 68 mg (67%); m.p. 197° to 198° (dec.); λ_{max} in nm ($\varepsilon_{m\underline{M}}$), in MeOH, 312 (8.6); at pH 11, 238 (13.2) and 291 (7.4); PMR data (Me_2SO-d_6): δ 8.12 (d, *J* 8 Hz, 1 H, H-6), 6.74 (d, *J* 3.5 Hz, 1 H, anomeric proton), and 6.22 (d, *J* 8 Hz, 1 H, H-5).

4-Chloro-1-(2,3,5-tri-O-benzoyl-β-D-ribofuranosyl)-2-pyrimidinone (3)

This material is obtained by a modification of the procedures given in Ref. 4. Freshly distilled thionyl chloride (1.19 g, 0.72 ml, 10 mmol) is added at room temperature to dry *N,N*-dimethylformamide (0.73 ml, 10 mmol) in a 100-ml, round-bottomed flask equipped with a reflux condenser and drying tube (all glassware dried for 3 hr at 120°). The mixture is stirred for 30 min at room temperature, 20 ml of dichloromethane (dried over phosphorus pentoxide is added, and the mixture is then stirred until it is homogeneous. 2',3',5'-Tri-*O*-benzoyluridine (5.5 g, 9.9 mmol) is added, and the solution is boiled under reflux for 14 hr.

---

[b]The use of a longer column length results in conversion (on the column) of 2-selenouridine into uridine.

A precipitate forms during the reaction. The reaction mixture is cooled for 1 hr at 0°, and the precipitate is collected by filtration and washed with cold dichloromethane (10 ml); yield 3.70 g (65%); m.p. 203° to 205°; TLC on SilicAR 7GF with 20:1 chloroform—methanol shows only one component.

1-(2,3,5-Tri-O-benzoyl-β-D-ribofuranosyl)-2-pyrimidinone-4-selone (4)

Absolute methanol (50 ml) (dried over magnesium) is added to a predried, 100-ml, three-necked flask equipped with a condenser (with drying tube), gas inlet, and stopper. Dry nitrogen is passed through the methanol for 30 min to purge the system of oxygen. Selenourea (4 g, 0.03 mmol) is then added, and the mixture is stirred until dissolution is complete. Compound *3* (1.47 g; 2.55 mmol) is added, and the mixture is boiled under reflux for 3 hr, with a continuous passage of nitrogen through the solution.[c] The mixture is allowed to cool to room temperature and filtered to remove selenium metal. The filtrate is evaporated *in vacuo* (< 35°), the residue is mixed with chloroform (30 ml), and the suspension is filtered[d] (the residue is unreacted selenourea). The filtrate and washings are combined and

---

[c]The solution turns red.
[d]This is repeated twice.

evaporated to dryness *in vacuo* (<35°), the residue is dissolved in absolute ethanol (20 ml), heptane (60 ml) is added, and the solution is kept for 18 hr at 10°. The crude product, 1.43 g (90%), is collected by filtration and recrystallized from the aforementioned solvent system, to afford a 78% yield of *4*; m.p. 120° to 121°; λ_{max} in nm ($\varepsilon_{m\underline{M}}$), in MeOH, 370 (30.9); at pH 1, 367 (28.5); and at pH 11, 339 (34.1).

4-Selenouridine (5)

To a solution of nucleoside *4* (0.7 g, 1.1 mmol) in dry methanol (40 ml) is added sufficient M sodium methoxide in methanol to bring the pH to an apparent value of ≃9 (universal pH paper).[e] The mixture is stirred for 1 hr at room temperature,[f] Dowex 50 (H^+) ion-exchange resin is added, and the suspension is stirred for 30 min. The resin is removed by filtration and washed with hot methanol (2 × 20 ml), and the combined filtrate and washings are evaporated to dryness under diminished pressure. The residue is dissolved in water (50 ml), and the solution is washed with chloroform (3 × 10 ml) to remove methyl benzoate and is lyophilized. The lyophilizate is continuously extracted in a Soxhlet apparatus with dichloromethane (250 ml) for 4 days to yield an

---

[e] Dissolution occurs very rapidly, to afford a solution having a yellow color with a red tinge.

[f] Thin-layer chromatography reveals no starting material after 45 min.

analytical sample of 5, yield 120 mg (35.6%); m.p. 150° to 151°; λ_{max} in nm (ε_{mM}), in MeOH, 263 (5.8) and 271 (21.5); at pH 1, 260 (4.6) and 368 (20.9); at pH 11, 237 (12.3) and 338 (16.3); pK_a 7.4 ± 0.1; PMR data (Me_2SO-d_6): δ 8.04 (d, *J* 7 Hz, 1 H, H-6), 6.74 (d, *J* 7 Hz, 1 H, H-5), and 5.82 (d, *J* 3.6 Hz, 1 H, anomeric proton).

REFERENCES

1. R. H. Hall, *The Modified Nucleosides in Nucleic Acids*, Columbia University Press, New York, 1971.

2. D. S. Wise and L. B. Townsend, *J. Heterocycl. Chem.*, 9, 1461 (1972).

3. H. G. Mautner, *J. Am. Chem. Soc.*, 78, 5292 (1956).

4. J. Zemlicka and F. Sorm, *Collect. Czech. Chem. Commun.*, 30, 2052 (1965); M. J. Robins and S. R. Naik, *Biochemistry*, 10, 3591 (1971).

[73] 1-β-D-ARABINOFURANOSYL-4-SELENOURACIL, 1-β-D-ARABINOFURANOSYL-2-SELENOURACIL, 4-SELENOURIDINE, 4-SELENOURACIL, 2-SELENOURIDINE, AND 2'-DEOXY-4-SELENOURIDINE

Replacement of the Amino Group of Pyrimidine Nucleosides

Chyng-Yann Shiue and Shih-Hsi Chu

Division of Biological and Medical Sciences, Brown University, Providence, Rhode Island 02912

INTRODUCTION

2- and 4-Thiouridine have been characterized as minor nucleoside components of transfer ribonucleic acid.[1,2] The preparation of these thiopyrimidine nucleosides, or the seleno analogs, involved multistip syntheses.[3] Ueda and co-workers[4] reported a facile method for synthesizing some 4-thiopyrimidine derivatives. The seleno analogs can be prepared similarly, as described here.[5]

PROCEDURE

1-β-D-Arabinofuranosyl-4-selenouracil[5] (5)

A solution of 1-β-D-arabinofuranosylcytosine (550 mg, 2 mmol) in pyridine (4 ml), 2-methoxyethanol (15 ml), and an excess of hydrogen selenide is kept overnight at room temperature and evaporated to dryness. The residue is dissolved in water, the solution is passed through a column of Dowex 50 (H^+), and the column is

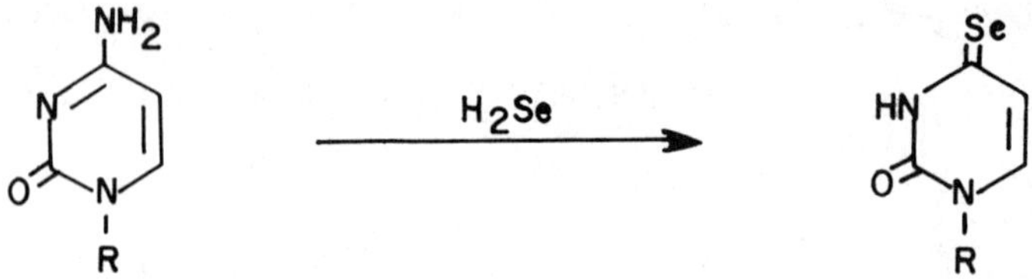

1 (243.2) 4 (311.7)

2 (243.2) 5 (311.7)

3 (227.2) 6 (295.7)

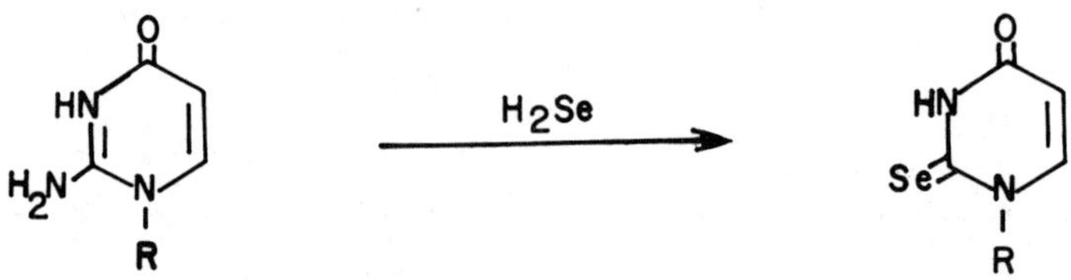

7 (243.2) 9 (311.7)

8 (243.2) 10 (311.7)

R=D-ribofuranosyl, D-arabinofuranosyl, or 2'-D-deoxyerythro-pentofuranosyl
(243.2) (311.7) (243.2) (311.7) (227.2) (295.7)

eluted with water. Evaporation of the eluate gives 130 mg (20%) of 5, m.p. 130° (dec.); λ_{max} in nm ($\varepsilon_{m\underline{M}}$), in H_2O 261 at (5.13) and 3.66 (16.48); at pH 1, 262 (3.798) and 366 (18.355); at pH 11, 240 (7.095) and 340 (14.915); pK_a 7.39; NMR data[a]: δ 7.69 (d, *J* 7 Hz, 1 H, H-6), 6.62 (d, *J* 7 Hz, 1 H, H-5), and 5.95 (d,

---

[a]Proton magnetic resonance spectra were recorded with a Varian A-60A spectrometer for solutions in Me_2SO-d_6 with Me_4Si as the internal standard.

J 4 Hz, 1 H, H-1').

1-β-D-Arabinofuranosyl-2-selenouracil[5] *(10)*

A mixture of 1-β-D-arabinofuranosylisocytosine[6] (170 mg, 0.7 mmol), pyridine (10 ml), *N,N*-dimethylformamide (10 ml), and hydrogen selenide (0.4 ml) is stirred for 4 days at room temperature, and evaporated to dryness. The residue is taken up in water, and the suspension is filtered. The filtrate is passed through a column of Dowex 50 (H^+), and the column is eluted with water. Evaporation of the appropriate fractions and drying of the residue *in vacuo* gives 60 mg (26%) of *10*, m.p. 130° (dec.); λ_{max} in nm ($\varepsilon_{m\underline{M}}$), in H_2O, 223 (14.31) and 308 (9.756); at pH 1, 224 (11.488) and 308 (7.599); at pH 11, 241 (13.871) and 292 (7.316); pK_a 7.53; NMR data[a]: δ 7.78 (d, *J* 7 Hz, 1 H, H-6), 6.08 (d, *J* 7 Hz, 1 H, H-5), and 6.88 (d, *J* 4 Hz, 1 H, H-1').

4-Selenouridine[5,7] *(4)*

A solution of cytidine (450 mg, 1.85 mmol) in *N,N*-dimethylformamide (20 ml), pyridine (20 ml), and hydrogen selenide (0.7 ml) is kept for 3 days at room temperature, and evaporated to dryness. The residue is dissolved in water, and the solution is passed through a column of Dowex 50 (H^+), which is then eluted with water. The eluate is evaporated to dryness, the residue is taken up in methanol, the suspension is filtered, and the filtrate is evaporated to dryness, and the residue dried *in vacuo* to give

252 mg (42%) of *3*. An analytical sample is obtained by purification on Avicel plates (1.000 mm), developed with water. The bands containing the compound are scraped off, extracted with water, and the extracts filtered. The filtrate is evaporated to dryness, and the residue is co-evaporated with ethanol, and dried *in vacuo*; m.p. 105° (dec.) [lit.[7] m.p. 150° to 151° (anhydrous)]; λ_{max} in nm ($\varepsilon_{\underline{mM}}$), in H_2O, 260 (4.532) and 366 (16.332); at pH 1, 260 (4.32) and 366 (16.36); at pH 11, 240 (7.40) and 340 (13.568); pK_a 7.60; NMR data[a]: δ 7.93 (d, *J* 7 Hz, 1 H, H-6), 6.63 (d, *J* 7 Hz, 1 H, H-5), and 5.72 (d, *J* 4 Hz, 1 H, H-1').

4-Selenouracil[5]

A solution of cytosine (300 mg, 2.8 mmol) in pyridine (10 ml), water (25 ml), and hydrogen selenide (0.7 ml) is kept for 5 days at 70° and evaporated to dryness. The residue is taken up in water and the suspension filtered. The filtrate is passed through a column of Dowex 50 (H^+), which is eluted with water. The eluate is evaporated, and the residue is dried *in vacuo* at 100° to give 50 mg (12%) of product, m.p. 235° (dec.); λ_{max} nm ($\varepsilon_{\underline{mM}}$), H_2O, 260 (7.069) and 368 (18.625); at pH 1, 260 (6.763) and 368 (16.708); at pH 11, 235 (7.722) and 348 (14.415), NMR data[a]: δ 7.41 (d, *J* 7 Hz, 1 H, H-6) and 6.49 (d, *J* 7 Hz, 1 H, H-5).

2-Selenouridine[5,7] (*7*)

A solution of isocytidine (450 mg, 1.85 mmol) in *N*,*N*-dimethylformamide (10 ml), pyridine (10 ml), and hydrogen selenide (0.6 ml) is kept for 3 days at room temperature, and evaporated to dryness. The residue is dissolved in water, and the solution is passed through a column of Dowex 50 (H^+), which is then eluted with water. Evaporation of the appropriate fractions gives 270 mg (47%) of *7*; this is suspended in a small volume of methanol, the suspension is filtered, and the solid is dried *in vacuo* to give an analytical sample, m.p. 170° (dec.) [lit.[7] m.p. 197° (dec.) (anhydrous)]; λ_{max} in nm ($\varepsilon_{\underline{mM}}$), in H_2O, 225 (16.150) and 308 (13.65); at pH 1, 223 (17.246) and 308 (14.856); at pH 11, 240 (13.988) and 293 (9.294); pK_a 7.57; NMR data[a]: δ 8.18 (d, *J* 7 Hz, 1 H, H-6), 6.17 (d, *J* 7 Hz, 1 H, H-5), and 6.72 (d, *J* 4 Hz, 1 H, H-1').

2'-Deoxy-4-selenouridine[5] *(6)*

A solution of 2'-deoxycytidine (300 mg, 1.32 mmol) in *N,N*-dimethylformamide (10 ml), pyridine (5 ml), and an excess of hydrogen selenide is kept for 2 days at room temperature and evaporated to dryness. The residue is dissolved in water and the solution passed through a column of Dowex 50 (H^+), which is eluted with water. Evaporation of the eluate gives 120 mg (31%) of *6*, m.p. 135° (dec.) λ_{max} in nm ($\varepsilon_{\underline{mM}}$), in H_2O, 260 (3.142) and 368 (11.783); at pH 1, 260 (2.686) and 368 (10.663); at pH 11, 239 (5.854) and 340 (8.960); NMR data[a]: δ 7.87 (d, *J* 7 Hz, 1 H, H-6),

6.58 (d, *J* 7 Hz, 1 H, H-5), and 6.02 (d, *J* 6 Hz, 1 H, H-1').

REFERENCES

1. J. A. Carbon, L. Hung, and D. S. Jones, *Proc. Natl. Acad. Sci. U.S.A.*, 53, 979 (1965).

2. M. N. Lipsett, *J. Biol. Chem.*, 240, 3975 (1965).

3. L. B. Townsend and C. C. Cheng, in A. C. Sartorelli and D. G. Johns (eds.), *Rational Design of Cytotoxic Agents: Pyrimidine Nucleoside Analogs, Handbook of Experimental Pharmacology*, Springer-Verlag, West Berlin, 1973.

4. T. Ueda, M. Imazawa, K. Miura, R. Iwata, and K. Odajima, *Tetrahedron Lett.*, 2507 (1971).

5. C. Y. Shiue and S. H. Chu, *J. Org. Chem.*, 40, 2971 (1975).

6. (a) D. M. Brown, A. R. Todd, and S. Varadarajan, *J. Chem. Soc.*, 2388 (1956); (b) I. L. Doerr and J. J. Fox, *J. Org. Chem.*, 32, 1462 (1967).

7. D. S. Wise and L. B. Townsend, *J. Heterocycl. Chem.*, 9, 1461 (1972).

[74] 2'-*O*-(TETRAHYDROPYRAN-2-YL)URIDINE

An Intermediate in the Synthesis of an Internucleotide Bond

J. Smrt

Institute of Organic Chemistry and Biochemistry, Czechoslovak Academy of Sciences, Prague, Czechoslovakia

INTRODUCTION

2'-*O*-(Tetrahydropyran-2-yl)uridine (*2*) is an intermediate in the preparation of 5'-*O*-[bis(*p*-methoxyphenyl)phenylmethyl]uridine.[1] The readily accessible[2] phosphate *1* is used as the starting material. By the present procedure, the preparative utilization of enzymic dephosphorylation is exemplified.[3] The procedure may be used as a general method for the unambiguous preparation of ribonucleoside 2'-*O*-(tetrahydropyran-2-yl) derivatives from the corresponding, known phosphates.[4,5] In an alternative preparation of compounds of the present type, ribonucleosides are used as the starting materials.[6]

PROCEDURE

2'-O-(Tetrahydropyran-2-yl)uridine (2)

The pH of a solution of 5'-*O*-acetyl-2'-*O*-(tetrahydropyran-2-yl)-uridine 3'-phosphate, calcium salt[2] (*1*) (5 g) and barium chloride (2 g) in water (200 ml) is adjusted to 9.5 by the addition of

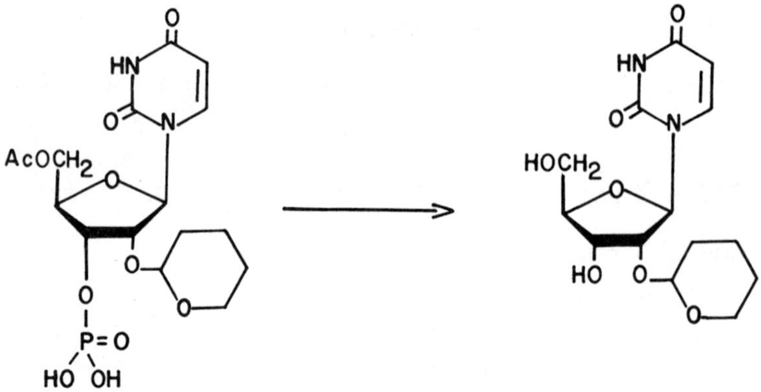

1 (450.3) → 2 (328.3)

aqueous ammonia. A solution of calf-intestine alkaline phosphatase[a] (100 mg) in water (10 ml) is added, and the mixture is kept at 37°. Aqueous ammonia is added at 30-min intervals to maintain the pH at 9.0 to 9.5. After precipitation of barium phosphate is complete, and the supernatant solution is clear (≃4 hr), concentrated aqueous ammonia (100 ml) is added. The mixture is kept for 20 hr at room temperature, treated with dodecyl alcohol (0.1 ml), and evaporated in a rotary evaporator at 40°/15 torr. A mixture of residue with ethanol (100 ml) is boiled for 10 min under reflux and filtered. The cake is washed with boiling ethanol (100 ml), and the filtrate and washings are combined and evaporated. The flask containing the residue is immersed to the neck

---

[a]Available from Calbiochem, San Diego, Calif.

in a water bath, and the acetamide is removed by sublimation for 4 hr at 40°/0.1 torr. The final residue (2.9 g; 89%) is a mixture of diastereoisomers of compound *2*; R_F 0.21 and 0.28 on ready-for-use sheets of Silufol UV_{254} silica gel (Kavalier Glassworks, Votice, Czechoslovakia) in 9:1 chloroform—methanol. The mixture may be used directly in the next step. The crystalline isomer, having a lower R_F value, may be obtained in a yield of 1.8 g by dissolving the mixture in methanol (10 ml) and adding ether (100 ml).

REFERENCES

1. J. Smrt and A. Malkievicz, *Collect. Czech. Chem. Commun.*, 38, 2962 (1973).

2. J. Smrt, in W. W. Zorbach and R. S. Tipson (eds.), *Synthetic Procedures in Nucleic Acid Chemistry*, Vol. 1, Interscience, New York, 1968, pp. 487 to 490.

3. J. Smrt and A. Holý, *Tetrahedron Lett.*, 981 (1967).

4. J. Smrt and F. Sorm, *Collect. Czech. Chem. Commun.*, 28, 61 (1963).

5. A. Holý, *Collect. Czech. Chem. Commun.*, 35, 3686 (1970).

6. B. E. Griffin, M. Jarman, and C. B. Reese, *Tetrahedron*, 24, 639 (1968).

[75] 2',3',5'-TRI-*O*-BENZOYL-5-NITROURIDINE

Nucleoside Synthesis by the $SnCl_4$-Catalyzed Silyl-Hilbert-Johnson Reaction

Ulrich Niedballa and Helmut Vorbrüggen

Schering A. G., 1000 Berlin 65, West Germany

INTRODUCTION

2',3',5'-Tri-*O*-benzoyl-5-nitrouridine (*4*) was first prepared in low yield by the reaction of 2,4-dimethoxy-5-nitropyrimidine,[1] and later in higher yield by the reaction of 5-nitrouracil[2] with 2,3,5-tri-*O*-benzoyl-β-D-ribofuranosyl chloride. The present synthesis of *4* is simpler, consisting in the reaction of 5-nitro-di-*O*-(trimethylsilyl)-uracil[2] with 1-*O*-acetyl-2,3,5-tri-*O*-benzoyl-β-D-ribofuranose (*3*) in the presence of stannic chloride.[3]

PROCEDURE

5-Nitro-2,4-bis(trimethylsilyloxy)pyrimidine (2)

A mixture of 5-nitrouracil (*1*)[a] (7.85 g, 50 mmol), hexamethyldisilazane[b] (42 ml, 200 mmol), and chlorotrimethylsilane (1 ml) is boiled under reflux for 1 hr, during which time *1* dissolves. The

---

[a]Aldrich Chem. Co.

[b]Standard quality, Dow Chem. Co.; Wacker Chemie, GmbH, 8000 München, West Germany.

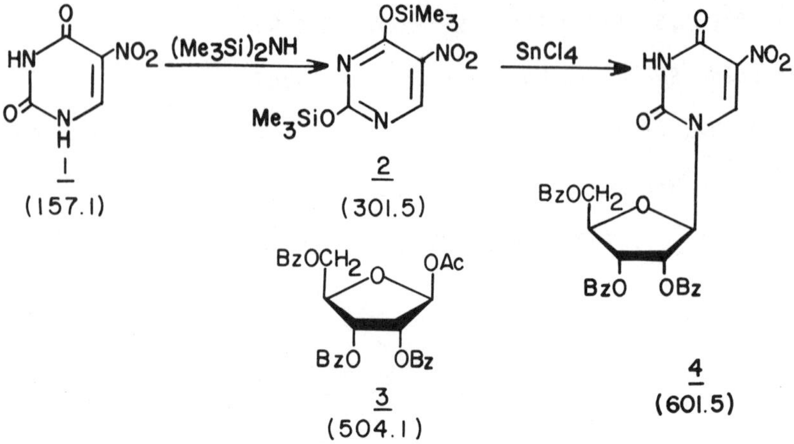

excess of hexamethyldisilazane (b.p. 127° to 128°) is removed by distillation, and the residue is distilled[c] at 110° to 111°/0.2 torr, with careful exclusion of moisture to give 14.3 g (95%) of *2* as a yellowish viscous liquid, which is dissolved in absolute 1,2-dichloroethane[d] (100 ml) to give a 475 m$\underline{M}$ standard solution.

1,2,3,4-Tetrahydro-5-nitro-1-(2,3,5-tri-O-benzoyl-β-D-ribofuranosyl)-2,4-pyrimidinedione (4)

To a solution of 1-*O*-acetyl-2,3,5-tri-*O*-benzoyl-β-D-ribofuranose[e] (*3*) (5.04 g, 10 mmol) in absolute 1,2-dichloroethane (75 ml) are added a standard solution of *2* (23 ml, 11 mmol) and a solution of stannic chloride (0.35 ml, 3 mmol) in absolute 1,2-dichloroethane (40 ml). After 1 hr the mixture is diluted with chloroform

---

[c]Evaporative bulb-to-bulb distillation by use of a Büchi Kugelrohrofen.

[d]Distilled from phosphorus pentaoxide.

(150 ml) and shaken with saturated sodium hydrogen carbonate solution (200 ml). The aqueous phase is washed with chloroform (3 × 100 ml), and the organic phases are combined, washed with water (3 × 100 ml), dried (sodium sulfate), and evaporated to give 6.9 g of crude *4*, which crystallizes from ethanol (200 ml) to give 5.79 g (96.3%) of *4* in four crops, m.p. 184° to 185° (lit.[2] m.p. 183° to 184°).

REFERENCES

1. M. Pryštaš and F. Šorm, *Collect. Czech. Chem. Commun.*, 31, 3990 (1966).

2. K. A. Watanabe and J. J. Fox, *J. Heterocycl. Chem.*, 6, 109 (1969).

3. U. Niedballa and H. Vorbrüggen, *J. Org. Chem.*, 39, 3654 (1974).

[76] 4-AMINO-1-β-D-RIBOFURANOSYL-*s*-TRIAZIN-2(1*H*)-ONE (5-AZACYTIDINE)

Direct Synthesis of a 5-Azapyrimidine Ribonucleoside by the Trimethylsilyl Procedure

Alois Pískala and Frantisek Sorm

Institute of Organic Chemistry and Biochemistry, Czechoslovak Academy of Sciences, 166 10 Prague 6, Czechoslovakia

INTRODUCTION

In view of its high biological activity, 5-azacytidine (*6*) is among the most potent cytostatics. The best results have thus far been obtained in applications against the acute leukemia of children.[1] The present procedure exemplifies the use of the general trimethylsilyl method in the field of 5-azapyrimidine nucleosides,[2] and also includes an improved preparation of 2,3,5-tri-*O*-benzoyl-D-ribosyl chloride. In the synthesis of 5-azacytidine (*6*), some additional modifications of the trimethylsilyl process have recently been used.[3-5]

PROCEDURE

2-(Trimethylsilylamino)-4-(trimethylsilyloxy)-s-triazine (2)

A mixture of 5-azacytosine[6] (*1*) (11.2 g, 0.1 mol), hexamethyldisilazane (35 ml), and powdered ammonium sulfate (0.2 g) is heated

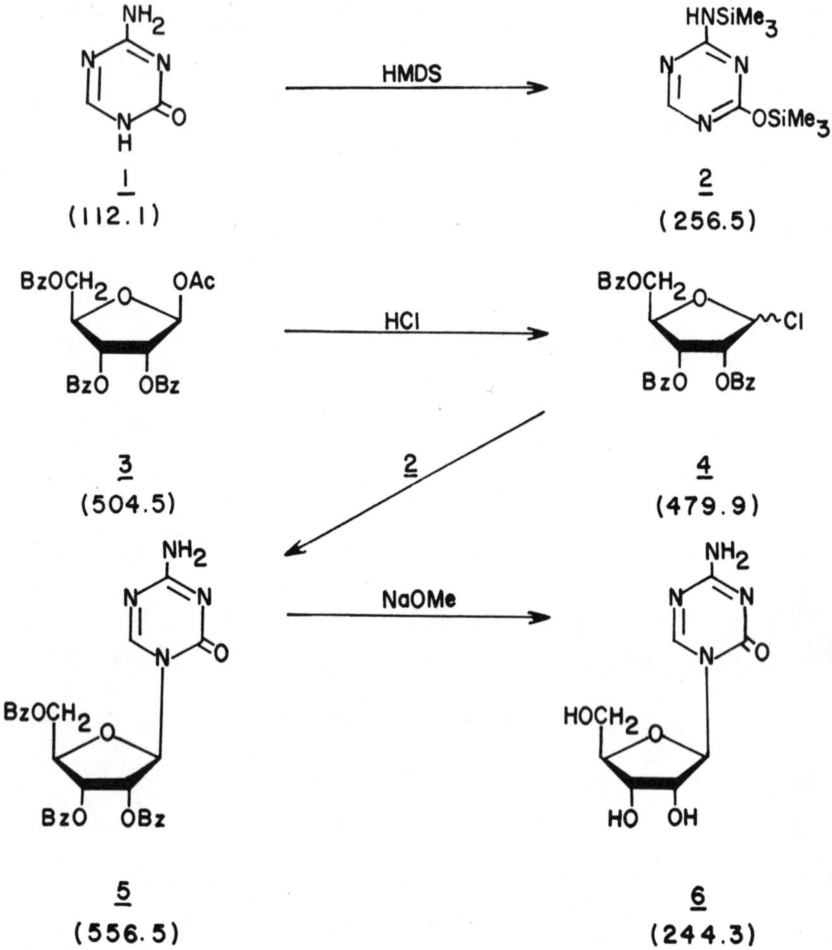

(oil bath, 160° to 170°) under a reflux condenser (potassium hydroxide tube) for 8 hr, the starting material dissolving in the course of 3 to 5 hr. The excess of hexamethyldisilazane is evaporated *in vacuo* (water pump) in a rotatory evaporator at 55° to 60° (bath temperature). The crystalline residue is thoroughly triturated with dry toluene (50 ml) and the solvent evaporated under diminished pressure. The residue is powdered and dried *in vacuo* (water pump) in a rotatory evaporator at 55° to 60° for 1 to 2 hr; yield 25.0 to 25.4 g (97% to 99%) of compound *2*, m.p. 128° to 130° (dec.; sealed capillary). An analytical sample (same m.p.) is obtained (loss, 2% to 5%) by sublimation at 100°/0.1 torr; λ_{max}^{MeCN} 224 nm, log ε (4.00), 246 (sh) nm (log ε 3.62); ν_{max}^{Nujol} 3165 (N-H), 1056 (O-Si), 912 (N-Si), and 847 cm^{-1} (C-Si).

2,3,5-Tri-O-benzoyl-D-ribosyl Chloride (4)

A solution of 1-*O*-acetyl-2,3,5-tri-*O*-benzoyl-β-D-ribose[a] (*3*) (20.16 g, 0.04 mol) in a mixture of dry 1,2-dichloroethane (80 ml) and acetyl chloride (1 ml) is saturated for 2 hr with dry gaseous hydrogen chloride under exclusion of atmospheric moisture and with external ice-cooling. The solution is kept (calcium chloride tube) overnight at room temperature, and evaporated under diminished pressure (water pump) in a rotatory evaporator at 35° to 40° (bath temperature). The almost colorless residue is coevaporated under diminished pressure at 35° to 40° with dry toluene (2 × 50 ml) to remove the remaining acetic acid. The crude chloride

---

[a]For its preparation, see E. F. Recondo and H. Rinderknecht, *Helv. Chim. Acta*, 42, 1171 (1959).

4 is used directly in the next step.[b]

4-*Amino*-1-(2,3,5-*tri*-O-*benzoyl*-β-D-*ribosyl*)-s-*triazin*-2(1H)-*one* (5)

A mixture of the crude chloride *4* (prepared from 20.16 g, 0.04 mol of compound *3*), dry acetonitrile (50 ml), and the trimethylsilyl derivative *2* (10.3 g, 0.04 mol) is stirred with an efficient magnetic stirrer for 30 min to 2 hr at room temperature in a tightly stoppered flask until the contents solidify. The mixture is kept overnight at room temperature, broken (in the flask) with a spatula into small pieces, and triturated with dry 1,2-dichloroethane (100 ml) under exclusion of atmospheric moisture. The strongly turbid solution resulting is run in a thin stream from a separatory funnel into 150 ml of a vigorously stirred, ice-cooled, 5% aqueous solution of sodium hydrogen carbonate, and stirring is continued for 10 min. The mixture is filtered through a layer of Celite, and the material on the filter is washed with chloroform.[c] The organic layer of the filtrate is separated, briefly (5 to

---

[b]When kept in a tightly stoppered flask, the chloride *4* may be stored in a refrigerator for several weeks without any substantial decomposition. It is, however, recommended that freshly prepared material be used. The crude product *4* usually contains only 2% to 4% of the starting material. When chloride *4* is prepared by the conventional procedure in ether at low temperature, the yields in the subsequent step of the present synthesis are 15% to 20% lower.

[c]The filtration is accompanied by the formation of a sticky layer, which should be systematically disturbed by means of a spatula. To avoid hydrolysis of the product, the filtration should be as brief as possible.

10 min) dried (anhydrous sodium sulfate) and filtered, and the filtrate is evaporated under diminished pressure in a rotatory evaporator at 35° to 40° (bath temperature). The resulting thick syrup is dissolved in boiling and dry ethanol (40 ml), and the solution is nucleated and kept in a stoppered flask for 2 days at room temperature to deposit the product.[d] The crystals are collected with suction, washed successively with ethanol and ether, and dried under diminished pressure; yield,[e] 14.5 to 15.6 g (65% to 70%) of compound 5, m.p. 182° to 184° (dec.). The product is sufficiently pure for use in the next step. Recrystallization from ethanol affords an analytical sample, m.p. 186° to 187° (dec.), $[\alpha]_{\underline{D}}^{25}$ -49.24° (*c* 0.526; $HCONMe_2$); ^{1}H-NMR data (Me_2SO-d_6; Me_4Si): δ 8.44 (s, 1, H-6), 6.05 (m, 3, H-1',2',3'), and 4.69 (m, 3, H-4',5',5').

4-Amino-1-β-D-ribofuranosyl-s-triazin-2(1H)-one (5-Azacytidine) (6)

A mixture of compound 5 (0.01 mol) and absolute methanol (45 ml) is heated to the boiling point, treated with M methanolic sodium methoxide (1 ml), and thoroughly swirled. The starting material rapidly dissolves, and the solution immediately begins to deposit

---

[d] The turbidity, if any, may be removed by moderate heating or by the addition of a small volume of benzene.

[e] When the yield is lower than stated, some additional product (5% to 7%) may by obtained by evaporation of the mother liquor and by successive crystallization of the residue from benzene—ether and then from ethanol.

the product. The mixture is kept in a stoppered vessel for 4 hr at room temperature, and then overnight in a refrigerator. The crystals are collected with suction, washed with ice-cold methanol, and dried under diminished pressure at room temperature; yield, 2.60 to 2.64 g (94% to 96%) of the methanol solvate of compound *6*, m.p. 234° to 235° (dec.). The methanol of solvation may be removed by drying to constant weight at 110°/0.1 torr. Recrystallization[f] from 1:1 water—acetone (25 ml per g of 5-azacytidine) affords (loss, 15% to 20%) the hydrate,[g] having the same m.p. as the methanol solvate. Drying at 110°/0.1 torr affords anhydrous *6*, m.p. 232° to 233° (dec.)[h] $[\alpha]_D^{25}$ +68.3 (5 min) ⟶ -16.0° (24 hr) (*c* 0.501; water); λ_{max}^{MeOH} 244 nm (log ε 3.84); CD (pH 7.1): λ 249 nm, [Θ] +11600 ⟶ λ 252 nm, [Θ] +5300 (24 hr); ^{1}H-NMR data (D_2O, DSS): δ 8.56 (s, 1, H-6), 5.76 (d, 1, H-1'), 4.37 (dd, 1, H-2'), 4.20 (m, 1, H-3'), 4.15 (m, 1, H-4'), 3.77 (dd, 1, H-5'$_B$) and 3.96 (dd, 1, H-5'$_C$), $J_{1',2'}$ 3.0 Hz; $J_{2',3'}$ 4.5 Hz; $J_{4',5'B}$ 2.4 Hz; $J_{4',5'C}$ 3.8 Hz; $J_{5'B,5'C}$ -13 Hz.

---

[f]The yellow color, if any, may be removed by the addition of a small amount of active charcoal prior to the filtration. The preparation of both the solution and the filtration should be effected as quickly as possible to avoid hydrolysis of 5-azacytidine. The crystallization is completed by keeping the mixture overnight at -25°.

[g]After being dried at room temperature under diminished pressure, the product contains from 5.6% to 6.2% of water.

[h]When protected against atmospheric moisture, the product is almost completely stable.

REFERENCES

1. For leading references, see A. Cihák, *Oncology*, 30, 405 (1974).

2. A. Piskala, P. Fiedler, and F. Sorm, *Nucleic Acid Res., Special Publication*, No. 1, s17 (1975).

3. A. Piskala and F. Sorm, German Patent 1,935,028 (1970); *C. A.*, 73, 45,793 (1970).

4. M. W. Winkley and R. K. Robins, *J. Org. Chem.*, 35, 491 (1970).

5. U. Niedballa and H. Vorbrüggen, *J. Org. Chem.*, 39, 3672 (1974).

6. A. Piskala and F. Sorm, in W. W. Zorbach and R. S. Tipson (eds.), *Synthetic Procedures in Nucleic Acid Chemistry*, Vol. 1, Wiley—Interscience, New York, 1968, pp. 107 to 108.

[77] ANOMERIC 4-AMINO-1-(2-DEOXY-D-*ERYTHRO*-PENTOFURANOSYL)-*s*-TRIAZIN-2(1*H*)-ONES (2'-DEOXY-5-AZACYTIDINE AND ITS α-D ANOMER)

Synthesis of 5-Azapyrimidine Deoxyribonucleosides via Acylglycosyl Isocyanates

Alois Pískala and Frantisek Sorm

Institute of Organic Chemistry and Biochemistry, Czechoslovak Academy of Sciences, 166 10 Prague 6, Czechoslovakia

INTRODUCTION

The activity of 2'-deoxy-5-azacytidine (*6b*) is of the same order of magnitude as that of the nucleoside antibiotic 5-azacytidine, but the mechanism of the biological action is different.[1] The present procedure[2,3] exemplifies the use of the "isocyanate process" in the synthesis of 5-azapyrimidine deoxyribonucleosides. As the reaction of chloride *1* with silver isocyanate affords the β anomer of the isocyanate *2* as the preponderant product, the present procedure is particularly advantageous in the preparation of the biologically active β-D anomer *6b*. 2'-Deoxy-5-azacytidine (*6b*) is also accessible by an alternative route[4] or by a direct glycosylation of trimethylsilylated 5-azacytosine.[5,6]

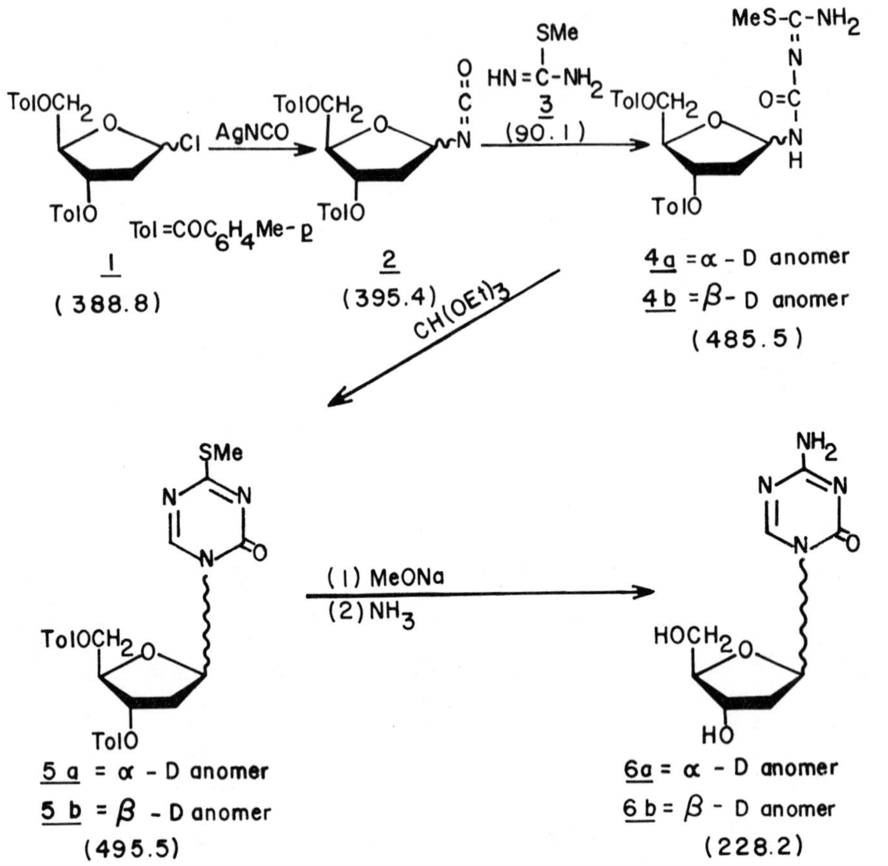

PROCEDURE

1-(2-Deoxy-3,5-di-O-p-toluoyl-β-D-erythro-pentofuranosyl)-4-methylisothiobiuret and its α-D Anomer (4b and 4a)

A solution of freshly prepared *S*-methylisothiourea[8] (4.5 g, 0.05 mol) in dry acetone (100 ml) is added at 0° with stirring and under exclusion of atmospheric moisture to a toluene solution of isocyanate *2* obtained from 19.4 g (0.05 mol) of chloride[7] *1* analogously to the preparation of 2,3,5-tri-*O*-acetyl-β-D-ribosyl isocyanate.[a] The mixture is kept for 30 min at room temperature and evaporated under diminished pressure at 35° to 40° (bath temperature). The residue is dissolved in chloroform (200 ml), and the solution is washed with water (3 × 50 ml), dried (anhydrous sodium sulfate), and evaporated under diminished pressure to a thick syrup, which is dissolved in dry acetone (50 ml). Dry, light petroleum is added to faint opalescence, and the inside walls of the vessel are scratched with a rough-tipped glass rod to induce crystallization. The compound is allowed to crystallize for 4 days, a second portion of light petroleum being added on the 2nd day to faint opalescence of the supernatant liquor. The crystals are collected with suction, and washed with ether; yield, 13.2 g (54%) of the β-D-anomer *4b*, m.p. 164° to 168° (dec.). This product may be used in the subsequent step without additional purification. By recrystallization from acetonitrile the melting point is raised to 170° to 172° (dec.); $[\alpha]_D^{25}$ -19.9° (*c* 0.47, *N*,*N*-dimethylformamide).

The mother liquors remaining after the isolation of the β-D-anomer *4b* are diluted with light petroleum to faint turbidity, and the mixture is kept for 6 days at room temperature, with

---

[a]When heated rapidly this sample of *S*-methylisothiourea melts at 87° to 89° to a clear melt that immediately resolidifies. A sample containing higher-melting, decomposition products should not be used in this reaction. For the preparation of glycosyl isocyanates, see A. Piskala and F. Sorm, *Collect. Czech. Chem. Commun.*, 29, 2060 (1964).

occasional stirring, to deposit crystals.[b] The solid is collected with suction and washed with ether; yield, 6.8 g (28%) of the α-D anomer *4a*, m.p. 125° to 128° (dec.)[c]; $[\alpha]_D^{25}$ +50.2° (*c* 0.72, *N*,*N*-dimethylformamide).

1-(2-Deoxy-3,5-di-O-p-toluoyl-α-D-erythro-pentofuranosyl)-4-(methylthio)-s-triazin-2(1H)-one (5a)

Dry nitrogen is slowly passed through a mixture of the isothiobiuret *4a* (4.86 g, 0.01 mol) and triethyl orthoformate (50 ml) in a flask equipped with a descending Liebig condenser. The mixture is heated (internal temperature, 120° to 124°; bath temperature, 130° to 135°) for 2 hr and then kept overnight at room temperature. The solid is collected with suction and washed with ether; yield, 3.56 g (72%) of the α-D anomer *5a*, m.p. 174° to 175° (dec.)[d]; $[\alpha]_D^{25}$ -79.7° (*c* 0.41, *N*,*N*-dimethylformamide); ^{1}H-NMR data ($CDCl_3$; Me_4Si): δ 8.30 (s, 1, H-6), 6.28 (q, 1, H-1'), 5.64 (m, 1, H-3'), 4.87 (m, 1, H-4'), 4.60 (m, 2, H-5'), 2.81 (m, 2, H-2'), and 2.45 (s, 3, SCH_3); $J_{1',2'}$ 5.0 Hz, $J_{1',2''}$ 10.5 Hz, $J_{3',4'}$

---

[b]The α-D anomer *4a* may also be obtained by evaporation of the solvent and crystallization of the residue from benzene.

[c]The α-D anomer *4a* is a new compound. The melting point does not change on recrystallization from absolute ethanol. Analysis calculated for $C_{24}H_{21}N_3O_6S$ (485.5): C, 59.36; H, 5.60; N, 8.65; S, 6.60. Found: C, 59.16; H, 5.62; N, 8.73; S, 6.54.

[d]The α-D anomer *5a* is a new compound. The melting point does not change on recrystallization from absolute ethanol. Analysis calculated for $C_{25}H_{25}N_3O_6S$ (495.5): C, 60.58; H, 5.08; N, 8.48; S, 6.47. Found: C, 60.50; H, 4.82; N, 8.23; S, 6.36.

4.5 Hz.

1-(2-Deoxy-3,5-di-O-p-toluoyl-β-D-erythro-pentofuranosyl)-4-(methylthio)-s-triazin-2(1H)-one (5b)

The isothiobiuret *4b* (4.86 g; 0.01 mol) is processed analogously to the compound in the preceding paragraph; yield, 3.25 g (67%) of the β-D anomer *5b*, m.p. 201° to 202° (dec.), $[\alpha]_D^{22}$ +15.5° (c 0.45, *N,N*-dimethylformamide); ^{1}H-NMR data ($CDCl_3$; Me_4Si): δ 8.40 (s, 1, H-6), 6.23 (t, 1, H-1'), 5.56 (m, 1, H-3'), 4.65 (m, 2, H-5'), 2.25° to 2.50° (m, 2, H-2'), and 2.45 (s, 3, SCH_3); $J_{1',2'}$ 5.5 Hz.

4-Amino-1-(2-deoxy-α-D-erythro-pentofuranosyl)-s-triazin-2(1H)-one (6a)

A mixture of finely powdered *5a* (2.48 g, 5 mmol), absolute methanol (50 ml), and M methanolic sodium methoxide (6.2 ml) is magnetically stirred in a stoppered vessel for 140 min at room temperature, and then applied to a column of Amberlite IRC-50 (H^+) ion-exchange resin (50 ml) packed in absolute methanol. The column is washed with absolute methanol (300 ml), and the effluent is evaporated under diminished pressure at 35° to 40° (bath temperature). The residue is dissolved in 7M methanolic ammonia (10 ml), and the solution is kept for 1 hr at room temperature and evaporated under diminished pressure. The residue is dissolved in absolute methanol (5 ml), and material is precipitated with dry ether (50 ml). The whole mixture is kept for 1 hr at room temperature and the supernatant liquor is then decanted. The residue is thoroughly triturated with absolute methanol (5 ml), and the slurry is kept overnight at -10°. The solid is collected with suction and washed with ice-cold methanol; yield, 0.73 g (64%) of the α-D anomer *6a*, m.p. 176° to 178° (dec.). After recrystallization from methanol, m.p. 181° to 182° (dec.), $[\alpha]_D^{22}$ -45.4 ⟶ -23.6° (48 hr; *c* 0.5, water); CD (H_2O): δ 247 nm, [θ] -16200 ⟶ λ 246 nm, [θ] -12200 (24 hr); $\lambda_{max}^{pH\ 7.0}$ 244 nm (log ε 3.86); ^{1}H-NMR data (D_2O; DSS): δ 8.48 (s, 1, H-6), 6.16 (q, 1, H-1'), 4.3 to 4.65 (m, 2, H-3',4'), 3.58 to 3.81 (m, 2, H-5'), and 2.00 to 3.12 (m, 2, H-2'); $J_{1',2'}$ 2.0 Hz, $J_{1',2''}$ 7.0 Hz.

*4-Amino-(2-deoxy-β-D-*erythro*-pentofuranosyl)-*s*-triazin-2(1H)-one (2'-Deoxy-5-azacytidine) (6b)*

A mixture of finely ground *5b* (2.48 g, 5 mmol), absolute methanol (50 ml), and M methanolic sodium methoxide (6.2 ml) is magnetically stirred in a stoppered flask for 165 min at room temperature. The solution is decationized (as described in the preceding paragraph) on a column of Amberlite IRC-50 (H^+) ion-exchange resin, and the effluent is evaporated under diminished pressure. The residue is dissolved in 7M methanolic ammonia (10 ml), and the solution is kept for 30 min in a stoppered flask at room temperature and then overnight at -10°. The solid is collected with suction and washed with ice-cold methanol; yield, 0.89 g (78%) of the β-D anomer *6b*, m.p. 196° to 198° (dec.). After recrystallization from methanol, m.p. 201° to 202° (dec.), $[\alpha]_D^{22}$ +68.5 (30 min) → +57.8° (6 hr)(*c* 0.5, water); CD (H_2O): δ 249 nm, [θ] +7840 → λ 247 nm, [θ] +2100 (24 hr); $\lambda_{max}^{pH\ 7.0}$ 244 nm (log ε 3.86); ^{1}H-NMR data (D_2O; DSS): δ 8.58 (s, 1, H-6), 6.29 (t, 1, H-1'), 3.80 to 3.94 (m, 2, H-5'), and 2.36 to 2.67 (m, 2. H-2'), $J_{1',2'}$ 6.5 Hz.

REFERENCES

1. For leading references, see J. Vesely and A. Cihák, *Oncology*, 28, 204 (1973).

2. F. Sorm and A. Piskala, Dutch Patent 6,414,959 (1965); *C. A.*, 64, 2154 (1966).

3. A. Piskala and F. Sorm, German Patent 1,922,702 (1969); *C. A.*, 72, 79427 (1970).

4. J. Pliml and F. Sorm, *Collect. Czech. Chem. Commun.*, 29, 2576 (1964).

5. M. W. Winkley and R. K. Robins, *J. Org. Chem.*, 35, 491 (1970).

6. U. Niedballa and H. Vorbrüggen, *J. Org. Chem.*, 39, 3672 (1974).

7. C. C. Bhat, in W. W. Zorbach and R. S. Tipson (eds.), *Synthetic Procedures in Nucleic Acid Chemistry*, Vol. 1, Wiley—Interscience, New York, 1968, pp. 521 to 522.

8. F. Arndt, *Ber.*, 54, 2336 (1921).

[78] 3-METHYL-6-AZAURIDINE [4-METHYL-2-β-D-RIBOFURANOSYL-AS-TRIAZIN-3,5-(2*H*,4*H*)-DIONE]

Alkylation of a Nucleoside Antimetabolite by Use of N,N-Dimethylformamide Dimethyl Acetal

Jiří Žemlička

Michigan Cancer Foundation and Department of Oncology, Wayne State University School of Medicine, Detroit, Michigan 48201

INTRODUCTION

6-Azauridine (*1*) is a nucleoside antimetabolite that exhibits cancerostatic and virostatic activity.[1] It is also active against psoriasis and as an abortive agent.[2] The alkylation of 6-azauridine (*1*) is of interest from the standpoint of structure—activity relationships.[1,3] 3-Methyl-6-azauridine (*2*) has been prepared by reaction of 2',3',5'-tri-*O*-benzoyl-6-azauridine with diazomethane, and subsequent debenzoylation.[4] A more direct and, incidentally, a simpler route consists in treating 6-azauridine (*1*) with *N,N*-dimethylformamide dimethyl acetal at room temperature to give *2* in 60% yield.[5]

PROCEDURE

6-Azauridine (*1*) (0.23 g, 0.98 mmol) and *N,N*-dimethylformamide dimethyl acetal (0.48 g, 4 mmol) are shaken in chloroform (5 ml)

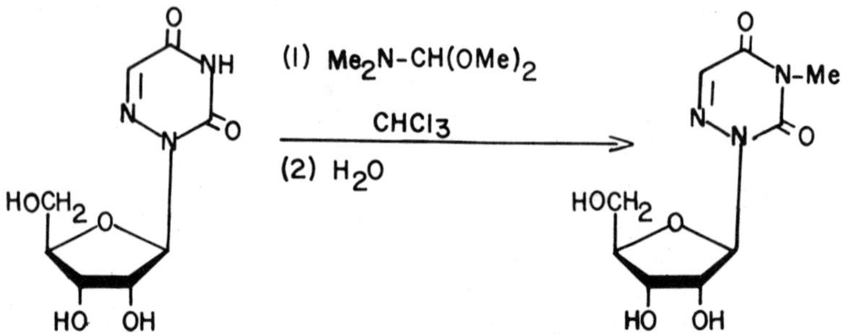

1
(245.2)

2
(259.3)

for 22 hr at room temperature. The mixture, which becomes homogeneous in ≃8 hr, is evaporated *in vacuo* (aspirator),[a] the residue is dissolved in water, the solution is evaporated,[a] benzene is added, and the product is removed by filtration. Recrystallization from ethanol affords pure *2*, wt. 0.15 g (61%), m.p. 125°, undepressed on admixture with an authentic sample[4] of *2*. It can be readily distinguished from 6-azauridine (*1*) by paper electrophoresis in 25m<u>M</u> disodium hydrogen phosphate (pH 7.5); compound *2* stays at the origin, whereas *1* moves toward the anode.

REFERENCES

1. J. Škoda, *Prog. Nucleic Acid Res. Molec. Biol.*, 2, 197 (1963).

2. J. Škoda, in A. C. Sartorelli and D. G. Johns (eds.), *Antineoplastic and Immunosuppressive Agents*, Springer-Verlag, New York, Vol. 2, 1975, p. 348.

---

[a]A rotary evaporator is conveniently used.

3. J. Žemlička, J. Smrt, and F. Šorm, *Collect. Czech. Chem. Commun.*, 27, 1462 (1962).

4. J. Prystaš, J. Gut, and F. Šorm, *Collect. Czech. Chem. Commun.*, 27, 1572 (1962).

5. J. Žemlička, *Collect. Czech. Chem. Commun.*, 35, 3572 (1970).

[79] 1-β-D-RIBOFURANOSYL-*s*-TRIAZINE-2,4-(1*H*,3*H*)-DIONE (5-AZAURIDINE)

Direct Synthesis of a 5-Azapyrimidine Ribonucleoside by the Fischer-Helferich Procedure

Alois Pískala and Frantisek Sorm

Institute of Organic Chemistry and Biochemistry, Czechoslovak Academy of Sciences, 166 10 Prague 6, Czechoslovakia

INTRODUCTION

5-Azauridine (5) is a pyrimidine antimetabolite the mechanism of whose biological action has been investigated.[1] The present procedure[2] exemplifies the use of the general Fischer—Helferich method for the preparation of 5-azauracil nucleosides that readily undergo hydrolytical degradation.[3] As the glycosylation of 5-azauracil silver salt[4,5] (*1*) occurs selectively at N-1 of the *s*-triazine ring, and as all operations are performed under anhydrous conditions, the present approach to the synthesis of 5-azauridine (5) and related nucleosides is of special importance. 5-Azauridine (5) is also accessible by the "isocyanate process," but in lower yield.[3]

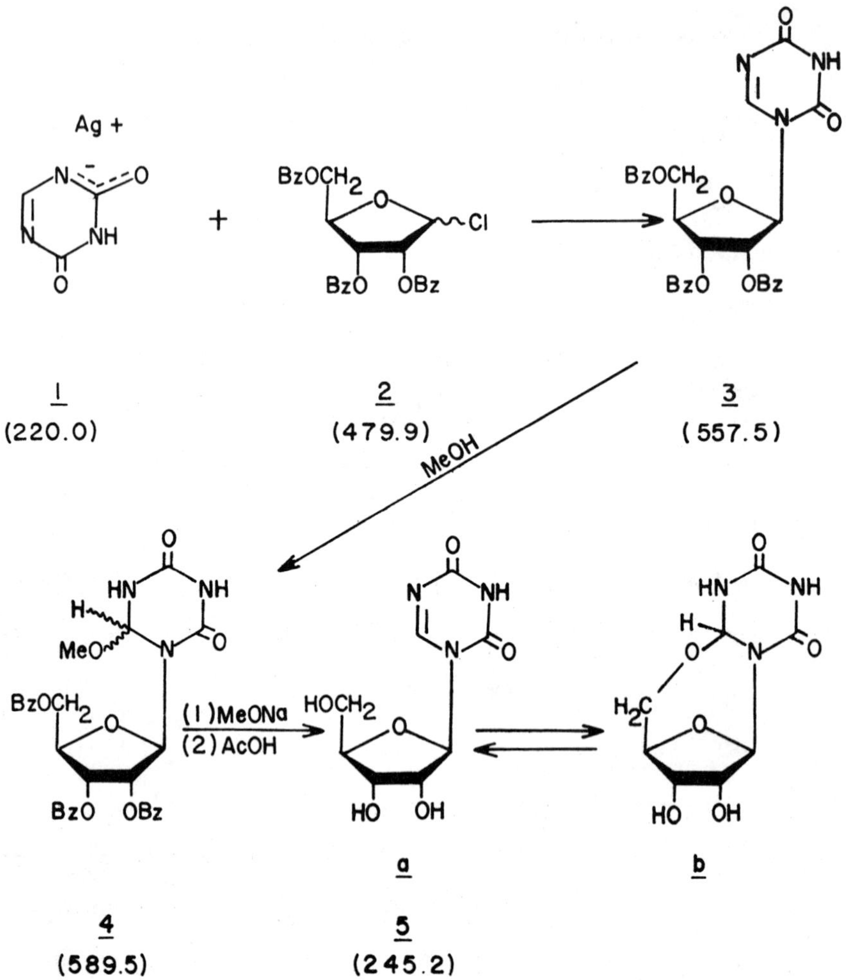

PROCEDURE

5-Azauracil Silver Salt (1)

A solution of 2.26 g (0.02 mol) of 5-azauracil[6] in water (50 ml) is treated with a solution of silver nitrate (3.4 g, 0.02 mol) in water (20 ml). The white precipitate resulting is collected by filtration, washed with water, and dried to constant weight under diminished pressure over potassium hydroxide pellets and concentrated sulfuric acid; yield, 4.40 g (100%) of *1*.

2',3',5'-Tri-O-benzoyl-5-azauridine (3)

A mixture of *1* (1.98 g, 0.009 mmol), dry toluene (100 ml), and the chloride[a] *2* [prepared from 5.04 g (0.01 mol) of 1-*O*-acetyl-2,3,5-tri-*O*-benzoyl-β-D-ribose[b]] is boiled under reflux for 4 hr with magnetic stirring and exclusion of atmospheric moisture. The hot mixture is filtered and the silver salts washed with hot toluene (3 × 20 ml). The filtrate and washings are combined, cooled to room temperature, and treated, with stirring, with dry, light petroleum (750 ml). The white, amorphous precipitate of crude tribenzoate *3* is allowed to settle, and the supernatant liquor is discarded. The residual slurry is dissolved in boiling absolute methanol (50 ml), and the solution is kept for 10 days at 0° to deposit crystals that are collected by filtration and washed with methanol; yield, 2.65 g (45%) of the covalent adduct[c] *4*, m.p. 152° to 158° (dec.). This product may be used directly in the preparation of 5-azauridine. By recrystallization from acetonitrile the m.p. is raised to 164° to 168° (dec.), $[\alpha]_D^{22}$ -48.9° (*c* 0.46,

---

[a]For a preparation, see Reference 7.

[b]For a preparation, see contribution #76 of this volume.

[c]The configuration of C-6 of the *s*-triazine ring in this unstable product has not been examined. The product is probably a mixture of the *R* and *S* stereoisomers.

$HCONMe_2$). On heating to constant weight at 130° to 140°/0.1 torr, the covalent adduct *4* may be quantitatively converted into the amorphous tribenzoate *3*.

1-β-D-Ribofuranosyl-s-triazine-2,4(1H,3H)-dione (5-Azauridine) (5)

A solution of the covalent adduct *4* (0.589 g, 1 mmol) in 0.1M methanolic sodium methoxide (25 ml) is kept in a stoppered flask for 24 hr at room temperature. The solution is acidified with glacial acetic acid (0.2 ml), and decationized with moist Dowex 50 W (H^+) ion-exchange resin (3 g) prewashed with water and then methanol. The resin is collected by filtration and washed with methanol. The filtrate and washings are combined and evaporated under diminished pressure at 25° to 30° (bath temperature). The crystalline residue is washed with ether and triturated with absolute methanol (3 ml), and the slurry is kept overnight at 0°. The solid is collected by filtration and washed with methanol; yield, 0.196 g (80%) of the title compound[d] 5, m.p. 207° to 208° (dec.),

---

[d]In the crystalline state, 5-azauridine (5) exists in the cyclic *b*-form. The configuration at C-6 of the *s*-triazine ring has not been examined, but the *S* configuration may be assumed for steric reasons. In the solid state, 5-azauridine is almost unlimitedly stable. An equilibrium between the *5a* and *5b* forms exists in nonaqueous solutions. In aqueous solutions 5-azauridine is rapidly hydrolyzed, with the formation of 1-formyl-5-β-D-ribofuranosylbiuret. 5-Azauridine is obtained by the present procedure in an analytically pure state. Owing to the unstability of 5-azauridine, the recrystallization from 5% aqueous acetic acid is accompanied by a considerable loss of material; the recrystallized compound, m.p. 210° to 211° (dec.) is obtained in only 10% yield.

$[\alpha]_{D}^{22}$ -30.5° (*c* 0.2, $HCONMe_2$); UV (pH 5.0)[e]: λ_{max} 240 nm (log ε 2.68); CD (pH 5.0)[f]: λ 250 nm, [Θ] +170; ^{1}H-NMR data (Me_2SO-d_6, Me_4Si): δ 8.37 (s, H-6 of the *a*-form), 5.48 (s, H-6 of the *b*-form).

REFERENCES

1. A. Cihák and F. Sorm, *Collect. Czech. Chem. Commun.*, 30, 2091 (1965).
2. A. Piskala, P. Fiedler, and F. Sorm, *Nucleic Acids Res.*, Special Publication, No. 1, s17, (1975).
3. A. Piskala and F. Sorm, *Collect. Czech. Chem. Commun.*, 29, 2060 (1964).
4. A. Piskala and J. Gut, *Collect. Czech. Chem. Commun.*, 26, 2519 (1961).
5. H. Biltz and R. Robl, *Ber.*, 53, 1967 (1920).
6. A. Piskala and J. Gut, in W. W. Zorbach and R. S. Tipson (eds.), *Synthetic Procedures in Nucleic Acid Chemistry*, Vol. 1, Wiley—Interscience, New York, 1968, pp. 105 to 106.
7. E. F. Recondo and H. Rinderknecht, *Helv. Chim. Acta*, 42, 1171 (1959).

---

[e]Measured 10 min after the preparation of the solution. After 1.5 hr at room temperature the solution does not exhibit any maximum at the wavelength stated.

[f]Measured immediately after the preparation of the solution.

[80] 2,4,6-TRIMETHOXY-1-β-D-RIBOFURANOSYLBENZENE

An Intermediate in the Synthesis of C-Nucleosides

L. Kalvoda and J. Farkaš

Institute of Organic Chemistry and Biochemistry, Czechoslovak Academy of Sciences, 16610 Prague, Czechoslovakia

INTRODUCTION

The procedure is an example of the synthesis of C-glycosyl derivatives by use of the Friedel-Crafts reaction.[1] Compound 5 is a versatile intermediate in the synthesis of C-nucleosides.[2,3]

PROCEDURE

2,4,6-Trimethoxy-1-(2,3,5-tri-O-benzoyl-β-D-ribofuranosyl)-benzene (4)

A solution of 1-*O*-acetyl-2,3,5-tri-*O*-benzoyl-β-D-ribofuranose (*1*) (10.0 g, 19.8 mmol) in benzene (50 ml) is saturated at 0° with hydrogen bromide and then kept for 30 min at room temperature and evaporated. The residue (*2*) is coevaporated at 40° with fresh benzene (2 × 50 ml). Benzene (50 ml), 1,3,5-trimethoxybenzene (*3*) (6.65 g, 39.6 mmol), and 10 g of zinc oxide (freshly dried for 1 hr at 400°) are added to the halide *2*, and the mixture is stirred for 20 hr at room temperature, filtered, and the filtrate concentrated under diminished pressure. The concentrate is

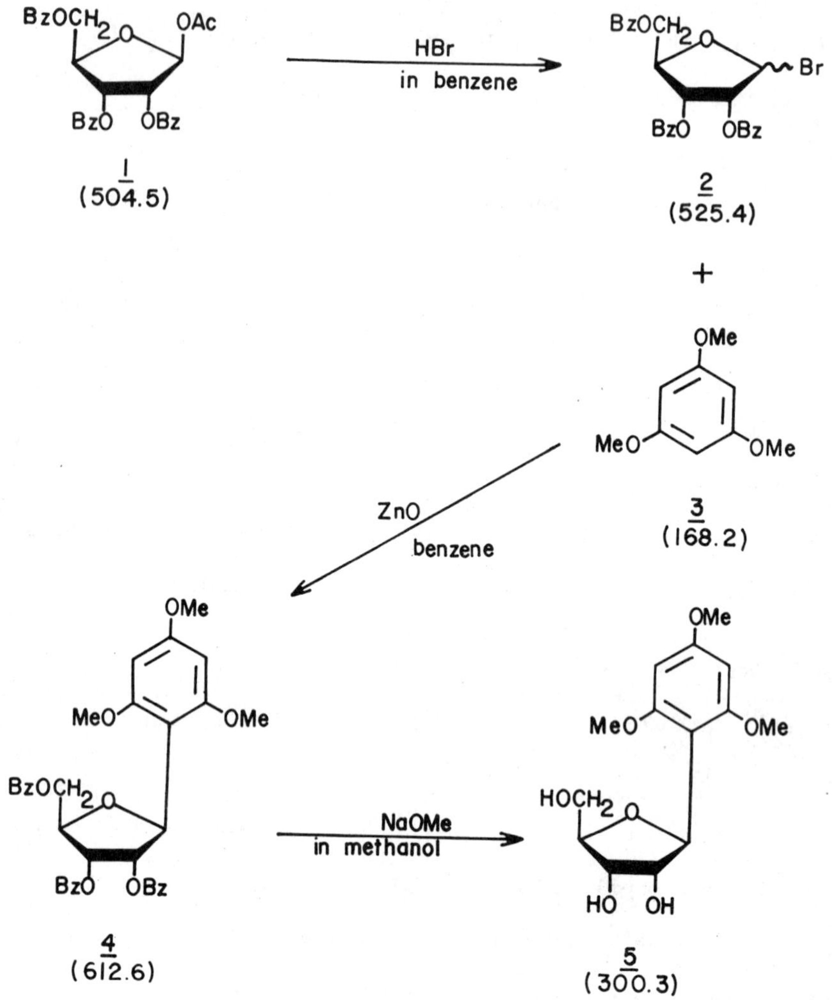

applied to a column (4 × 14 cm) of silica gel, and successively eluted with benzene (200 ml) and 1:19 ethyl acetate—benzene (200 ml). The eluates are combined and evaporated, the residue is dissolved in ether (30 ml), and the solution is treated with cyclohexane until slightly turbid. The solution is nucleated with a crystal of compound *4* and kept for 2 days at room temperature and then for 1 day at -15° to deposit crystals that are collected with suction and washed with 1:1 ether—light petroleum; yield, 4.4 g (36%) of compound *4*, m.p. 108° to 110°, $[\alpha]_{D}^{25}$ -46.3° (*c* 0.5, chloroform).

2,4,6-Trimethoxy-1-β-D-ribofuranosylbenzene (5)

Methanolic sodium methoxide (0.6 ml of M solution) is added to a suspension of 4.4 g of crystalline[a] compound *4* in absolute methanol (50 ml). The mixture is kept for 20 hr at room temperature, rendered neutral with acetic acid, and evaporated. Water (20 ml) is added to the residue, and the mixture is washed with ether (2 × 20 ml). The aqueous layer is concentrated to ≃7 ml, and the concentrate is extracted with ethyl acetate (2 × 30 ml). The extract is dried, concentrated, and allowed to deposit crystals, which are collected with suction and washed with ether; yield 2.05 g (95%) of compound *5*, m.p. 102° to 103°; $[\alpha]_{D}^{25}$ -37.8° (*c* 0.5, water).

REFERENCES

1. L. Kalvoda, *Collect. Czech. Chem. Commun.*, 38, 1679 (1973).
2. L. Kalvoda, J. Farkaš, and F. Šorm, *Tetrahedron Lett.*, 2297 (1970).

---

[a]In the preparations on a larger scale crude *4* may be used, as the crystallization of compound *4* is rather difficult; the overall yield (34%) of compound *5* (based on the sugar derivative *1*) is then almost the same as with crystalline *4*.

3. J. Farkaš, E. Flegelová, and F. Šorm, *Tetrahedron Lett.*, 2279 (1972).

[81] 3-(2,3,5-TRI-*O*-BENZOYL-β-D-RIBOFURANOSYL)-5-TRICHLOROMETHYL—1,2,4-OXADIAZOLE

An Improved Synthesis of 2,5-Anhydro-3,4,6-tri-O-benzoyl-D-allonamidoxime, a Versatile Intermediate

Ganapathi R. Revankar and Roland K. Robins

ICN Pharmaceuticals, Inc., Nucleic Acid Research Institute, Irvine, California 92715

INTRODUCTION

2,5-Anhydro-3,4,6-tri-*O*-benzoyl-D-allonamidoxime (*2*) is an important intermediate for the synthesis of *C*-glycosyl nucleosides of 1,2,4-oxadiazoles.[1] Although *2* is obtained in good yield via low-temperature chlorination of 2,5-anhydro-3,4,6-tri-*O*-benzoyl-D-allose oxime[2] followed by treatment with ammonia at 0°, the direct preparation of *2* from 2,3,5-tri-*O*-benzoyl-β-D-ribofuranosyl cyanide[3] gave a rather low yield (34%) and involved chromatographic purification.[1] The improved synthesis of *2* described herein is relatively simple and gives better yields. Ring closure of *2* with trichloroacetyl chloride provided the title compound (*3*), in which the reactive trichloromethyl group is available for various nucleophilic substitution-reactions.

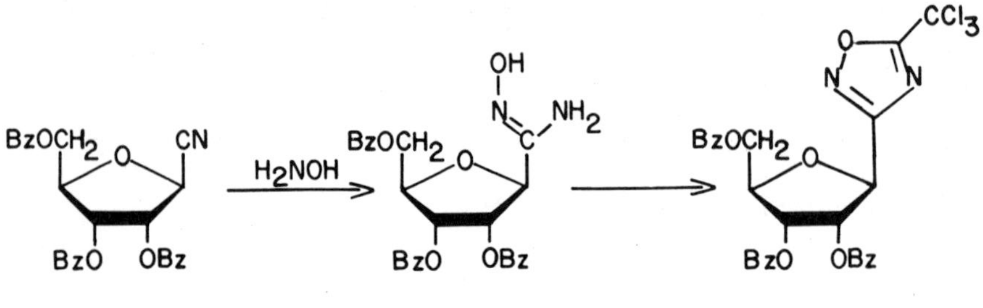

1 (471.5) 2 (504.5) 3 (631.9)

PROCEDURE

2,5-Anhydro-3,4,6-tri-O-benzoyl-D-allonamidoxime (2)

To a solution of hydroxylamine hydrochloride (1.042 g, 15 mmol) in water (10 ml) is added anhydrous sodium carbonate (0.796 g, 7.5 mmol) with stirring. The solution is stirred for 5 min and then 2,3,5-tri-*O*-benzoyl-β-D-ribofuranosyl cyanide[3] (*1*, 4.714 g, 0.10 mol) in ethanol (10 ml) is added. The mixture is heated under gentle reflux for 1 hr and then refrigerated (0° to 4°) overnight. The white precipitate that separates is collected by filtration and successively washed with cold ethanol (2 × 5 ml) and cold water (5 ml). A solution of the residue in ethyl acetate (60 ml) is successively washed with 0.4M hydrochloric acid (25 ml) and cold water (3 × 25 ml), dried (anhydrous sodium sulfate), and evaporated[a] to yield 4.3 g (85.3%) of analytically pure, colorless foam, m.p. 78° to 80°; $[\alpha]_D^{25}$ -17.5° (c 0.6,

[a]The evaporation is carried out on a rotary evaporator under reduced pressure with bath temperature below 30°.

chloroform); λ_{max}^{MeOH} in nm 228 (ε_{mM} 41.90), 273 (4.20), and 280 sh (3.60).

3-(2,3,5-Tri-O-benzoyl-β-D-ribofuranosyl)-5-(trichloromethyl)-1,2,4-oxadiazole (3)

To a solution of *2* (4.20 g, 8.3 mmol) in dry benzene[b] (50 ml) containing dry pyridine (2.5 ml) is added freshly distilled trichloroacetyl chloride (1.67 g, 9.1 mmol), and the mixture is boiled under reflux for 1 hr and then refrigerated (0° to 4°) overnight. The mixture is evaporated[a] to dryness, and a solution of the residue in chloroform (200 ml) is successively washed with saturated aqueous sodium hydrogen carbonate (2 × 50 ml) and water (2 × 50 ml), dried (anhydrous sodium sulfate), and evaporated to dryness; the residual gum (5.2 g) is chromatographed on a column[c] (300 g, 4 × 60 cm) of silica gel prepacked in chloroform. The column is eluted with 17:3 (v/v) chloroform—acetone, and the fractions containing the product are pooled and evaporated to yield 4.80 g (91.2%) of analytically pure, colorless foam having no definite m.p.; $[\alpha]_{D}^{25}$ -38.5° (*c* 1, chloroform); λ_{max} in nm (pH 1 and 7) 245 (ε_{mM} 38.70), 275 (19.60), and 282 sh (19.00), $\lambda_{max}^{pH\ 11}$ in nm 233 (ε_{mM} 38.10), 273 sh (8.00), and 282 sh (6.70); ^{1}H NMR data ($CDCl_3$): δ 8.0 (m, 6, aromatic), 7.45 (m, 9,

---

[b] Dried over sodium.

[c] ICN-Woelm silica gel (70 to 230 mesh) is used.

aromatic), 6.05 (m, 2, H-2',3'), 5.54 (d, *J* 4.5 Hz, H-1'), and 4.75 (m, 3, H-4',5',5').

Analysis calculated for $C_{29}H_{21}Cl_3N_2O_8$ (631.85): C, 55.12; H, 3.35; Cl, 16.83; N, 4.43. Found: C, 55.26; H, 3.42; Cl, 17.01; N, 4.25.

REFERENCES

1. D. B. Repke, H. P. Albrecht, and J. G. Moffatt, *J. Org. Chem.*, 40, 2481 (1975).

2. H. P. Albrecht, D. B. Repke, and J. G. Moffatt, *J. Org. Chem.*, 40, 2143 (1975).

3. M. Bobek and J. Farkas, *Collect. Czech. Chem. Commun.*, 34, 247 (1969).

[82] 4-AMINO-1-β-D-RIBOFURANOSYLPYRAZOLE-3-CARBOXAMIDE

Synthesis of Pyrazole Nucleosides by the Fusion Method Without a Catalyst

I. A. Korbukh, O. V. Budanova, and Maria N. Preobrazhenskaya

Cancer Research Center of the U.S.S.R. Academy of Medical Sciences, 115478 Moscow, U.S.S.R.

INTRODUCTION

4-Amino-1-β-D-ribofuranosyl-pyrazole-3-carboxamide (*5*) is of interest as an analog of the D-ribosyl derivative of 5-aminoimidazole-4-carboxamide. Its synthesis is achieved via the useful intermediate 3-(ethoxycarbonyl)-4-nitropyrazole[1] (*1*), which after ribosylation gives a mixture of 5-(ethoxycarbonyl)-4-nitro-1-(2,3,5-tri-*O*-acetyl-β-D-ribosyl)pyrazole (*2*) and 3-(ethoxycarbonyl)-4-nitro-1-(2,3,5-tri-*O*-acetyl-β-D-ribosyl)pyrazole (*3*). The latter is amidated and the product catalytically reduced to yield 4-amino-1-β-D-ribofuranosylpyrazole-3-carboxamide[2] (*5*).

PROCEDURE

5-(Ethoxycarbonyl)-4-nitro-1-(2,3,5-tri-O-acetyl-β-D-ribosyl)-pyrazole (2) and 3-(Ethoxycarbonyl)-4-nitro-1-(2,3,5-tri-O-acetyl-β-D-ribosyl)pyrazole (3)

3-(Ethoxycarbonyl)-4-nitro-pyrazole (*1*) (1.85 g, 0.01 mol) and

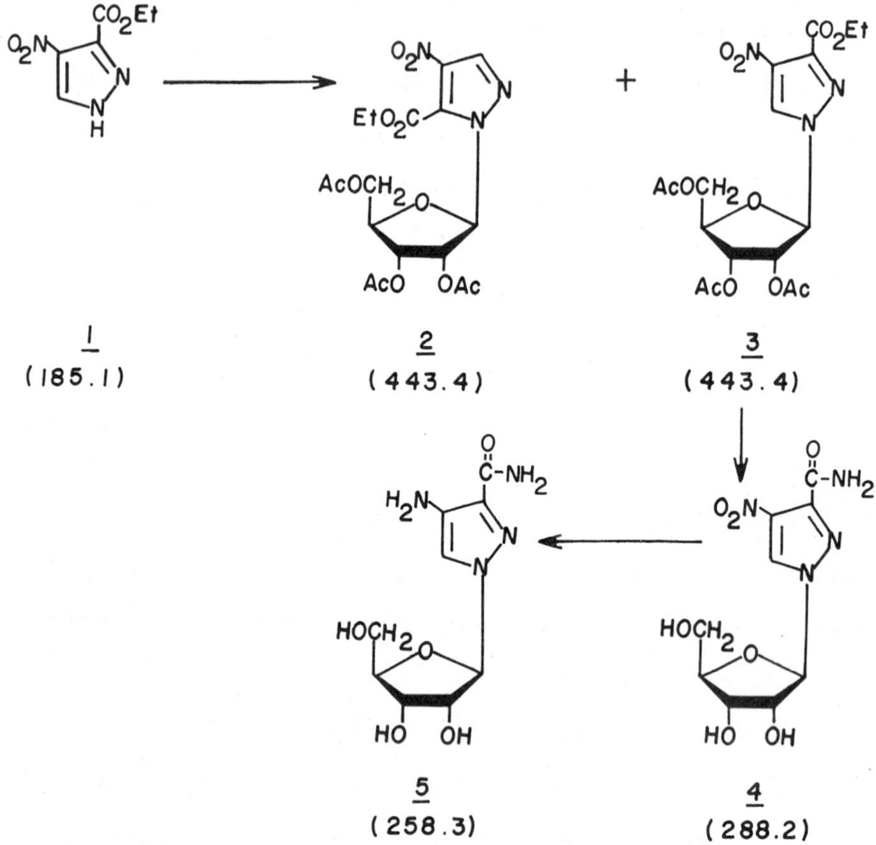

1,2,3,5-tetra-*O*-acetyl-β-D-ribose (3.18 g, 0.01 mol) are thoroughly mixed in a mortar and then placed in a round-bottomed flask connected to a rotary evaporator. The mixture is heated at 150° until dissolution occurs (≈5 min), and heating with rotation is continued *in vacuo* for 3 hr.[a] The resulting dark syrup is

---

[a]An efficient water-aspirator may be used as the vacuum source (10 to 20 torr is ideal). Use of an oil bath of sufficient depth to permit complete submersion of the reaction flask prevents the accumulation of unfused reactants on the walls of the flask.

dissolved in the minimal volume of chloroform (≃3 ml), and the solution is applied to a column (3 × 70 cm)[b] of silica gel prepacked in chloroform. The column is eluted with chloroform, 10-ml fractions being collected. The course of elution is monitored with a Uvicord UV spectrometer and by TLC.[c] Fractions 14 to 25 are combined and evaporated to give 1.36 g (31%) of *2*, m.p. 82.5° to 83.5°, $[\alpha]_{\underline{\underline{D}}}^{22}$ -100° (*c* 0.5, chloroform); R_F 0.52 (99:1 chloroform—methanol); λ_{max}^{MeOH} in nm ($\varepsilon_{\underline{mM}}$), 205 (9.6), 260 (8.0); PMR data[d] ($CDCl_3$): δ 8.25 (1, s, H-3), 6.33 (1, d, $J_{1',2'}$ 3.0 Hz, H-1'), 6.04 (1, q, $J_{2',3'}$ 5.0 Hz, H-2'), 5.74 (1, t, $J_{3',4'}$ 5.0 Hz, H-3'), 4.67 to 4.07 (3, m, H-4',5',5"), 4.57 (2, q, CH_2-Et), 2.16 (3, s, CH_3COO), 2.13 (3, s, CH_3COO), 2.05 (3, s, CH_3COO), 1.44 (3, t, CH_3-Et); (Me_2SO-d_6): δ 8.54 (1, s, H-3), 6.21 (1, d, $J_{1',2'}$ 2.0 Hz, H-1'), 5.83 (1, q, $J_{2',3'}$ 5.2 Hz, H-2'), 5.50 (1, q, $J_{3',4'}$ 6.5 Hz, H-3'), 4.57 to 3.93 (3, m, H-4',5',5"), 4.42 (2, q, CH_2-Et), 2.09 (3, s, CH_3COO), 2.04 (3, s, CH_3COO), 1.90 (3, s, CH_3COO), 1.28 (3, t, CH_3-Et). Fractions 37 to 66 provide 1.52 g (34%) of *3*. $[\alpha]_{\underline{\underline{D}}}^{22}$ -46° (*c* 1.0, chloroform); R_F 0.32 (99:1 chloroform—methanol); λ_{max}^{MeOH} in nm ($\varepsilon_{\underline{mM}}$), 207 (10.4), 264

---

[b]The column is prepacked with silica gel L 40/100 (Chemapol, Czechoslovakia).

[c]Thin-layer chromatography is conducted on ready-for-use Silufol UV_{254} (Kavalier Glassworks, Czechoslovakia).

[d]Nuclear magnetic resonance data are obtained on a JEOL JNM-MH-100 spectrometer.

(6.0); PMR data ($CDCl_3$): δ 8.63 (1, s, H-5), 6.03 (1, d, $J_{1',2'}$ 3.0 Hz, H-1'), 5.83 (1, q, $J_{2',3'}$ 5.0 Hz, H-2'), 5.56 (1, q, $J_{3',4'}$ 5.7 Hz, H-3'), 4.50 to 4.20 (3, m, H-4',5',5"), 4.59 (2, q, CH_2-Et), 2.19 (3, s, CH_3COO), 2.22 (6, s, CH_3COO), 1.48 (3, t, CH_3-Et); (Me_2SO-d_6): δ 9.34 (1, s, H-5), 6.32 (1, d, $J_{1',2'}$ 3.2 Hz, H-1'), 5.80 (1, q, $J_{2',3'}$ 5.4 Hz, H-2'), 5.59 (1, t, $J_{3',4'}$ 5.4 Hz, H-3'), 4.60 to 4.16 (3, m, H-4',5',5"), 4.46 (2, q, CH_2-Et), 2.17 (3, s, CH_3COO), 2.15 (3, s, CH_3COO), 2.08 (3, s, CH_3COO), 1.33 (3, t, CH_3-Et).

4-Nitro-1-β-D-ribofuranosylpyrazole-3-carboxamide (4)

Methanol (24 ml) presaturated with ammonia at 0° is added to *2* (1.7 g, 4 mmol) and the mixture is kept in a sealed flask for 5 days at room temperature. The solvent is removed *in vacuo* to provide 0.96 g of a syrup, which is crystallized by addition of methanol (≃5 ml), to give 0.54 g (49%) of solid *4*; m.p. 147.5° to 148.5°, $[\alpha]_D^{22}$ -67° (*c* 0.8 methanol); R_F 0.21 in (4:1 chloroform—methanol); λ_{max}^{MeOH} in nm (ε_{mM}); 212 (7.1) and 268 (6.0); PMR data (CD_3OD): δ 9.14 (1, s, H-5), 5.92 (1, d, $J_{1',2'}$ 3.5 Hz, H-1'), 4.72 to 3.68 (5, m, H-2',3',4',5',5"). The filtrates from several preparations are combined and evaporated to give additional quantities of *4* after crystallization.

4-Amino-1-β-D-ribofuranosylpyrazole-3-carboxamide (5)

To a solution of *4* (1.10 g, 4 mmol) in dry methanol (≃100 ml) is added 10% palladium-on-carbon catalyst (0.6 g), and the suspension is stirred under hydrogen at room temperature until the uptake of hydrogen has ceased (≈2.5 hr). The catalyst is removed by filtration, and the filtrate is evaporated *in vacuo* at 25° to give 0.95 g (96%) of **5**, m.p. 157° to 158°, $[\alpha]_D^{22}$ -31° (*c* 0.5, methanol); R_F 0.09 (4:1 chloroform—methanol); λ_{max}^{MeOH} in nm (ε_{mM}), 214 (7.8) and 287 (3.2); PMR data (CD_3OD): δ 7.52 (1, s, H-5), 5.72 (1, d, $J_{1',2'}$ 3.7 Hz, H-1'), 4.63 to 3.60 (5, m, H-2',3',4',5',5").

REFERENCES

1. R. K. Robins, F. W. Furcht, A. D. Grauer, J. W. Jones, *J. Am. Chem. Soc.*, **78**, 2418 (1956).

2. I. A. Korbukh, O. V. Budanova, N. G. Yakunina, V. I. Seraya, M. N. Preobrazhenskaya, *Zh. Org. Khim.*, **12**, 1560 (1976).

[83] DIMETHYL 3-(2,3,5-TRI-*O*-BENZYL-β-D-RIBOFURANOSYL)PYRAZOLE-4,5-DICARBOXYLATE

Generation of a 1-Diazo Sugar to Afford C-Nucleoside Precursors

Edward M. Acton and Kenneth J. Ryan

Life Sciences Division, Stanford Research Institute, Menlo Park, California 94025

INTRODUCTION

2,5-Anhydro-1-deoxy-1-diazo sugars are uniquely useful in 1,3-dipolar additions to construct *C*-nucleosides having the furanose moiety already present.[1-4] As precursors to the diazo sugars, nitrosoamides[3,4] are preferred over nitrosoureas.[1,2] Nitrosoamides are formed and converted almost quantitatively, but ureas undergo nitrosation at both N-1 and N-3 to give unstable 1-nitrosoureas that produce isocyanates in addition to the desired 3-nitrosoureas. Generation of the diazo sugar requires strong aqueous alkali; thus, base-stable protecting groups (such as benzyl) are needed on the sugar, so that the diazo compound can be retained in the organic layer and cleansed of alkali and any water-soluble impurities. Diester *4* is a precursor to the formycins[2,3]; D-*arabino*formycins have been synthesized similarly.[4]

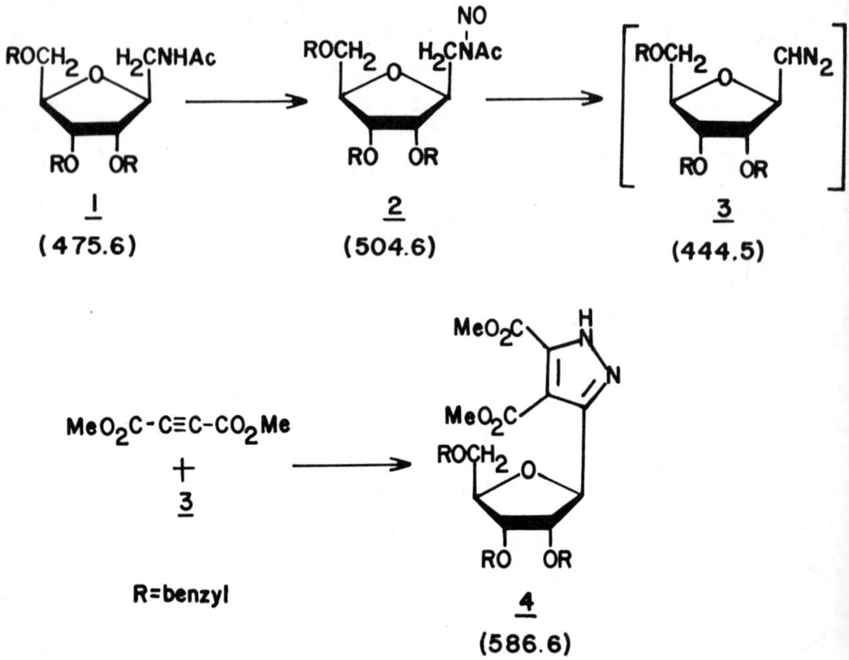

PROCEDURE

1-Acetamido-2,5-anhydro-3,4,6-tri-O-benzyl-1-deoxy-D-allitol (1)

Acetic anhydride (17 ml) is added in portions to a stirred solution of 1-amino-2,5-anhydro-D-allitol[5] (17 g, 0.10 mol) in water (100 ml). After 10 min the anhydride has completely dissolved, and the solution is evaporated under diminished pressure to leave syrupy 1-acetamido-2,5-anhydro-1-deoxy-D-allitol (21 g, 98%); λ_{max}^{film} 6.05 μm (C=O). A stirred solution of the acetamide (1.86 g, 9.1 mmol) in dimethyl sulfoxide (20 ml) mixed with powdered potassium hydroxide (85%; 1.6 g, 25 mmol) is cooled to 15° and treated dropwise with benzyl chloride (3.8 g, 30 mmol).[a] Stirring is continued for 2 hr at 15°, and the mixture is poured into ice and water (150 ml), further stirred for 1 hr, and extracted with three 50-ml portions of benzene. The extracts are dried with anhydrous magnesium sulfate and evaporated *in vacuo*. The residual syrup (3.30 g, 76%) is purified by crystallization[b] from 1:1 ether—cyclohexane (30 ml) to give 2.10 g (48%) of *1*, m.p. 65° to 68°; λ_{max}^{Nujol} 3.00 (NH), 6.03 (C=O), and 6.41 μm (Amide II); PMR data ($CDCl_3$, internal Me_4Si): δ 1.58 s (NAc); TLC homogeneous on silica gel in 1:1 ethyl acetate—benzene, R_F 0.2. An analytical sample melted at 67° to 68.5°. Alternatively, *1* can be prepared by acetylation of 1-amino-2,5-anhydro-3,4,6-tri-*O*-benzyl-1-deoxy-D-allitol.[6,7]

2,5-Anhydro-3,4,6-tri-O-benzyl-1-deoxy-1-(nitrosoacetamido)-D-allitol (2)

To a solution of *1* (10.0 g, 21.0 mmol) in 1:1 carbon tetrachloride glacial acetic acid (120 ml) is added anhydrous sodium acetate (10 g) as a buffer for the nitric acid to be liberated. The stirred mixture is cooled[c] to 3°, and treated with enough (6 to

---

[a] If the benzylation is conducted at room temperature or with sodium hydride[5] as the base, significant amounts of the *N*-benzylacetamide (R_F 0.6) are produced.

[b] Nucleating crystals can be obtained after an initial chromatographic purification on a column of silica gel in 1:1 ethyl acetate—benzene.

[c] Even if there is partial freezing, the mixture becomes stirrable

8 ml) liquid dinitrogen tetraoxide[d] (nitrogen dioxide, Matheson) to give a permanent green color. After being stirred for 1.5 hr at 3° the mixture is poured into ice water (1 l), stirred for 30 min, and extracted with two 200-ml portions of dichloromethane. The extracts are washed with saturated aqueous sodium hydrogen carbonate (400 ml), dried (anhydrous magnesium sulfate), and evaporated *in vacuo* to a syrup (11 g); λ_{max}^{film} 5.75 (C=O of nitrosoamide) and 6.6 μm (NO), with no absorption near 3.0 or 6.0 μ to suggest unreacted *1*; PMR data ($CDCl_3$, internal Me_4Si): δ 2.56 s (CH_3CO-N-NO), with no signal near δ 1.6 to suggest the presence of unreacted *1*. The nitrosoamide is used immediately, or stored at 0°.

2,5-Anhydro-3,4,6-tri-O-benzyl-1-deoxy-1-diazo-D-allitol (*3*)

A solution of syrupy nitrosoamide *2* (11 g, from 21.0 mmol of amide *1*) in ether (50 ml) is mixed with an ice-cold solution of potassium hydroxide (12 g) in water (26 ml) and stirred[e] to provide intimate mixing that almost gives an emulsion. To monitor the reaction, a few drops of the ether layer are allowed to evaporate on a sodium chloride plate; the infrared spectrum of the

---

after addition of the dinitrogen tetraoxide.

[d]Poured from an inverted lecture-bottle and measured in a graduated cylinder; an excess does not interfere with the reaction.

[e]Use of a magnetic stirrer is adequate; on a larger scale a mechanical stirrer is required.

resultant film shows the appearance, in successive aliquots, of a strong band at 4.83 μm (CHN_2) and gradual disappearance of the band at 5.75 μm (a weak band at 5.72 μm always remained). After being stirred for 45 min at 3° the mixture is diluted with ether (100 ml) and water (200 ml) and the layers separated. The ether layer is washed with water (100 ml), and dried rapidly (in <5 min) first by swirling with potassium hydroxide pellets and decantation and then with anhydrous magnesium sulfate and filtration. The yellow filtrate (220 ml) is used immediately.

Dimethyl 3-(2,3,5-tri-O-benzyl-β-D-ribofuranosyl)pyrazole-4,5-dicarboxylate (4)

The yellow solution of 1-diazo sugar *3* (from 21.0 mmol of amide *1*) is added dropwise to a chilled, stirred solution of dimethyl acetylenedicarboxylate (3.30 g; 23.0 mmol) in ether (30 ml) at 3°, with almost immediate discharge of the yellow color. Stirring is continued for 10 min, and the solution is evaporated to dryness under reduced pressure. The residual syrup is dried for 2 to 5 hr *in vacuo* at 65° (mainly to ensure absence of dimethyl acetylenedicarboxylate, δ 3.78 for CO_2Me) to yield 12.5 g (100%, based on *1*) of *4*, $[\alpha]_D^{21}$ +125.5° (*c* 0.5, chloroform), lit.[2] +110.4°; λ_{max}^{film} 3.1 (NH) and 5.73 μm (broad, C=O) (lit.[2] 2.92, 5.73, and 5.74 μm); PMR data ($CDCl_3$, internal Me_4Si): δ 7.33 s (broadened, 7.45 to 7.15; Ph), 5.71 s (H-1'), 3.88 s and 3.82 s (two CO_2Me); TLC on silica gel (chloroform), R_F 0.3, with trace impurities at R_F 0.0, 0.4, and 0.8 (spots detected under UV light and by charring with sulfuric acid). A purity of >90% was estimated from PMR spectroscopy by comparing signals for Ph and H-1'; the purity was not measurably improved by column chromatography[2] (or as for the D-*arabino* isomer[4]), except that the trace impurity having R_F 0.8 was removed.

REFERENCES

1. E. M. Acton, K. J. Ryan, and L. Goodman, *Chem. Commun.*, 313 (1970).

2. M. Bobek, J. Farkas, and F. Sorm, *Tetrahedron Lett.*, 4611 (1970); J. Farkas and F. Sorm, *Collect. Czech. Chem. Commun.*, 37, 2798 (1972).

3. E. M. Acton, K. J. Ryan, D. W. Henry, and L. Goodman, *Chem. Commun.*, 986 (1971).

4. E. M. Acton, A. N. Fujiwara, L. Goodman, and D. W. Henry, *Carbohydr. Res.*, 33, 135 (1974).

5. M. Bobek and J. Farkas, *Collect. Czech. Chem. Commun.*, 34, 1684 (1969).

6. T. Ogawa, Y. Kikuchi, M. Matsui, H. Ohrui, H. Kuzuhara, S. Emoto, *Agri. Biol. Chem.*, 35, 1825 (1971).

7. M. W. Winkley, *Carbohydr. Res.*, 31, 245 (1973).

[84] 1-β-D-RIBOFURANOSYLPYRIDIN-2-ONE

Nucleoside Synthesis by the $SnCl_4$ Catalyzed Silyl-Hilbert-Johnson Method

Ulrich Niedballa and Helmut Vorbrüggen

Schering A. G., 1000 Berlin 65, West Germany

INTRODUCTION

Compounds *4* and *5* were first prepared via the *O*-glycoside followed by *O,N*-rearrangement,[1] but are much more readily accessible in higher yield via the $SnCl_4$-catalyzed silyl Hilbert-Johnson reaction.[2]

PROCEDURE

2-Trimethylsilyloxypyridine (2)

A mixture of pyrid-2-one[a] (9.51 g, 100 mmol) and hexamethyldisilazane (HMDS)[b] (80 ml, 400 mmol) was refluxed for 1 hr in an oil bath at 140° with magnetic stirring and evolution of NH_3 whereupon *1* had dissolved. The excess of HMDS was removed by distillation (b.p. 127° to 128°) and the residue distilled at 64° to

---

[a] Sublimed; if only crude material containing NaCl is available, a 10% to 20% excess of *1* should be employed.

[b] Standard quality, Dow Chem. Co.; Wacker Chemie, GmbH, 8000 München 22, West Germany; redistilled, b.p. 127° to 128°.

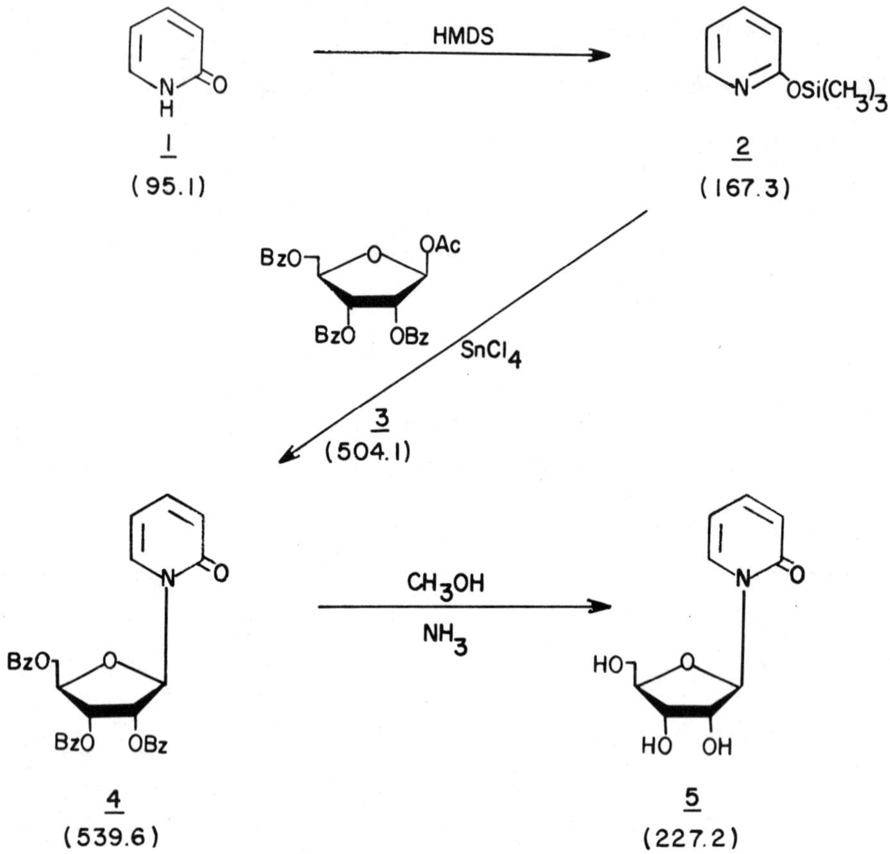

65°/12 mm[c] under dry N_2 or Ar to give 15.9 g (95%) of *2*, which was dissolved in absolute 1,2-dichloroethane[d] to give a 1.085 m standard solution.

---

[c]Evaporative bulb-to-bulb distillation using a Büchi Kugelrohrofen.

[d]Distilled from P_2O_5.

1-(2',3',5'-Tri-O-benzoyl-β-D-ribofuranosyl)-1,2-dihydropyridin-2-one (4)

To 1-*O*-acetyl-2,3,5-tri-*O*-benzoyl-β-D-ribofuranose (*3*)[e] (5.04 g, 10 mmol) and a standard solution of *2* (10.1 ml, 11 mmol) in absolute 1,2-dichloroethane[d] a solution of $SnCl_4$ (1.41 ml, 12 mmol) in absolute 1,2-dichloroethane (50 ml) was added under N_2 or Ar. After 4 hr the reaction was practically complete according to TLC.[f] The reaction mixture was poured on 200 ml $CHCl_3$ and 300 ml saturated $NaHCO_3$-solution, the phases separated and the aqueous phase extracted with 2 × 100 ml $CHCl_3$. If emulsions formed, the crude mixture was filtered over a layer of Celite and the Celite layer carefully washed with $CHCl_3$. The combined organic phase was washed with saturated NaCl-solution (100 ml), dried (Na_2SO_4) and evaporated to give 6.1 g colorless oil, which crystallized from CCl_4 (90 ml) to give in two crops 4.56 g (84.6%) of *4*, m.p. 140° to 142° (lit.[1,2] m.p. 139° to 142°).

1-β-D-Ribofuranosyl-1,2-dihydropyridin-2-one (5)

Compound *4* (2.7 g, 5 mmol) was dissolved in methanolic ammonia (150 ml) and stirred for 24 hr at 24° and evaporated. The residue was dissolved in H_2O (100 ml) and the aqueous solution

---

[e]Pharma Waldhof GmbH, 4000 Düsseldorf 11, West Germany. The compound should be pulverized and dried, m.p. 131° to 132°.

[f]Toluene—acetic acid—H_2O: 5:5:1. For the practical use of two phase systems, see the work by Vorbrüggen and Strehlke.[3]

extracted with ether (3 × 75 ml) to remove the benzamide. After evaporation the residue was recrystallized from ethanol/isopropanol (15 + 5 ml) to give in two crops 0.9 g (80%) of 5, m.p. 147° to 150° (lit.[1] 151° to 152°).

REFERENCES

1. H. Pischel and G. Wagner, *Arch. Pharm.*, 300, 602 (1967).

2. U. Niedballa and H. Vorbrüggen, *J. Org. Chem.*, 39, 3668 (1974).

3. H. Vorbrüggen and P. Strehlke, *Chem. Ber.*, 106, 3039 (1973).

Nucleic Acid Chemistry

AUTHOR INDEX

This index covers Volumes 1 and 2. Therefore, references to pages 1 through 484 will be found in Volume 1 and pages 485 through 1067 in Volume 2. Underlined numbers indicate authorship of a contribution in Volume 1 or 2 of this series. The other reference numbers indicate the pages on which literature references appear and the numbers in parentheses are the reference numbers in that contribution.

B

N

S

W

SUBJECT INDEX

The number following the keyword refers to contribution number rather than page number. Contributions 1 through 84 will be found in Volume 1 and contributions 85 through 177 in Volume 2.

E